INTELLIGENT AUTONOMOUS SYSTEMS 7

Intelligent Autonomous Systems 7

Edited by

Maria Gini

Department of Computer Science and Engineering,
University of Minnesota, USA

Wei-Min Shen

Information Sciences Institute,
University of Southern California, USA

Carme Torras

Institut de Robòtica i Informàtica Industrial (CSIC-UPC),
Barcelona, Spain

Hideo Yuasa

Department of Precision Engineering,
University of Tokyo, Japan

IOS
Press

Ohmsha

Amsterdam • Berlin • Oxford • Tokyo • Washington, DC

ISBN 1 58603 239 9 (IOS Press)
ISBN 4 274 90496 2 C3055 (Ohmsha)

Publisher
IOS Press
Nieuwe Hemweg 6B
1013 BG Amsterdam
The Netherlands
fax: +31 20 620 3419
e-mail: order@iospress.nl

Distributor in the UK and Ireland
IOS Press/Lavis Marketing
73 Lime Walk
Headington
Oxford OX3 7AD
England
fax: +44 1865 75 0079

Distributor in the USA and Canada
IOS Press, Inc.
5795-G Burke Centre Parkway
Burke, VA 22015
USA
fax: +1 703 323 3668
e-mail: iosbooks@iospress.com

Distributor in Germany, Austria and Switzerland
IOS Press/LSL.de
Gerichtsweg 28
D-04103 Leipzig
Germany
fax: +49 341 995 4255

Distributor in Japan
Ohmsha, Ltd.
3-1 Kanda Nishiki-cho
Chiyoda-ku, Tokyo 101
Japan
fax: +81 3 3233 2426

Preface

The goal of the 7th International Conference on Intelligent Autonomous Systems (IAS-7) is to exchange and stimulate research ideas that make future robots and systems more intelligent and autonomous. This conference emphasizes that intelligence should and could be best illustrated by systems that can directly sense and act in their own environment without demanding detailed supervision from humans. As more and more robots are entering our daily life, researchers in IAS are facing increasingly many challenges in how to construct, control, and interact with such systems.

The IAS-7 conference provides an exciting environment for researchers to present and discuss their novel theoretical results, implementations, and applications. A total of 71 technical papers were submitted by authors from 15 countries, representing more than 40 universities and research institutions. All the submissions were rigorously reviewed by the Program Committee. Of those submissions 40 were accepted as full papers and 9 as short papers. The overall outcome of the revision process is an excellent selection of papers that showcase the research of autonomous systems today. In addition, Dr. Karsten Berns, from the Institute for Industrial Applications of Informatics and Microsystems of the University of Karlsruhe organized an invited session on "Biologically motivated robots", and Prof. Heinz Wörn, of the Institute for Process Control and Robotics, University of Karlsruhe organized an invited session on "Innovative ways of using manipulators".

IAS-7 builds on the success of its predecessors. The first IAS conference was held in Amsterdam in 1986, the second in Amsterdam in 1989, the third in Pittsburgh in 1993, the fourth in Karlsruhe in 1995, the fifth in Sapporo in 1998, and the sixth in Venice in 2000. We are confident that IAS-7 will continue the tradition established by the previous IAS conferences of being a key venue for presenting cutting-edge research in intelligent autonomous systems.

IAS-7 is held in the beautiful city of Marina del Rey in California, USA. Marina del Rey, overlooking the world's largest man-made yacht harbor, is one of the most desirable areas in the Los Angeles basin. The full diversity of the area's cultural and recreational activities are within easy reach.

We would like to take this opportunity to thank everyone involved with the organization of IAS-7. First, we would like to thank the members of the Program Committee, who did a thorough and conscientious job in reviewing a large number of papers. Wei-Min Shen in addition to his duties as program co-chair coordinated superbly the local arrangements. Liz Hall did an excellent job to keep the finances of the conference in order. Paul Rybski helped with the publicity of the conference and managed the Web site. The members of the Steering Committee provided invaluable help and support through the process of organizing the conference. Finally, we would like to thank our sponsors. Their enthusiastic support is a good indication of the vitality of the conference. It has been with the support of all these people and several others not mentioned here that the conference organization was possible. We sincerely appreciate all the hard work they put in to make this conference a success.

We warmly welcome all representatives from industry, government, and academia to join us in Marina del Rey in March 2002!

Maria Gini, Wei-Min Shen, Carme Torras, and Hideo Yuasa

IAS-7 Conference Organization

General Chair
Maria Gini
Department of Computer Science and Engineering, University of Minnesota, USA

Program Co-Chairs
Wei-Min Shen
Information Sciences Institute, University of Southern California, USA

Carme Torras
Institut de Robòtica i Informàtica Industrial (CSIC-UPC), Barcelona, Spain

Hideo Yuasa
Department of Precision Engineering, University of Tokyo, Japan

Steering Committee Chair
Franz Groen, University of Amsterdam, The Netherlands

Steering Committee Members
Enrico Pagello, University of Padua and LADSEB-CNR, Italy
Rüdiger Dillmann, University of Karlsruhe, Germany
Tamio Arai, University of Tokyo, Japan
Anthony Stentz, Carnegie Mellon University, USA

Local Arrangements Chair
Wei-Min Shen, ISI/University of Southern California, USA

Publicity Chair
Paul Rybski, University of Minnesota, USA

Sponsored by
The Intelligent Autonomous System Society

Supported by
The Air Force Office of Scientific Research
The Army Research Office
The Office of Naval Research
The Information Sciences Institute, University of Southern California
The University of Minnesota

The views, opinions, and/or findings contained in this report are those of the author(s) and should not be construed as an official Department of the Air Force/Army/Navy position, policy, or decision, unless so designated by other documentation.

Program Committee Members

Marcelo H. Ang, National University of Singapore, Singapore
Ron Arkin, Georgia Tech, USA
Tamio Arai, University of Tokyo, Japan
Minoru Asada, Osaka University, Japan
Hajime Asama, RIKEN, Japan
Riccardo Cassinis, University of Brescia, Italy
Enric Celaya, IRI (CSIC-UPC), Spain
Raja Chatila, LAAS, France
Hyung Suck Cho, KAIST, Korea
Henrik Christensen, Royal Institute of Technology, Sweden
Jim Crowley, INRIA Rhone-Alpes, France
Kerstin Dautenhahn, University of Hertfordshire, UK
Dario Floreano, Swiss Federal Institute of Technology, Switzerland
John Hallam, University of Edinburgh, UK
Takakazu Ishimatsu, Nagasaki University, Japan
Koji Ito, Tokyo Institute of Technology, Japan
Ray Jarvis, Monash University, Australia
Gal Kaminka, Carnegie Mellon University, USA
Pradeep Khosla, Carnegie Mellon University, USA
Kurt Konolige, SRI International, USA
Ben Krose, University of Amsterdam, The Netherlands
Jean-Claude Latombe, Stanford University, USA
James J. Little, University of British Columbia, Canada
Ramon Lopez de Mantaras, IIIA, Barcelona, Spain
Zhi-Wei Luo, RIKEN, Japan
Satoshi Murata, Tokyo Institute of Technology, Japan
Robin Murphy, University of Southern Florida, USA
Ulrich Nehmzow, University of Manchester, UK
Anibal Ollero, University of Sevilla, Spain
Jun Ota, University of Tokyo, Japan
Alan Schultz, Navy Research Lab, USA
Reid Simmons, Carnegie Mellon University, USA
Peter Stone, AT&T Research, USA
Gaurav Sukhatme, University of Southern California, USA
Keiji Suzuki, Hokkaido University, Japan
R. Lal Tummala, Michigan State University, USA
Kanji Ueda, Kobe University, Japan
Peter Will, ISI/University of Southern California, USA
Mark Yim, Xerox Park, USA

Contents

Intelligent Autonomous Systems 7
M. Gini et al. (Eds.)
IOS Press, 2002

Reactive Reflex based Control for a Four-Legged Walking Machine *

J. Albiez, T. Luksch, W. Ilg, K. Berns, R. Dillmann
Forschungszentrum Informatik an der Universität Karlsruhe (FZI)
Interactive Diagnosis and Servicesystems
Haid-und-Neu-Str. 10-14, 76131 Karlsruhe, Germany
albiez@fzi.de

December 20, 2001

Abstract

This paper presents methods and first experiments of a reflex based posture control for a four legged walking machine. Starting with a description of the existing control architecture we introduce the concepts of reflexes and their integration into the system. The reflex units are implemented using fuzzy control. All reflexes needed for a statically stable gait over irregular terrain are described and initial ideas for reflexes scheduling are introduced.

1 Introduction

As a fundamental requirement for walking on irregular terrain a reactive control should not only plan the body trajectory but also perform sensor-based adaptation of the motion. Problems in implementing such a reactive control consist in the sensoric equipment and the real-time interpretation of its signals as well as in the necessity of adaptation to changing environments.

Biological research of the last years has identified some key elements being used in nature for adapting the motion. These range from the geometrical structure of legs [13] and the dynamic properties of muscles [10] to the neural networks used for walking by insects [3]. The results of this research suggest a transfer of these principles to legged robots. Due to the high complexity of real walking machines and the impracticality of mimicking especially nature's activators and sensors, up to now only some of the ideas have been transferred into the control architectures of real robots. In [8] and [7] a neuro-oscillator based pattern generator is introduced. The adaptation to the terrain is solved by directly influencing the activation of the oscillators neurons. [4] proposes a

*This research is funded by the Deutsche Forschungsgemeinschaft (DFG), grants DI-10-1, DI-10-2 and DI-10-3

reflex based gait generation system, triggered by the input of a camera system mounted on the robot.

In the following, we first describe the four-legged walking machine BISAM addressing its special features and the aimed movements. Next we present an overview of the control architecture and the integration of the adaptive control component as well as the models used to process the information coming from the BISAMs sensors. In section 4 the methods for implementing the reactive control based on a reflex system and the first experiments are presented. The paper closes with a conclusion and an outlook on further developments.

2   The Walking Machine BISAM

BISAM (Biologically InSpired wAlking Machine) consists of the main body, four equal legs and a head (fig 1). The main body is composed of four segments which

Figure 1: The quadrupedal walking machine BISAM. Due to the five active degrees of freedom in the body and the ability to rotate the shoulder and hip, BISAM implements key elements of mammal-like locomotion.

are connected by five rotary joints. Each leg consists of four segments, that are connected by three parallel rotary joints and attached to the body by a fourth. The joints are all driven by DC motors and ball screw gears. The height of the robot is 70 cm, its weight is about 23 kg. 21 joint angle encoders, four three dimensional foot sensors and two inclinometers mounted on the central body provide the necessary sensoric input. A more detailed description of BISAM can be found in [1]. Research on BISAM aims at the implementation of mammal-like movement and different gaits like statically stable walking and dynamic trotting with continuous gait transitions. Due to this target, BISAM is developed with joints in the shoulder and in the hip, a mammal-like leg-construction and small foot contact areas. These features have strong impact on the appliable methods for measuring stability and control. For example, caused by BISAMs small feet the ZMP-Criterion [12] is not fully adequate to describe the aspired movements.

The control design has to consider the high number of 21 active joins and especially the five joints in the body. One common way to reduce the model complexity is to combine joins and legs by the approach of the virtual leg, as used in many walking machines [11], [8], [14].This approach poses problems when modelling BISAM's body joints and lead to a strong reduction in the flexibility of the walking behaviour [9]. A second way is to reduce the mechanical complexity of the robot so it is possible to create an exact mathematical model of the robot [2].

In considering the described problems we opt for a strategy using a planned trajectory for the CoG based on a simplified model (with all degrees of freedom). Reflexes will correct the guidance of CoG as well as the posture of the machine in respect to its kinematic constraints based on the sensoric input.

3 Control Architecture

The classical approach to robot control is to determine the joint trajectories by inverse kinematics, paramerized body motion and foot trajectories. Based on this control architecture a statically stable walk (b = 0.8) and a dynamically stable trot (b =0.6) as well as the parameterised gaits extended trot [14] and intermittent trot [15] have been implemented for flat terrain [5][6].

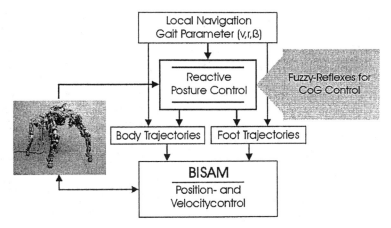

Figure 2: Overview of the control architecture with the reactive component for posture control.

To make these locomotion behaviour more robust against disturbances and applicable on irregular terrain, the neccessity of a reactive posture control is identified [7] and posture control mechanisms are integrated into the control architecture (see fig 2).

3.1 CoG Modelling and Stability Margin

For developing a efficient reactive control a model for the CoG based on the available sensor information is required. In this paper we use the sensor based CoG Model SCM [6] [5], which provides a two component scheme to locate the CoG in respect to the central coordinate frame of the robot shown in eq. (1) and (2) and Fig 3a.

$$SSM_x = \frac{F_{FL} + F_{FR} - (F_{RL} + F_{RR})}{\sum_{F_{XY} \in \mathcal{F}} F_{XY}} \tag{1}$$

$$SSM_y = \frac{F_{FR} + F_{RR} - (F_{FL} + F_{RL})}{\sum_{F_{XY} \in \mathcal{F}} F_{XY}} \qquad (2)$$

The SCM is also used as basis for calculating the stability margin (see [12] and Fig 3b) for statically stable movements when surpassing obstacles.

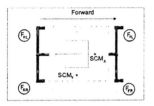 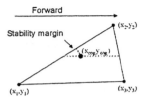

Figure 3: (left) Illustration of the parameters SCM_x and SCM_y for the sensor based measurement of the effective CoG based on foot sensor information.
(right) Visualisation of the stability margin for statically stable movement, which is termined by the minimum distance of the CoG to the margin of the support area.

4   Reactive Posture Control

The idea of the proposed reactive control is to use small reflex-like units which take care of a certain stability aspect by watching only a limited part of the sensor bandwidth. Their output is superimposed on the basic trajectory generation system. These small reflex-units can be placed on any level inside the control hierarchy, depending on their functionality. In this paper reflexes for controlling the posture of BISAM while standing and walking (b = 0.8) are presented.

4.1   Structure of the Reflexes

To implement the reflexes a fuzzy approach has been chosen due to the following to reasons: Fuzzy controllers provide a rapid and easy way to adjust the reflex output making good use of the experience gained by BISAM's operators. Furthermore there are several methods to optimise fuzzy system with machine learning which can be applied in future works.

Each reflex consists of two main parts: A sensor input pre/post-processing unit and a fuzzy controller (FC) (fig.4a). The pre/post-processing unit has the following tasks: Scale and filter the sensor inputs (eg. inclinometers) so that they meet the requirements of the FC concerning scale and range; calculate a criterion r of how much the current position of the machine meets the target of the reflex; adapt the output vector u' of the FC to the needs of the lower layers of the control architecture. The fuzzy controller is implemented as a classical Fuzzy-PD-Controller (see fig 4b).

Beside the input from the pre-/post-processing unit it has a special fuzzy on/off input i which basically scales the output of the fuzzy controller, and an activity output a defined as the norm of the FCs output vector. The FC is designed to minimize the

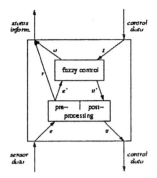

		Control Deviation				
		NL	NS	Z	PS	PL
Derived Control Deviation	NL	LL	LL	LL	LS	Z
	NS	LL	LL	LS	Z	RS
	Z	LL	LS	Z	RS	RL
	PS	LS	Z	RS	RL	RL
	PL	Z	RS	RL	RL	RL

Figure 4: (left)The internal structure of a Fuzzy-Reflex Unit. The sensor data e is processed by the reflex and an action u is generated. a and i are used to control the reflex from higher levels. (right) The inference matrix of the Fuzzy-PD controller. There are five fuzzy sets (negative large and small, zero, positive small and large) for both input variables and five sets (left large and small, zero, right and large) for the output variable.

criterion r by generating posture adjustments which are added to the cyclic movement of BISAM.

4.2   Implemented Reflexes

Various experiments [6] and considerations on the insights from biological research as in [13] and [3] have shown the necessity for the following key reflexes:

- **EvenForceXY** The purpose of this reflex is to evenly distribute the forces at the feet touching the ground. The SCM_{xy} values are used as measurement criterion of the load distribution and as input values e for the FC. $r = \|SCM\|$ is the absolute value of the SCM vector. As reflex output u offset values shifting the central body in the xy-plane are generated by the FC.

- **EvenForceZ** Similar to EvenForceXY this reflex tries to achieve the same goal by individually adapting the length of all four legs in z direction.

- **BodyHeight** This reflex tries to keep the body on a certain height level. It uses only the internal representation of the machine, and therefore the actual height h of BISAM is calculated from the average length in the z direction of all legs touching the ground

- **BodyInclination** The inclinometers mounted on BISAM's body are used by this reflex as input e to keep the body parallel to the ground. This results in r to be defined as:

$$r = |\phi - \phi_0| + |\theta - \theta_0| \tag{3}$$

where ϕ and θ are the pitch and yaw angles. The rotation of the body (the output u) is implemented by equally stretching the legs on one side while bending the legs on the other side.

- **LegRelieve** Before BISAM can take a step when walking in stride the leg which is about to swing (l_{swing}) has to be relieved which is insured by this reflex. The force in l_{swing} is used as criterion r, the output u of the reflex is a shift of the body in the y direction and an appropriate bending of the leg opposite to l_{swing}.

4.3 Reflex Scheduling

Unfortunately just turning on all the reflexes won't suffice to assure a statically stable walk as some reflexes work against each other. One such reflex conflict occurs while trying to relieve the leg that is about to swing: the LegRelieve reflex will move the machine body to one side and bend the leg opposite the swing leg, whereas the Even-ForceXY, EvenForceZ and BodyInclination reflexes will try to move the body back to the most stable posture but with $\frac{1}{4}$ of the load on the swing leg. To resolve these conflicts a reflex weighting mechanism depending on the current stance phase has been developed (Fig 5left). Fig 5right visualises the resulting reflex activities while swinging with a single leg.

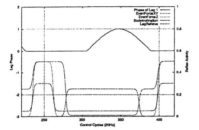 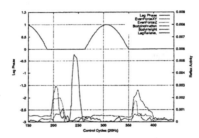

Figure 5: (left) The activation of the different reflexes using the input i of the reflex units. Note that while in the intermittent phase (cycle 20 to 310), the reflexes for stabilizing the machine are activated. In the swing phase (cycle 310 to 380) the *LegRelieve* reflex is the significant one, only *EvenForceXY* is switched on a bit to ensure the stability of the robot in case of disturbances.
(right) The activation of the different reflexes using the input i of the reflex units. Note that while in the intermittent phase (cycle 20 to 310), the reflexes for stabilizing the machine are activated. In the swing phase (cycle 310 to 380) the *LegRelieve* reflex is the significant one, only *EvenForceXY* is switched on a bit to ensure the stability of the robot in case of disturbances.

4.4 Experiments

To verify the functioning of the reflexes in uncertain environment the following experiment setup has been used among others. BISAM is placed on a platform of about 10cm in height with one leg reaching over the platforms rim (Fig 6). When toppling down the platform (at about $t \sim 35$ in Fig 6) the machines stability will decrease until the lower leg touches the ground ($t \sim 65$). Without any further posture control the fourth leg won't reach the ground and the stability will remain on a low level (solid line), whereas it will dramatically increase when the reflexes are activated (dotted line at about $t \sim 70$).

 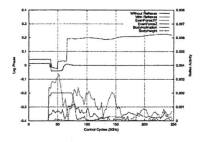

Figure 6: (Left) BISAM stepping down a platform. (Right) Comparing stability with and without activated fuzzy reflexes. The stability margin and the reflex activity are plotted against time. The numbers denote certain points of the experiment: 1 - The leg is over the rim; 2 - BISAM is tipping forward; 3 - The swinging leg hits the ground; 4 - Regaining stability. Without activated reflexes stability is significantly lower (stability margin < 0) in all states, whereas the machine tries to shift the CoG to a better position already while falling if the reflexes are activated. With activated posture control the final stability margin is by far higher than without.

5 Conclusion and Outlook

As experiments showed, a reactive control substantially increases the robustness and thus the range of use of the robot. The proposed posture control mechanisms have been used to surpass a broad range of obstacles without modelling them. The first implementation of reflex scheduling proofed that a time and state dependent system for activation is feasible.

In future experiments the reflex architecture will be expanded with additional layers (Evasion reflexes, local navigation with external sensors) and with control mechanisms combining the reflexes on different layers to reflex chains. The cyclic movement system will be revised and reflexes integrated for stabilisation while in trot or in a climbing-free-gait. Therefore a more robust system to schedule the reflex outputs to avoid conflicts between oppositional reflexes with respect to the actual state of the walking machine and the desired movement has to be designed. Furthermore it is planned to optimise the fuzzy controllers, by transferring them to radial basis function networks and train them with reinforcement learning methods.

References

[1] K. Berns, W. Ilg, M. Deck, J. Albiez, and R. Dillmann. Mechanical construction and computer architecture of the four-legged walking machine BISAM. *IEEE Transactions on Mechatronics*, 4(1):1–7, Mrz 1999.

[2] M. Buehler, A. Cocosco, K. Yamazaki, and R. Battaglia. Stable open loop walking in quadruped robots with stick legs. In *Proceedings of the IEEE International Conference on Robotics and Automation*, pages 2348–2354, Detroit, Mai 1999.

[3] H. Cruse and C. Bartling. Movement of joint angles in the legs of a walking insect, carausius morosus. *J. Insect Physiol.*, (41):761–771, 1995.

[4] K. Hosoda, T. Miyashita, and M. Asada. Emergence of quadruped walk by a combination of reflexes. In *Procceedings of the International Symposium on adaptive Motion of Animals and Machines*, Montreal, August 2000.

[5] W. Ilg. *Eine biologisch motivierte adaptive Bewegungssteuerung fr eine vierbeinige Laufmaschine.* Infix-Verlag, 2001. Dissertationsschrift (in german).

[6] W. Ilg, J. Albiez, and R. Dillmann. Adaptive posturecontrol for a four-legged walking machine using some principles of mammalian locomotion. In *Int. Symposium on Adaptive Motion of Annimals and Machines*, 2000.

[7] H. Kimura and H. Nakamura. Biologically inspired dynamic walking on irregular terrain - adaptation based on vision. In *International Symposium on Adaptive Motion of Animals and Machines*, Montreal, August 2000.

[8] H. Kimura, I. Shimoyama, and H. Miura. Dynamics in the dynamic walk of a quadruped robot. *Advanced Robotics*, 4(3):283–301, 1990.

[9] O. Matsumoto, W. Ilg, K. Berns, and R. Dillmann. Dynamical stable control of the four-legged walking machine bisam in trot motion using force sensors. In *Intelligent Autonomous Systems 6*, 2000.

[10] K.G. Pearson. Proprioceptive regulation of locomotion. *Current Opinions in Neurobiology*, 5(6):768–791, 1995.

[11] Marc H. Raibert. *Legged Robots That Balance.* MIT Press, Cambridge, MA, 1986.

[12] M. Vukobratovic, B. Borovac, D. Surla, and D. Stokic. *Biped Locomotion.* Springer–Verlag, Heidelberg, Berlin, New York, 1990.

[13] H. Witte, R. Hackert, K. Lilje, N. Schilling, D. Voges, G. Klauer, W. Ilg, J. Albiez, A. Seyfarth, D. Germann, M. Hiller, R. Dillmann, and M.S. Fischer. Transfer of biological priciples into the construction of quadruped walking machines. In *Second International Workshop On Robot Motion And Control*, Bukowy Dworek, Poland, 2001.

[14] K. Yoneda and S. Hirose. Dynamic and Static Fusion Gait of a Quadruped Walking Vehicle on a Winding Path. In *Proceedings of the IEEE/RSJ International Conference on Intelligent Robots and Systems*, pages 143–148, Nizza, Mai 1992.

[15] K. Yoneda, H. Iiyama, and S. Hirose. Intermittent trot gait of a quadruped walking machine dynamic stability control of an omnidirectional walk. In *Proceedings IEEE International Conference on Robotics and Automation*, pages 3002–3007, Minneapolis, April 1996.

Intelligent Autonomous Systems 7
M. Gini et al. (Eds.)
IOS Press, 2002

9

Smooth Task Switching through Behaviour Competition

Philipp Althaus and Henrik I. Christensen

Centre for Autonomous Systems, Numerical Analysis and Computer Science
Royal Institute of Technology (KTH), SE-10044 Stockholm, Sweden
[philipp, hic]@nada.kth.se

Abstract. Navigation in large-scale environments is composed of different local tasks. To achieve smooth switching between these tasks and thus a continuous control signal, usually a precise map of the environment and an exact pose estimate of the robot are needed. Both are hard to fulfill for experiments in real-world settings. We present a system that shows how one can relax the need for accurate metric models of the environment while at the same time achieving smooth task switching. To facilitate this scheme the dynamical system approach [10] is used, which incorporates behaviour coordination through competition in a dynamic framework. Feature detectors use sonar data to provide means for local navigation. This ability combined with a simple topological map constitutes a complete navigation system for large-scale office environments. Experiments showed that a Scout robot using this scheme is able to successfully navigate through our whole institute. Through the use of the dynamic behaviour coordination, switching between the navigational tasks occurs in a smooth manner leading to continuous control of the platform.

1 Introduction

A complete navigation system of an autonomous robot is traditionally of a hybrid nature. It consists of a reactive part, which allows to deal with unforeseen local events. Then, there is a deliberative part, which takes care of global issues like path planning. See [2] for an overview of different architectures. The deliberative part is making use of a map to allow navigation in large-scale structures. To the implementation of this map and its use there are two conceptually different approaches: Using either an exact geometric map or a qualitative map (e.g. topological map).

An exact geometric map (a grid map [7], for example) allows to determine an appropriate control action at any robot position, which leads to smooth control and trajectories. However, this precise map first has to be obtained and the exact robot position has to be known at all times. These conditions are hard to fulfill and might be computationally very expensive. In contrast, the use of a topological map requires only an approximate guess of the robots pose; also the small-scale geometry does not have to be known precisely. Through sensory information from the robot it is determined when the robot has reached a node in the map. This information is enough to switch between the necessary behaviours to deal with the navigation task at hand. The Polly system [4], for example, shows how successful navigation can be achieved using this simple type of qualitative map. However, through discrete switching between tasks the control signals are not continuous. Furthermore, also the trajectory of the robot is not smooth anymore.

In this paper we propose a unified control scheme combining the advantages of each of the two approaches above. We present a navigation system for office environments using feature detectors which extract representations from sensory data. This geometric place recognition facilitates safe local navigation. To capture the large-scale properties of the environment a simple topological map is used. Nevertheless, through smooth task switching in a dynamical system framework, continuity in the control signal can be achieved. The dynamical system approach to behaviour based robotics introduced by Schöner and Dose [10] provides this necessary framework. It utilizes a competition scheme among individual behaviours. So far, this competition has only been exploited in simulation work [6] using simplified small-scale environments. The system proposed in this paper, however, acts in a large-scale real-world environment using a topological map containing the global information necessary for navigation. This information about the whereabouts of the robot is still of a discrete nature; nevertheless, it is only used as parameters in the dynamic coordination among the behaviours. This dynamic system then ensures smooth switching between tasks. In this way the discrete nature of a qualitative map is combined with the need for continuous navigation in a unified control scheme. In essence, the system presented shows that one can relax the need for continuous pose estimation while maintaining a possibility for smooth behaviour switching.

2   The Topological Map and its Use

The topological map contains qualitative information about the large-scale structure of the environment. This information is reflected in nodes and edges that connect these nodes. The nodes stand for important places in the environment. There has to be one in front of each door, at each corridor crossing and at other places of interest (e.g. goal locations and charging station). Each node has a location in a fixed coordinate system. The edges that connect these nodes can be of three different types: room, corridor, door. Due to the limited amount and simplicity of information in the topological map, it is a matter of minutes to construct a new one for a previously unknown office environment. Figure 1 shows the topological map of our institute. To plan a path between two arbitrary nodes, the system does an exhaustive search through the map. This is a feasible solution for environments of the size of an office area.

At execution of the plan it is assumed that the initial position and orientation of the robot are known (e.g. charging station). From there odometry is used to determine the robots location. This introduces errors in the estimation of the exact position of the robot, but is totally sufficient to determine if the system is in the vicinity of a node. This information is enough to evoke the appropriate behaviours, which in turn are capable of dealing with the small-scale structures of the environment (see section 4.2). Note that the error in the position estimate would grow bigger than desired on long trials over a great distance. To avoid this deficiency the pose estimate is corrected based on detected features (see section 3).

3   Extracting Geometric Representations from Raw Sensor Data

In our experiments we used the Scout robot from Nomadic Technologies (Figure 1). The platform has a cylindrical shape with a diameter of 38 cm and moves at a speed of up to $1 \frac{m}{s}$. The robot is equipped with a ring of 16 evenly spaced ultrasonic sensors. The perception and geometric reconstruction of obstacles, walls and doors is solely based on

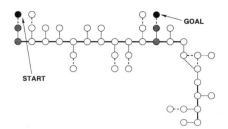 

Figure 1: The topological map of our institute and the Scout robot used in the experiments: The circles in the map depict nodes, which have an exact location in a coordinate system. Edges are of three different types: corridor (thick line), room (dashed line), and door (thin line). Additional nodes for goal points and starting positions can be added arbitrarily. The nodes denoted with "start" and "goal" correspond to the initial (charging station) and final location of the experiment described in section 5. The nodes in gray are the ones used to execute this plan.

the information provided by these sensors. Each sonar has a beam width of 25° and a detection range of 6 to 255 inches. The robot possesses a two wheel differential drive located at the geometric center which allows omni-directional steering at zero turning radius. The wheels are equipped with encoders to obtain odometric data.

To navigate along a corridor two behaviours have been designed: CORRIDOR FOLLOWING and WALL AVOIDANCE. They are based on the orientation of the corridor and the distance to its walls. To obtain this information the 200 most recent sonar readings are kept in a FIFO buffer. A Hough transform [3] is invoked on the sonar data every few seconds in order to extract the pair of parallel lines (one on either side of the robot) that coincide with the largest number of sonar echos. No assumptions on the width or direction of the corridor are made.

The behaviour DOOR PASSING guides the robot safely through an arbitrary, large enough opening in a wall. In order to find a door, when the robot finds itself in a corridor, the direction to the detected corridor wall is used. The 25 most recent sonar readings, that lie in the direction of this wall and not more than 50 cm behind it, are kept in a FIFO buffer. The largest angular segment (from the robots point of view) that does not contain any sonar reading is determined. If this segment is greater than 15° we consider a door to be detected and its direction is defined as the centre of the free segment. This process is invoked at every control cycle of the robot. Note that this door detector is very crude, due to the simplicity of the sensors used. Especially half blocked doors, with passages that are to small to pass, will sometimes still be detected as doors. However, situations like this are resolved by the coordination among a door passing and an obstacle avoidance behaviour (see the experimental results in section 5). Further, if the robot is in a room the same strategy to detect a door is applied. However, first the wall at which the door is located has to be extracted. In order to do this a Hough transform is invoked on the 100 most recent sonar echos.

Due to the limited angular resolution of sonar sensors, the geometric representation of obstacles (used by the behaviour OBSTACLE AVOIDANCE) is rather simple and closely linked to the actual perception of the robot. Out of the 50 most recent sonar readings that do not belong to detected walls, the ones in the frontal half plane of the current robot heading are considered. Obstacles are reconstructed from these detected echos in ascending order of their distance to the robot. The echo closest to the robot defines the first obstacle whose orientation in the robot frame is given by the axis of the sensor that received the echo. A new obstacle is recorded for every subsequent echo whose orientation differs by an angle of at least 22.5° from any previously identified obstacle. New obstacles are added in an incremental fashion until the sonar buffer contains no further echos. Obstacle reconstruction is invoked at every control cycle of the robot. Notice, that our representation only considers the distance to an obstacle but ignores its shape or size. Despite its simplicity the chosen representation is powerful enough to successfully navigate in cluttered areas [1].

Each of the above detectors keeps a certain number of the most recent sonar readings in a FIFO buffer. While collecting these readings the robot is driving a short distance. Odometry is used, to calculate the location of sonar readings taken at different robot positions, which introduces further uncertainty in the sonar data. These errors, however, are comparatively small and hardly influence the performance of the behaviours.

The behaviour GO TO aligns the robots heading with the direction of a goal point. We do not use any detector for this goal point yet; its location is only defined by a node in the topological map (section 2). Also to determine which of the above detectors should be invoked the map is used. The information about the exact location of its nodes and the robots position estimate (from odometry) determine if the robot finds itself in a corridor or in a room and/or close to a door. The information from the detected features, in turn, is used to update the position estimate. Otherwise, the error in this estimate may grow bigger than desired using only odometry. Hence, in a corridor the robots computed orientation and position are adjusted relative to the corridor walls each time the Hough transform is invoked (i.e. every few seconds). In addition, every time a door is passed orientation and position relative to the door posts can be updated accordingly.

4 Behaviour Coordination

The dynamical system approach has been used to design the individual behaviours and their interaction. The conceptual framework of this approach is based on the theory of nonlinear dynamical systems [8]. In section 4.1 we only provide a brief outline of this framework and refer the interested reader to [11] for a more detailed description. Section 4.2 describes the design of our system, especially the dynamic coordination of the individual behaviours. Both, the behaviours and their coordination, anchor their parameters in the features extracted from the sensory data and in the topological map (introduced in section 2).

4.1 Dynamical System Approach

A behaviour b emerges from the time evolution of the *behavioural variables* described by the vector $\vec{x}$. In a navigation task for example the robots heading and velocity may constitute the set of behavioural variables. In the dynamical system described by

$$\dot{\vec{x}} = \vec{f}_b(\vec{x}) \tag{1}$$

the function $\vec{f}_b$ can be interpreted as a *force* acting on the behavioural variables. This force is designed such that the desired values of $\vec{x}$ (e.g. direction of a target) form an attractor and undesired values (e.g. direction of an obstacle) form a repellor in the dynamics of the behavioural variables. The function $\vec{f}_b$ depends on the relative pose between the robot and its environment. However, the dynamics of $\vec{x}$ takes place at a much faster time scale than the gradual changes that emerge in $\vec{f_b}$ as a result of the robots motion. This property assures that the dynamic variables remain close to the attractor state at all times. Multiple behaviours are aggregated by weighted addition of the individual contributions $\vec{f_b}$.

$$\dot{\vec{x}} = \sum_b |w_b| \vec{f_b}(\vec{x}) + \text{noise} \tag{2}$$

The weights $w_b \in [-1, 1]$ define the strength of each behaviour and are computed based on the perceived context of operation. The noise has a small amplitude and merely assures that the dynamics escapes unstable fix-points (repellors). Coordination among behaviours is modeled by means of an additional competitive dynamics that controls the weights w_b, which evolve in the following fashion:

$$\tau_b \dot{w}_b = \alpha_b(w_b - w_b^3) - \sum_{b' \neq b} \gamma_{b',b} w_{b'}^2 w_b + \text{noise} \tag{3}$$

The first term constitutes a pitchfork bifurcation, i.e. the dynamics possesses stable fix-points at

$$w_b = \begin{cases} \pm 1 & \text{if } \alpha_b > 0 \\ 0 & \text{if } \alpha_b < 0 \end{cases} \tag{4}$$

The factors $\alpha_b \in [-1, 1]$ are called *competitive advantages*. They determine to which degree a behaviour is appropriate and desirable in the present context. The second term in equation 3 captures the competitive dynamics in that an active behaviour b' of higher priority suppresses the activation of another conflicting behaviour b. Hence, the factors $\gamma_{b',b} \in [0, 1]$ are called *competitive interactions*. For $|w_{b'}| \sim 1$ and $\gamma_{b',b} > \alpha_b$, the point $w_b = 0$ becomes the new stable fix-point of behaviour b, despite a positive competitive advantage $\alpha_b > 0$. A detailed analysis of how the stability of fix-points varies across different values of competitive advantages and interactions is given in [6]. The time constant τ_b determines the rate at which the behaviours are switched on and off. Similar to the behavioural dynamics, the noise term helps the system to escape unstable fix-points.

4.2 System Design

We chose the robot heading ϕ as the behavioural variable of the dynamical system as it offers the advantage that the behaviours can be naturally expressed in this variable. Furthermore, the commanded turn rate $\dot{\phi}$ can be directly applied as a control action to the robot. The translational velocity is regulated by an external control loop, which reduces the robot speed based on two values: 1) the proximity of nearby obstacles, for safety reasons, 2) a high turn rate $\dot{\phi}$, to assure that the robots heading remains close to an attractor state at all times (see section 4.1).

As mentioned in section 3, a total of five behaviours have been designed: CORRI-DOR FOLLOWING, WALL AVOIDANCE, DOOR PASSING, OBSTACLE AVOIDANCE, and GO TO. The design of these behaviours and their functional form $f_b(\phi)$ (equation 1) is motivated and discussed in our previous work [1]. These behaviours or combinations of them are able to deal successfully with the environment on the small scale. In [1] we showed analytically, for example, that the system can reliably distinguish between those passages that are too narrow to pass and gaps that are wide enough to traverse safely.

The overall dynamics of the system is obtained from the weighted summation of individual behaviours based on equation 2:

$$\dot{\phi} = \sum_b |w_b| f_b(\phi) + \text{noise} \tag{5}$$

with $b \in \{goto, corr, wall, door, obst\}$. For the coordination of the behaviours the competitive advantages α_b, the competitive interactions $\gamma_{b',b}$, and the time constants τ_b in equation 3 have to be chosen appropriately.

The competitive advantages reflect the relevance and applicability of a behaviour in a particular context. Obviously, GO TO should be activated whenever the agent finds itself in a room and is supposed to approach a goal (e.g. the charging station or the node in front of the next door) and turned off otherwise. For $\alpha_{goto} \in (0,1]$ behaviour GO TO is switched on. To have the possibility for any competitive interaction $\gamma_{b,goto} \in [0,1]$ to be greater or smaller than α_{goto}, we chose a value of 0.5 for the competitive advantage.

$$\alpha_{goto} = \begin{cases} 0.5 & \text{if in a room} \\ -0.5 & \text{otherwise} \end{cases} \tag{6}$$

Equivalently CORRIDOR FOLLOWING and WALL AVOIDANCE are relevant if the robot finds itself in a corridor.

$$\alpha_{corr} = \alpha_{wall} = \begin{cases} 0.5 & \text{if in corridor} \\ -0.5 & \text{otherwise} \end{cases} \tag{7}$$

The competitive advantage of DOOR PASSING is tightly coupled to the perception. It is set to a positive value as soon as the door we want to pass is detected.

$$\alpha_{door} = \begin{cases} 0.5 & \text{if door detected} \\ -0.5 & \text{otherwise} \end{cases} \tag{8}$$

The relevance of OBSTACLE AVOIDANCE depends on the number and proximity of the obstacles currently surrounding the robot. The competitive advantage of OBSTACLE AVOIDANCE is related to the obstacle density $\rho = \sum_i e^{-d_i}$, where d_i denotes the distance to obstacle i, and is computed according to

$$\alpha_{obst} = \tanh(\rho - \rho_0) \tag{9}$$

The constant ρ_0 determines the density at which obstacle avoidance becomes relevant (i.e. $\alpha_{obst} > 0$ if $\rho > \rho_0$). The tangent hyperbolic assures that the magnitude of α_{obst} is limited to the interval $[-1,1]$.

The competitive interaction $\gamma_{b',b}$ reflects to which degree an active behaviour b' suppresses another behaviour b. In fact, there are situations where behaviours would interfere with each other in an undesirable, counterproductive manner. A door that is half blocked by an obstacle might still be detected as a door, although the gap to pass is actually too narrow. Hence we want OBSTACLE AVOIDANCE to suppress DOOR PASSING in the presence of a high obstacle density. Furthermore, if two obstacles lie close to each other, the dynamics of ϕ generates a weak repellor in the middle of them (this has been shown in [1]). This repellor, however, could be dominated by an attractor of another behaviour. This would inevitably lead to collision. Consequently, OBSTACLE AVOIDANCE ought to suppress GO TO and CORRIDOR FOLLOWING as well, if the obstacle density exceeds a critical threshold ρ_c. This type of prioritization is achieved by appropriately choosing the competitive interactions:

$$\gamma_{obst,goto} = \gamma_{obst,corr} = \gamma_{obst,door} = \frac{1}{2}(1 + \tanh(\rho - \rho_c)) \qquad (10)$$

The constant ρ_c determines the density at which obstacle avoidance suppresses an other active behaviour b ($\gamma_{obst,b} > 0.5 = \alpha_b$ if $\rho > \rho_c$). The functional form of the term is chosen such that $\gamma_{obst,b} \in [0,1]$. All other competitive interactions $\gamma_{b',b}$ are set to zero, since there exist no potential conflicts among any other pair of behaviours.

The time constants τ_b determine the time scale at which the behaviours are switched on and off respectively. τ_{obst} is chosen very small, such that the robot reacts almost immediately if a new obstacle is perceived. The same holds for τ_{wall}. When a door is detected the robot should turn towards it before driving out of detection range again. Consequently τ_{door} is also chosen to be small. The dynamics of w_{goto} and w_{corr} evolve at a slower rate $\tau_{goto} = \tau_{corr} \gg \tau_{obst}$. Once OBSTACLE AVOIDANCE becomes less relevant, e.g. after the robot has circumnavigated an obstacle, the other behaviours switch on gradually rather than causing jitter among themselves and OBSTACLE AVOIDANCE.

5 Experiment

Figure 2 shows the trajectory of the robot during a typical task: driving from the charging station in the living room to a goal point in the manipulator lab. During this task, the robot covered a distance of about 50 meters. In the middle of the corridor the way was blocked by people. Therefore, the robot was circling for about a minute before the passage was cleared and it was able to proceed (see [1] for details on behaviour switching in blocked corridors). Through this circling the error in the position estimate, obtained from odometry only, would be too large. However, by correcting this estimate on the basis of the detected corridor (section 3) the error was kept small and the robot continued successfully to the goal point.

Figure 3 visualizes more details on the trajectory from following a corridor to passing a door and reaching a goal point. The absolute values of the different weights $|w_b|$ (equation 5) are plotted over time. First, at position A the vicinity of the next node in the topological map was reached and the door was detected. DOOR PASSING was gradually switched on and guided the robot towards the door. CORRIDOR FOLLOWING was turned off on a slower time scale than WALL AVOIDANCE. When the robot reached

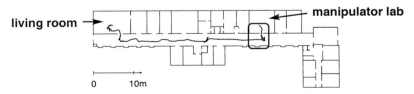

Figure 2: The trajectory from a trial in our institute: The robot started at the charging station in the living room and proceeded through the corridor to a goal point in the manipulator lab. The bold rectangle can be seen enlarged in Figure 3.

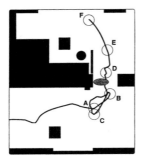
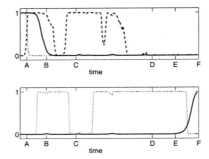

Figure 3: Details on the task switching from following a corridor to reaching a goal point in a room. The left image shows an outline of the trajectory of the robot. The gray ellipse denotes a person that was leaving the room, when the robot was at location B. On the right hand side, the absolute values of the weights (see equation 5) are plotted over time: $|w_{corr}|$ (upper plot, solid curve), $|w_{wall}|$ (upper plot, dotted curve), $|w_{door}|$ (upper plot, dashed curve), $|w_{obst}|$ (lower plot, dotted curve), and $|w_{goto}|$ (lower plot, solid curve). The situations A-F in the left image correspond to the time instances labeled in the plots.

point B the door was blocked by a person. However, OBSTACLE AVOIDANCE competed with DOOR PASSING and finally deactivated the latter due to a high obstacle density (equation 10). Consequently, the robot turned away from the door. Thereafter, the door was detected again, obstacle density has decreased, and therefore DOOR PASSING was gradually activated and the robot turned towards the door again (C). Subsequently, the robot passed the door (D). Due to the high obstacle density DOOR PASSING was switched off again and OBSTACLE AVOIDANCE guided the robot through the opening. At point E the vicinity of the next node was reached. Thus, GO TO was gradually turned on leading to a smooth switch in tasks again. Eventually, the goal point was reached (F) and the trial was completed.

Note that in Figure 3 not the actual trajectory of the robot is plotted, but the position estimate obtained from the odometry of the robot. The sharp bend right after passing the door (D) is therefore not a turn that the robot actually made. It is simply an artifact of updating the position estimate after having passed a door (see section 3).

6   Discussion

Feature detectors were used to extract environmental representations from the sensory data obtained by the sonars of the Scout robot; namely corridors, doors, walls, and obstacles. This geometric information facilitates the single behaviours to perform individual tasks: navigating safely along corridors, through doors and around obstacles. To determine which one of these tasks is appropriate in a given context, a simple topological map and odometric data from the robot were used. To combine the continuous control of the different behaviours with the discrete information from the qualitative map we deployed the dynamical system approach to behaviour based robotics. In this way smooth switching between the navigational tasks could be achieved. In essence, this comprises a unified control scheme that relaxes the need for accurate metric models of large-scale environments, through embedded place recognition in the behaviours and behavioural competition.

This framework has been successfully tested in the premises of our institute (70×20

meters). An example of an experiment has been presented, which shows that the system is capable of conducting long trials through a large-scale real-world office environment. It can also be seen that the competition among the behaviours is able to deal with more complex situations like half-blocked doors.

There are other approaches to navigate in large-scale environments. Many of them (Xavier [5], for example) need more detailed models of the environment and sophisticated algorithms (e.g. Markov decision process models) to determine an appropriate control action for the robot at all times. An other approach [9] uses, as we do, the superposition of different local behaviours (motor schemas). However, discrete context changes lead to discrete changes in control. In our system this information from the topological map can be merged with the continuous nature of the individual behaviours using the dynamical system approach. This leads to a unified control scheme.

The use of sonars as the the only sensors restricts our system in different ways. The representations of the environment are rather simple, which can lead to problems (e.g. if two doors are right next to each other). Future research in this project will be directed towards integration of more accurate sensors (e.g. laser), to get a more reliable representation of the environment. With these more sophisticated sensing capabilities we will also try to solve the problem of global localization (neglected in this paper) using just a simple topological map.

Acknowledgment
This research has been sponsored by the Swedish Foundation for Strategic Research.

References

[1] P. Althaus, H. I. Christensen, and F. Hoffmann. Using the dynamical system approach to navigate in realistic real-world environments. In *Proceedings of the IEEE/RSJ International Conference on Intelligent Robots and Systems*, pages 1023–1029, 2001.

[2] R. C. Arkin. *Behavior-Based Robotics*. MIT Press, Cambridge, MA, 1998.

[3] J. Forsberg, U. Larsson, and Å. Wernersson. Mobile robot navigation using the range-weighted hough transform. *IEEE Robotics & Automation Magazine*, 2(1):18–26, 1995.

[4] I. Horswill. The Polly system. In D. Kortenkamp, R. P. Bonasso, and R. Murphy, editors, *Artificial Intelligence and Mobile Robots: Case studies of successful robot systems*, chapter 5, pages 125–139. AAAI, Menlo Park, CA, 1998.

[5] S. Koenig and R. G. Simmons. Xavier: A robot navigation architecture based on partially observable markov decision process models. In D. Kortenkamp, R. P. Bonasso, and R. Murphy, editors, *Artificial Intelligence and Mobile Robots: Case studies of successful robot systems*, chapter 4, pages 91–122. AAAI, Menlo Park, CA, 1998.

[6] E. W. Large, H. I. Christensen, and R. Bajcsy. Scaling the dynamic approach to path planning and control: Competition among behavioral constraints. *The International Journal of Robotics Research*, 18(1):37–58, 1999.

[7] H. P. Moravec. Sensor fusion in certainty grids for mobile robots. *AI Magazine*, 9(2):61–74, 1988.

[8] L. Perko. *Differential Equations and Dynamical Systems*. Springer, New York, 1991.

[9] A. Saffiotti, K. Konolige, and E. Ruspini. A multi-valued logic approach to integrating planning and control. *Artificial Intelligence*, 76(1–2):481–526, 1995.

[10] G. Schöner and M. Dose. A dynamical systems approach to task-level system integration used to plan and control autonomous vehicle motion. *Robotics and Autonomous Systems*, 10:253–267, 1992.

[11] G. Schöner, M. Dose, and C. Engels. Dynamics of behavior: theory and applications for autonomous robot architectures. *Robotics and Autonomous Systems*, 16(2–4):213–245, 1995.

Intelligent Autonomous Systems 7
M. Gini et al. (Eds.)
IOS Press, 2002

Application of Mobile Code to Development of Cooperative Multirobot Systems

Francesco Amigoni and Marco Somalvico
Politecnico di Milano Artificial Intelligence and Robotics Project
Dipartimento di Elettronica e Informazione
Politecnico di Milano, Milan, Italy
amigoni@elet.polimi.it, somalvic@elet.polimi.it

Abstract. Multirobot systems address an increasing number of different applications. However, a general assessed methodology for designing and developing the cooperation structure of these systems is still lacking. In this paper we present one of such methodologies, called *dynamic agency*, which is strongly based on a novel distributed production system, called *DCLIPS*, that acts an infrastructure for inferential code mobility. The adoption of mobile code promotes ease and flexibility in developing and managing the cooperation structure of multirobot systems. We have experimentally tested our methodology by building a multirobot system that maps unknown environments.

1 Introduction

The present market and technology yield the possibility to build systems composed of several robots that cooperate in order to reach a global goal [4]. Usually, the cooperation structures of such systems are made in an ad-hoc fashion to uniquely address the particular applications at hand [6]. The modern technology of mobile code systems [8] can provide the required flexibility for building in a general way the cooperative components of multirobot systems. In mobile code systems, the code and possibly the state of an executing program can migrate over hosts at run-time.

The purpose of this paper is to illustrate a novel distributed production system, called *DCLIPS*, that enables code mobility over robots. This code mobility supports *dynamic agency* [1]: an architecture and methodology for the development of a uniform cooperative framework among different specific operative robots, to drive them toward the solution of a global problem. The code that can be moved among robots by means of DCLIPS environment is composed of rules and facts, namely it is *inferential code* executable by the production systems located on robots.

This paper is organized as follows. The next section illustrates DCLIPS both at the programmer's level and at the implementation level. In Section 3, we outline the importance of DCLIPS for robotic applications by describing the experimental activities we have carried on with cooperating exploring mobile robots. In Section 4 we relate our work to the literature on distributed production systems and mobile code systems for robotic tasks. Finally, Section 5 concludes the paper.

2 The DCLIPS Environment

The DCLIPS environment is a collection of communicating independent production systems that are located on different nodes of a network and that exchange knowledge by a global shared memory. Each production system is based on the well-known CLIPS [5] environment. DCLIPS is a distributed production system [10] that supports both the migration of inferential code (facts and rules) and the communication among the composing production systems.

2.1 Programming with DCLIPS

From the programer's point of view, DCLIPS is composed of a number of *agent environments* that exchange knowledge via a shared memory area called *knowledge space (KS* for short), as shown in Fig. 1(a). Each agent environment is a modified CLIPS environment composed of a *facts base* (also called working memory), a *rules base* (also called production memory), and an *inference engine* that matches through pattern matching the preconditions of the rules with facts, selects a rule to fire among those with satisfied preconditions, and executes the actions of the selected rule. An agent environment of DCLIPS extends a CLIPS environment by offering a number of new commands to exchange knowledge with other agent environments through KS. These commands are discussed below.

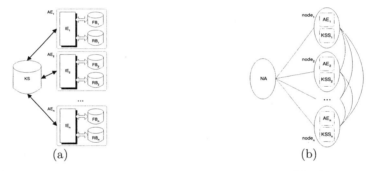

Figure 1: The structure of DCLIPS as it appears to the programmer (a): AE_i are the agent environments, FB_i are the facts bases, RB_i are the rules bases, IE_i are the inference engines, double arrows represent CLIPS communication, and single arrows represent DCLIPS communication. The internal architecture of DCLIPS (b): AE_i are the agent environments, KSS_i are the knowledge space servers, and NA is the network agent

Registration to knowledge space. na-join registers an agent environment to access the KS with a unique name. Conversely, na-leave un-registers the agent environment from accessing the KS.

Knowledge insertion in knowledge space. ks-put inserts a fact in the KS; the fact can be ordered or unordered (in this case the corresponding descriptor, the deftemplate, must be specified). ks-share inserts a construct in the KS; constructs can be rules or functions (procedural knowledge is considered as a particular subclass of rules).

Knowledge extraction from knowledge space. ks-read and ks-get retrieve the facts in the KS that match the specified pattern (when retrieving unordered facts, the corresponding deftemplate must be specified). The retrieved facts are inserted in the local facts base. ks-read leaves the retrieved fact in the KS, whereas ks-get removes the retrieved fact from KS. Command ks-use retrieves a construct in the KS and inserts it in the local rules base. The retrieved construct is left in the KS.

Error handling. `ks-errorp` returns the symbol TRUE if any error occurred during the execution of last command. `ks-error-msg` returns a string that describes the last occurred error.

2.2   The Implementation of DCLIPS

The main issue of the DCLIPS implementation is that the KS itself is distributed. At the implementation level, the DCLIPS architecture is composed of a set of distributed and communicating *nodes* (see Fig. 1(b)). Each node has two parts: the agent environment (illustrated in previous section) and the *knowledge space server* that manages the portion of KS assigned to the node. As better illustrated in Section 3, the nodes are located on robots.

The DCLIPS architecture is completed by a *network agent* that can be considered as a special node that locates the other nodes (and thus the agent environments) on the network. It maintains a database in which, for each agent environment, the physical address of the corresponding node is registered. The database is kept consistent by dynamically updating it as agent environments (and the corresponding nodes) join and leave the system. In particular, the network agent offers two services. The first one is registration/un-registration to/from the system; it is exploited for executing `ks-join` and `ks-leave` commands. The second one is multiple (single) query: it returns all the registered nodes (the first node) that match a given criterion. The query services are not available to the programmer and are part of the implementation of the DCLIPS commands that access the KS. The network agent allows mutually exclusive access to its functions. The use of a centralized name server (the network agent) has the disadvantage of creating a bottleneck in the system, but it is the simplest solution for naming services.

The agent environment of a node extends a CLIPS environment. It interprets in the usual way the standard commands of CLIPS and, for DCLIPS commands, it performs the following algorithm.

(1) Gather the parameters of the command (e.g., gather the pattern that describes a fact for `ks-read`)
(2) Query the network agent to obtain the list of currently registered nodes
(3) Send to the knowledge space servers of registered nodes the requests to execute the operations connected to the command (in the case of `ks-read`, send requests to read a fact described by the pattern)
(4) Collect the results (in the case of `ks-read`, collect the read facts)
(5) Return the result of the execution of the command (in the case of `ks-read`, return the first read fact)

The knowledge space server of the node i manages the local portion KS_i of KS by offering the following services.

– Allow the node i, and only node i, to write knowledge in KS_i (exploited for executing `ks-put` and `ks-share` commands).

– Allow every node to read or extract knowledge from KS_i (exploited for executing `ks-get`, `ks-read`, and `ks-use` commands). When a request to read or extract a fact arrives, the knowledge space server of node i tries to match the requested pattern with the facts present in KS_i. If a matching fact is retrieved, then it is returned, otherwise a null element is returned. In a request to use a construct, both the name and the type of the searched construct are specified. In this case, the knowledge space server of node i searches KS_i for an element with the key (name,type).

- Ensure the atomicity of accesses (read and write operations) to KS_i. The knowledge space server of node i gives exclusive access to KS_i to a single requesting node at a time. It also maintains a queue of pending requests to access KS_i.

The union of the contents of the portions KS_i of the KS constitutes the global knowledge shared by the whole system. The main disadvantage of our distributed solution for KS is that, when agent environment A wants to send knowledge k directly to agent environment B, A inserts k in KS_A. Then, B has to search all KS_i in order to find k. An alternative method (not yet experimentally tested) could be that A puts k in KS_B and B looks first in its local portion of KS. In our solution, the combined activity of agent environments and of knowledge space servers results in the transparency of the distribution of KS for the programmer of DCLIPS applications.

All the described modules have been coded according to a structure organized on several layers for ensuring the maximum flexibility in the use of DCLIPS. Apart from some interface modules between CLIPS and DCLIPS kernel that are written in C, the modules that form the DCLIPS kernel are written in Java. The integration of the two kinds of modules exploits JNI (Java Native Interface [11]). The interface modules have been developed in such a way that the KS can be accessed also by nodes that run procedural code, such as the perception units of robots, which can in this way access the KS for inserting knowledge about the perceived data. The pattern matching implementation in knowledge space servers exploits JESS [7]. The kernel of DCLIPS is a federation of distributed objects that are located in the nodes and that communicate via RMI. The main of these objects are the following ones. KSServer objects, the servers of knowledge space. KSClient objects, the clients of knowledge space that allow to interact with remote (i.e., located on other nodes) KSServer objects. NAServer object, the server that offers the services of network agent. NAClient objects, the clients of network agent that interact remotely with the network agent. The NAServer object implements the network agent previously described. A knowledge space server is implemented by a KSServer object. Finally, the implementation of an agent environment includes a KSClient object and a NAClient object that are connected with the modified CLIPS environment. As their names suggest, these objects interact together mainly in a client-server fashion.

3 Experimental Activity

We have defined a general methodology and architecture for developing multiagent systems, of which multirobot systems are a subclass, called *dynamic agency* [1]. The dynamic agency approach is based on the use of mobile code systems and on the conception of each agent of a multirobot system as composed of two parts (Fig. 2 left). The first part, called *op semiagent*, is a (usually preexisting) robot devoted to perform specific operative functions. The second part of an agent, called *co semiagent*, is a set of software modules designed to perform uniform cooperative functions. According to the dynamic agency approach, the co semiagents are the result of the evolutionary metamorphosis of a single mobile code that travels, visits, and evolves on the op semiagents. Hence, op semiagents host an incoming mobile code in order to put it into execution, by decoding it and assigning it the required resources. The mobile code is called *MIA* (from Mobile Intelligent Agent). The op semiagent is composed of the robot, called *FIA* (from Fixed Intelligent Agent) and of the hosting component, called *MFI* (MIA-FIA Interface). The activities of MIA are designed to install the co semiagents on the op semiagents to control them in order to cooperatively reach a global goal. In this way, the operative functions provided by the op semiagents can be exploited in the high-level cooperation framework set up by the co semiagents that exchange knowledge, coordinate activities, and divide and assign goals. This means that MFI must be designed in such a way that a co semiagent and the corresponding FIA, which are usually developed by different designers, can communicate.

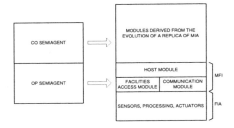

Figure 2: The architectural structure of an agent in a dynamic agency

DCLIPS provides the infrastructure for code mobility [2] required by dynamic agency: it assists the design, implementation, deployment, and execution of inferential mobile code.

We have developed a system composed of mobile robots for mapping unknown environments. This application is recognized as very important for multirobot systems; however, to date our effort has been more on showing the role of DCLIPS as infrastructure for developing cooperative multirobot systems than on the quality of resulting map. For this reason, we have implemented very basic robotic solutions. The multirobot system has been composed starting from four FIAs: three mobile robots and a computer. The mobile robots have three-dimensional vision systems and encoders as sensors and with driving wheels as actuators.

We now detail the structure of MFIs of the four FIAs of our system (refer to Fig. 2 right). Each MFI includes a *communication module* based on wireless IEEE 802.11 medium and protocol; the communication modules offer also a TCP/IP stack to send and receive messages to and from other op semiagents. The *facilities access modules* of robotic FIAs offer access to the following operative functions: moving the robot (forward, backward, and turn), getting the set of extracted segments from the images taken by the vision system, getting the robot position from the encoders, activating the path planning process (given a map, the starting, and the ending points), getting some information about the robot (e.g., its dimensions, the dimensions of its vision field, its battery level, its maximum moving velocity and precision, and so on). The *host modules* of (MFIs of) op semiagents are nodes of DCLIPS (see Section 2.2).

The MIA that constructs the co semiagents is composed of facts and rules and it is initially sent in the KS from a connected computer by the designer of the multirobot system. From there, it is detected and locally copied by all the op semiagents, providing code mobility. In particular, the host module (recall it is DCLIPS node) of an op semiagent i has initial sets of facts F_i and rules R_i that monitor the KS to detect and copy a MIA (an example of rule in R_i is shown in Fig. 3 left). After this step, the facts base and the rules base of the host module of op semiagent i are $FB_i = F_i \cup F_{\text{MIA}}$ and $RB_i = R_i \cup R_{\text{MIA}}$, respectively. F_{MIA} and R_{MIA} are the facts and rules of MIA that are organized in several modules whose sequential activation carries on the metamorphic evolution that brings to the development of the cooperation structure of the multirobot system. The switching between a module and the following one is performed by the former by putting an activation fact in the local fact base. In what follows, due to space limitations, we only list the phases (each one performed by a module of MIA) without giving their full explanation (see [1] for a more detailed illustration).

Figure 3: Some of the rules of the MIA that builds an exploring multirobot system

Each copy of MIA extracts (from a file) the knowledge about the corresponding robot (op semiagent): how to activate its functions, its physical dimensions, and so on. A rule for sharing (through KS) the results of extraction is shown in Fig. 3 center. Then, a hierarchical structure is formed according to the extracted knowledge: the computer agent acts as *coordinator* and is the root of the hierarchical organization of which robotic agents, the *explorers*, are the leaves. A rule for assigning roles to agents is shown in Fig. 3 right.

The robots are placed in an environment to be mapped and their initial positions are determined (by reading a file or by auto-localization techniques). The agents negotiate the areas worth to be explored until all explorers have been assigned a goal. All the required communication is performed via KS. Explorer agents extract segments, representing the map of the area to explore, from the images taken by vision system at the assigned location. Then the partial maps of the environment obtained by explorers (namely, sets of segments that are exchanged as facts through KS) are integrated (on the basis of robots' positions) by the coordinator to obtain the final global map. At this point a new mapping step starts with negotiation of interesting areas.

With the described system, we have been able to map part of the hall of our Department building. The travel of MIA on op semiagents takes some minutes, because of the dimensions of MIA code and of the limited bandwidth (up to $2Mb/s$) offered by wireless communication. However, the travel of MIA is performed only at the beginning of the construction of the system. A mapping step takes several minutes since the transfer of the sets of segments is very time consuming and, in the present version, the negotiation process lasts until every explorer has been assigned an area to explore. During the experimental activity, DCLIPS has proved to be a reliable infrastructure for multirobot systems.

The main advantage of dynamic agency approach over other multirobot methodologies is the *flexibility* that promotes an easy reuse of existing robots for different purposes. For example, we employed the same robots used in exploration task for the task of sweeping a known environment by simply sending a new MIA that sets up a new cooperation framework exploiting in a different way the operative functions of robots.

Moreover, multirobot systems can be automatically reconfigured. For example, a supervisor component of our exploring multirobot system might decide to recruit a new small robot to map a newly discovered room accessible by a narrow doorway. As another example, we can conceive a multirobot system that automatically switches robots from the task of exploring to the task of sweeping the floor of an environment as it is more and more mapped. Work in currently undergoing to implement these possibilities.

4 Related Works

Distributed production systems [10] have been widely studied and employed to build expert systems and to implement sense-and-act models in cognitive science and in robotics. The KS of DCLIPS can be thought a *tuple space* (as defined in the Linda programming language [9]) with a pattern matching mechanism. The KS represents also an example of the well-known *blackboard* approach: a shared memory paradigm widely used in distributed artificial intelligence [3]. Another recent system based on a Linda-like tuple space is LIME [14].

The investigation of the relations between robotics and mobile code systems has been addressed only by a very few papers. Mobile code is mainly used to implement teleoperation systems for robots [16]. For example, [13] describes a system in which a mobile robot moves in a museum according to the (asynchronous) commands issued by the user through a mobile agent (is a particular kind of mobile code that can autonomously decide when and where to move [15]). An unusual application of mobile code to mobile physical objects involves the tile-shaped computers called *Tiles* [12]. In conclusion, to the best of our knowledge, the dynamic agency approach represents the first attempt to spread a uniform cooperation structure over specific robots by means of mobile code systems.

5 Conclusions

In this paper we have presented a significant application of mobile code to the flexible development of the cooperation structure for multirobot systems. The proposed dynamic agency approach is based on the decoupling of operative and cooperative parts of each agent and on the implementation of the latter as an evolved mobile code. This promotes the possibility to easily change or modify the cooperative parts also during the activity of the multiagent system. The experimental activity has established the validity of the dynamic agency approach and the usefulness and reliability of the DCLIPS as an infrastructure for mobile code.

Future research is required to improve DCLIPS environment to ease the development and the installation of MFI on robotic FIAs. We are planning to further test DCLIPS in multiagent systems involving both robotic agents and software agents. More generally, future work will address the refinement of the dynamic agency approach as a general methodology for multiagent systems.

Acknowledgement

The authors are glad to thank the contribution of Gianmarco Antonini, Dario De Palma, Emanuele Prina, and Luigi Spalla to carry on this research.

References

[1] F. Amigoni and M. Somalvico. Dynamic agencies and multi-robot systems. In T. Lueth, R. Dillmann, P. Dario, and H. Worn, editors, *Distributed Autonomous Robotic Systems 3*, pages 215–224. Springer-Verlag, Berlin Heidelberg, Germany, 1998.

[2] Y. Aridor and M. Oshima. Infrastructure for mobile agents: Requirements and design. In K. Rothermel and F. Hohl, editors, *Mobile Agents, Second International Workshop, MA'98, Stuttgart, Germany, September 9-11, 1998*, pages 38–49. Springer, 1998.

[3] A. H. Bond and L. Gasser. *Readings in Distributed Artificial Intelligence*. Morgan Kaufmann, San Mateo, USA, 1988.

[4] Y. U. Cao, A. S. Fukunaga, and A. B. Kahng. Cooperative mobile robotics: Antecedents and directions. *Autonomous Robots*, 4:1–23, 1997.

[5] CLIPS. http://www.ghg.net/clips/clips.html.

[6] G. Dudek, M. Jenkin, E. Milios, and D. Wilkes. A taxonomy for multiagent robotics. *Autonomous Robots*, 3:375–397, 1996.

[7] E. Friedman-Hill. *Java Expert System Shell*. Sandia National Laboratories. herzberg.ca.sandia.gov/jess/.

[8] A. Fuggetta, G. P. Picco, and G. Vigna. Understanding code mobility. *IEEE Transactions on Software Engineering*, 24(5):342–361, 1998.

[9] D. Gelernter. Generative communication in linda. *ACM Transactions on Programming Languages and Systems*, 7(1):80–112, January 1985.

[10] T. Ishida. *Parallel, Distributed and Multiagent Production Systems*. Springer-Verlag, 1994. Lecture Notes in Artificial Intelligence 878.

[11] JavaSoft. *Java Native Interface Specification*, 1997. www.javasoft.com.

[12] K. Kramer. Mit tiles project: Moveable objects, mobile code, 1998. MIT MS Thesis, http://el.www.media.mit.edu/projects/tiles/msthesis-formatted.html.

[13] K. Matsubara, K. Kato, T. Maekawa, S. Maeyama, S. Yuta, and A. Harada. Asynchronous robot teleoperation via the internet: Mobile agent approach. In *Proceedings of SSGRR International Conference on Advances in Infrastrucure for Electronic Business, Science, and Education on the Internet*, 2000.

[14] Amy L. Murphy, Gian Pietro Picco, and Gruia-Catalin Roman. LIME: A Middleware for Physical and Logical Mobility. In Forouzan Golshani, Partha Dasgupta, and Wei Zhao, editors, *Proceedings of the 21st International Conference on Distributed Computing Systems (ICDCS-21)*, pages 524–533, May 2001.

[15] G. P. Picco. Mobile agents: An introduction. *Journal of Microprocessors and Microsystems*, 25(2):65–74, April 2001.

[16] J. Rning and K. Kangas. Interacting with personal robots and smart environments. In *Proceedings of the SPIE Conference 3837: Intelligent Robots and Computer Vision XVIII: Algorithms, Techniques, and Active Vision*, Boston, Massachusetts, September 19-22 1999.

Intelligent Autonomous Systems 7
M. Gini et al. (Eds.)
IOS Press, 2002

OccuBot VI - An Intelligent Robot System for Seat Testing Applications

Frank Beeh and Heinz Wörn
Institute for Process Control and Robotics
Department of Computer Science
University of Karlsruhe
Kaiserstrasse 12, D-76128 Karlsruhe
{beeh,woern}@ira.uka.de

Abstract. In this paper a new system for testing applications by using a standard robot manipulator as basic part is introduced. The robot
control is extended by force control algorithms and new movement profiles. The force control is able to compensate the static forces caused by
the weight of the dummy. The new defined movements allow the execution of independent sinusoidial movements in every cartesian dimension.
Furthermore it's possible to imitate human motions by executing free
paths specified by cartesian points including time and load.

1 Introduction

State of the art in seat testing are specialized hydraulic machines that are used for
wear tests. Most of them are constructed for the execution of only one test. These
machines have very restricted mobility and accuracy. This causes major restrictions on
the possible test since they're limited to simple one or two-dimensional sinusiodial or
linear movements.

Although the load is the most important factor in all tests, most of the systems
are not equiped with sensors nor have an active control mechanism. The only way to
handle this is to adjust the minimum resp. maximum position for reaching the desired
forces. But this approach has a big drawback. Since the seat wears during the test,
the reaction force is decreasing, leading to wrong forces applied during the test and to
a falsification of the test results.

These major limitations in the test equipment restrict the tests to a very simple cases
that are far away from the normal stress in real life. In normal test these inaccuracies
are handeled by executing much more cycles and causing additional stress on the seat.
But there are some cases where this approach fails and the company suffers great loss in
both money and reputation when the seat breaks during real use. The need for better
tests is therefor growing [2].

It's necessary to develop systems that are able to execute more realistic tests. These
systems have the following two major requirements. The ability to execute free programmable movements in the six-dimensional cartesian space and a six-dimensional
force/torque control to maintain the correct load on the seat. A standard robot can
comply partly with the first requirement, but the standard movements (PTP, linear
and circular) are not sufficient especially for executing timely correct movements. Furthermore there's no easy way to integrate a force/torque control. Our approach shows
how to enhance the robot and its control to fulfill the task.

2 System description

The system setup was already presented in [1]. The main part of the system is an industrial robot with a maximum load of 150 kg which is sufficient for this application. Attached to its flange is a six-dimensional force/torque sensor that can measure up to 5000 N resp. 400 Nm and a standard rigid seat test dummy. Figure 1 shows this setup. Since the standard robot movements are not sufficient for this application, a new inter-

Figure 1: OccuBot VI

polator that is able to execute sinusoidial movements in six cartesian dimensions with different frequencies, phases, amplitudes and offsets was developed. These movements are mostly used to execute the ordinary tests. Furthermore a new kind of spline movement was implemented that allows to reach user specified cartesian points in specified time as long as the velocity and acceleration limits of the robot are not violated. These so called free movements enable the user to specify completely new forms of tests by using time indexed points to specify paths that are executed in the correct time. They can be used to imitate human movements. Both kinds of movement consist of cartesian points with specified time and load. In case of the sinusoidial profile the extreme points in every cartesian dimension are used for this task, while the user has to specify them directly for the free movements. These points are connected by the interpolator using half sine waves resp. splines.

The second important topic in this approach is the force/torque control. First approaches with standard online impedance control failed. This was caused by the high dynamic of the cushion of the seat. The seat consists basically of a hard metal frame with springs that is filled with foam and covered with cloth. This can be regarded as a spring-damper system while the viscous character of the foam causes an additional time dependance. As a result, the reaction force of the seat is highly dependend on the velocity and the history of the movement. All these effects made a online impedance control during each cycle impossible since the control forced the robot to deviate heavily from the original path while breaching the time stamps.

A second completely new approach was a "cycle-spanning" control. The load of a defined point is controlled by modifying the cartesian position using an offset. The offsets are computed in every cycle by comparing the desired and the measured load of each point. This computed offset is used to change the position of the point in the next cycle. Figure 2 shows this for a simple example of a one dimensional sine wave. This control scheme can be regarded as a discrete impedance control while the discrete times are the cycle number of the respective test cycle (e.g time = 1 for cycle one, time = 2 for cycle two, etc.). This approach shows an excellent behaviour. It takes some cycles until the desired loads are found but their number is very small (<100) compared with the total number (>10,000) of the total test. The small corrections that are necessary for compensating the wear of the seat are handeled easily while the drift of the offsets can be used to get an impression of the gradual decay of the seat.

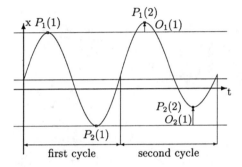

Figure 2: Example for the effect of the "cycle-spanning" control for a one-dimensional sine wave. The original points $P_1(1)$ and $P_2(1)$ of the first cycle are modified by the two offsets $O_1(2)$ and $O_2(2)$ to get the new points $P_1(2)$ and $P_2(2)$ of the second cycle. The offsets are computed using an impedance controller for the loads.

3   Experimental results

To generate a more realistic test, a person was tracked by a camera system while sitting into a seat and standing up. In the same time the load distributions on the cushion and the back are recorded using two sensing mats. The data was used to generate a OccuBot path by specifying time-indexed points with the appropriate loads. This path was the base for the new ingress/egress test that simulates realistically the strains that occur when a person is sitting down and standing up again. This movement stresses especially the side bolsters of the seat. Until now the jounce and squirm test (a rotation in the seat while going up an down) was used for this task, but since the movement was very unrealistic, the results have been very different from the stresses noticed in real life. The new ingress/egress test is now used by some manufacturers and it has shown already its usefullness by finding weak parts in the design in the early prototype phase.

Furthermore, the system is also capable to handle almost all of the normal sinusoidial tests as long as they're not high frequency. OccuBot is flexible enough to execute multiple different tests on different seats.

4   Conclusion and future work

OccuBot is a high efficient testing tool which is highly flexible and capable to fulfill the demands of the new tests (imitation of human movements) while being able to execute the old ones as well.

The major restriction of the system is the used rigid test dummy. It allows only movements of the complete body while additional movements for example of the legs are not executable yet. In the next step it's planned to use additional linear drives for simple movements of the legs. A flexible dummy called LifeForm (figure 3) with an internal human-like skeleton was already created by the Lear Corporation. The skeleton is covered by a flesh-like synthetic rubber. This dummy stresses the seat much more like a human than the old hardcover dummies. The main task is now to develop a kinematic structure that is capable of executing almost the same movements like a person during entering and leaving a seat.

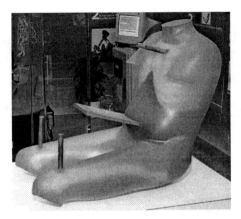

Figure 3: LifeForm: A new flexible dummy with internal skeleton.

Another very important task is to establish new realistic tests like the ingress/egress as standard. This will allow to improve the significance of the tests for the prediction of wear in real life. There's no research done in this area yet and there's no systematic comparison between the wear during real usage and during the tests. Since this is extremely important to create better tests, it's necessary to start research in this area as fast as possible.

References

[1] Frank Beeh, Thomas Längle, Heinz Wörn, Terry O'Bannon, John Lentz, and Bernd Schnoor. Occubot VI - Industrieroboter als intelligentes Sitztestsystem. In *Robotik 2000*, VDI Bericht 1552, pages 23–27. VDI, 2000.

[2] R. Stewart, T. O'Bannon, M. Müller, F. Beeh, B. Schnoor, and J. Lentz. Creating the next generation ingress/egress robot. In *SAE Technical Paper Series*, number 1999-01-0628. Warrendale, PA.

Intelligent Autonomous Systems 7
M. Gini et al. (Eds.)
IOS Press, 2002

Biological Inspired Walking Machines

K. Berns
Forschungszentum Informatik an der Universität Karlsruhe (FZI)
Interactive Diagnosis and Service Systems
Haid-und-Neu-Str. 10-14, 76131 Karlsruhe, Germany
berns@fzi.de

January 11, 2002

Abstract

In the last decade a lot of research is done in the fascinating area of biological inspired walking machines. This paper should summarise the research results and evaluate the potential for new applications. Based on a definition of this class of robots the requirements for its technical realisation are set up. An overview of the research results will be given due to the subtask mechatronic system, control architecture and adaptive behaviour control. At the end of the paper several applications are listed for which biological inspired walking machines would be an optimal solution.

1 Biological inspired walking machines

Today several scientific publications in the area of robotics exist, in which the idea of transferring principles from nature to robotic systems is described. The governments of different countries sponsor research programs, in which special transfer aspects are considered like the mechanical concept, the sensor system, the adaptive control or cognitive behaviours. Examples are the *Controlled Biological System Program* sponsored by the DARPA, USA or the German program *Autonomous Walking*.From the robotics point of view the main focus is not to *create a new generation of robots that imitate life*[1] but to use principles from nature to improve robotic systems.

In the following the term biological inspired walking machine, which belong to the above describe robots is introduced. This term can be defined as:

> Biological inspired walking machines contain all walking robots for which at least one of the system components like mechanical design, actuator system, control architecture or adaptive control is strongly influence by design principle in nature. The development aims to reach similar locomotion capability as comparable animals under rough environmental conditions.

---

[1] Taubes and Gray, Biologist and Engineers Create a new Generation of Robots that Imitate Life, Science 2000 288: 80-83

Examples of these kind of machines are an 8-legged ambulatory vehicle that is based on the lobster [2], the cockroach like machine Robot III Case Western University, Ohio USA [11], the Humanoid robot Asimo [3], and the biped machine of the University of Munich, Germany [10]. Based on the research results of the neuroethology of the stick insect Carausius Morosus[5] in Germany several machines are developed like Lauron [2], Airbug, [3] both from FZI, Karlsruhe, Tarry from the University of Duisburg, [4] and Max of the University of Munich [13].

Figure 1: (left) Lauron III, the stick insect like walking machine developed for walking in rough terrain (right) Johnnie [9], the biped robot, which allows jogging including ballistic phases.

2 System requirements

Based on the above given definition of biological inspired walking machines the following system requirement must be fulfilled:

Autonomy The machine must be an autonomous mobile robot.Basic intelligent must exist because wireless communication from outside can not be guaranteed under rough environmental conditions.

Energy autarky The energy supply must be on-board.

Flexibility The number of active and passive degree of freedom must be high so that the locomotion apparatus can adapt to ground. Bending forces or tenseness forces should be low to save energy and to allow smooth movements.

Reliability The robot must be reliable on all system levels starting from the computer architecture up the the behaviour control.

Adaptivity The control of locomotion must be implemented in a way that disturbances which arise e.g. from ground conditions can be managed.

---

[2]Northwestern University, MA, USA see http://www.dac.neu.edu/msc/burp.html

[3]Honda, Tokyo, Japan see http://ne.nikkeibp.co.jp/english/2000/11/1120asimo_d-ce.html

[4]see http://www.mechanik.uni-duisburg.de/publikationen/

For the realisation of biological inspired walking machines it is essential that knowledge from different research areas are used (see fig.2). Starting from research results of biologist concerning morphology or neuroethology of animals first an adequate mechatronic system must be designed and a sensor system which determines the inner state of the machine and detects the environment must be set up.

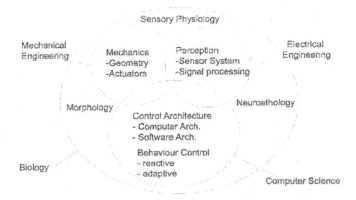

Figure 2: Interdisciplinary solution for the development of biological inspired walking machines

In the following the state of the art of the different system components for biological inspired walking machines are discussed.

3   Advanced mechatronic system

The physiology of the locomotion of creatures (drives and mechanics) are very different compared to walking machines constructed so far. It is very hard to transfer these concepts to walking machines. Mechanical design parameters from nature which are used for building biological inspired walking machines are the leg order, the proportion of the different leg segments (body parts),and the number of degrees of freedom.Because the machines are normally scaled up compared to their natural models the angle velocity is much smaller. In the case of the four legged machine BISAM [14] the biologically inspired aspects of the mechanics are the leg and the flexible shoulder and hip construction. Biologist had found out that most important for the locomotion of mammals is the hip, shoulder and spiral motion. Also a leg was constructed which is arranged in a rectangular zig-zag manner.

The actuators which are normally used to build up walking machine are not powerful enough to allow on one hand side fast movements and on the other side to generate necessary torques to move the machine with a payload.Even if control concepts with active compliance is implemented it is hard to cover impact problems. Therefore, actuators with passive compliance which behave similar than muscles in natures should be for the development of walking machines. In literature there are several types of artificial muscles. Fluidic muscles with high forces, high contraction velocities and a good energy efficiency could be a solution for the actuator problem. E.g. in [6] a robot primate and in [12] a spider like machine is developed using different types of pneumatic muscles.

4 Sensor system

Observations from nature have shown that the perception of creatures is based on a high amount of receptor cells (e.g. several thousands on the leg of the stick insect). A receptor cell delivers an analogue value or an on/of, value. In micro-mechanics research similar sensor components have been built. Up to now these sensors are only used in special robot systems. Walking machines should have at least the following internal and external sensors.

To the class of internal sensors belong shaft encoders (absolute or incremental), foot sensors (tactile or force sensors), inclinometers and gyroscopes to measure the state of the central body. Additionally, current and voltage in case of electrical motors and pressure in case of fluidic actuators must be determine. An indirect measurement of the torque and a protection against thermal overload can be performed with a known motor current.

For the control of the walking process in rough terrain it is essential to measure the distance to different objects and to determine soil properties which means to distinguish whether the ground is slippery, rather flat or strongly uneven and to find suitable areas for the foot points.Commercial sensors systems like laser scanner a normally to big and haevy to install them on small walking machines. Stereo camera system combined with a laser which generated a structured light image seems to be a promising solution. The main advantage of such a sensor is that the calculation effort for the detection of an object is very low [8].

5 Control architecture

Because biological inspired robots have huge number of actuators and sensors it is necessary for the reduction of the complexity problem to use a decentralised control architecture. The control architecture can be divided into computer architecture and software architecture. In fig. 3 a typical control architecture is shown which is applied to the insect-like walking machine Lauron.

In this example the computer architecture consists of three levels the micro-controller level, the PC level and the PC network level. The micro-controllers are directly coupled with a special power electronic card that allows the control of 4 motors. The micro-controllers are connected via Can-bus with an internal industrial PC.All sensors are directly coupled with the micro-controller. The sensors deliver both, analogue signals which are converted by the internal AD-converter of the C-167 and digital signals provided by joint encoders counted by the micro-controller. The second area provides the PC level with services including the management of the communication link to PC.

To handle the real-time control requirements a modular control architecture has been developed. Linux as well as Real Time Linux are used as operating systems. The selection has been performed because of the high number of available devices and the availability of source codes. The standard Linux kernel as a task of the RT-Linux kernel runs with a lower priority. For the efficient implementation of the different control levels the object oriented module concept MCA [4] has been implemented which enables a rapid development and the exchange of control algorithms on different control levels.

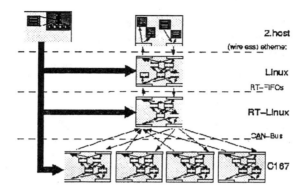

Figure 3: The control architecture of the stick insect like machine Lauron.

6   Adaptive behaviour control

Considering the features of neural networks like fault tolerance, the ability to process noisy sensor data and the possibility to learn from examples they seem to be an appropriate tool for implementing adaptive control mechanisms. In combination with Reinforcement Learning (RL), these approaches are very interesting because of their potential of online adaptability to changing environments. The main disadvantage is the huge amount of trial which are necessary to learn online a special control behaviour. Therefore, present research deals with the integration of a priori knowledge into the learning concept. Similar as observed in nature neuro-oscillators are used to predefined rhythmic movements for the leg and gait control [7].In combination with the neuro-oscillators the learning process can be used to learn and optimise the coupling weights and the influences of sensor information. For complex control behaviour it is still an open question how to determine the right weight.

Therefore, other approaches focused more on the realisation of adaptive control using reflexes with are coupled with traditional control methods. In [1] reflexes are implemented using fuzzy controller to adapt the locomotion of the mammal-like robot BISAM to rough terrain (see fig. 4).

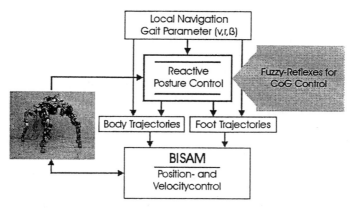

Figure 4: Overview of the adaptive control concept using reactive components for posture control of the four-legged machine BISAM

7 Potential applications and evaluation

Based on the advantages of biological inspired walking machines applications can be found in all areas that require a locomotion in very rough terrain like:

- transportation of heavy loads along badly rugged country;

- emergency rescue works in the destruction zones with a possibility of utilisation of separate legs as manipulators;

- repair inspection works in rooms suitable for mankind;

- moving on poor grounds, for example, moving in a forest, in the mountains;

- planatory exploration and submarine works;

- rehabilitation of people with restrictions of support movement apparatus.

However, up to now commercial and industrial use of walking is still in the status of developing prototypes. E.g. the Plustech company in Finland has developed a six-legged machine for forestry which was just sold a few times. The Odetics Inc. company in the US has built a small series six-legged machine for maintenance work in nuclear power stations. This leads to the question "what research problems have to be done in future"? Main problems when building these machines arise in the case of mechanics from effective drive system, and the construction of very light but robust machines as well as a special foot construction, which absorb the power stroke energy. The use of fluidic muscels offer a promising solution but also in this case a lot of improvements have to be done. Perception problems must be solved according to the interpretation of a huge amount of noisy and uncompleted sensor data and in the field of control from the fact that the locomotion behaviour can not completely determined a priori. In this area both new sensor systems as well as new methods for the interpretation of the measurements must be set up. In case of adaptive behaviour control several interesting approaches can be found in literature.These approaches are normally applied to simple control problems. It is still unclear how they can be used for complex control scenarios.Worldwide more and more projects have focused on the transfer of principles found in nature to develop powerful machines. The understanding of locomotion in nature will lead us to new concepts for walking machines, which will solve the above mention application problems.

References

[1] J. Albiez, W. Ilg, T. Luksch, K. Berns, and R. Dillmann. Learning reactive posture control on the four-legged walking machine bisam. In *International Conference on Intelligent Robots and Systems (IROS)*, Hawaii, USA, November 2001.

[2] K. Berns B. Gassmann, K.U. Scholl. Locomotion of lauron iii in rough terrain. In *International Conference on Advanced Intelligent Mechatronics*, Como, Italy, 2001.

[3] K. Berns, J. Albiez, V. Kepplin, and Hillenbrand. Airbug - insect-like machine actuated by fluidic muscle. In *CLAWAR 2001 - Climbing and Walking Robots and the Support Technologies for Mobile Machines*, september 24-26 2001.

[4] K. Berns, K.-U. Scholl, W. Ilg, and R. Dillmann. Aspects of flexible control for mobile robots. In *ICRA 2000*, 2000.

[5] H. Cruse, J. Dean, U. Mller, and J. Schmitz. The stick insect as a walking robot. In *Proceedings of the 5th International Conference on Advanced Robotics: Robots in unstructured Environment (ICAR '91)*, volume 2, pages 936–940, Pisa/Italien, 1991.

[6] S. T. Davis and D.G. Caldwell. The biomimetic design of a robot primate using pneumatic muscle actuators. In *4th International Conference on Climbing and Walking Robots (CLAWAR)*, pages 197–204, Karlsruhe, September 2001. Professional Engineering Publishing Limited.

[7] L. Jalics, H. Hemami, and Y. F. Zheng. Pattern generation using coupled oscillators for robotic and biorobotic adaptive periodic movement. In *Proceedings IEEE International Conference on Robotics and Automation*, pages 179–184, 1997.

[8] V. Kepplin, K. Berns, and R. Dillmann. Terrain and obstacle detection for walking machines using a stereo camera system. In *24th Annual Conference of the IEEE Industrial Electronics Society (IECON 98)*, 1998.

[9] K. Lffler, M. Gienger, and F. Pfeiffer. On theory and realization of a biped jogging robot. In *Proc. of the 10th International Workshop on Robotics in Alpe-Adria-Danube Region (RAAD)*, 2001.

[10] K. Lffler, M.Gienger, and F. Pfeiffer. A biped jogging robot - theory and realization. In *Proc. of the 3rd International Conference on Climbing and Walking Robots (CLAWAR)*, pages 51–58, 2000.

[11] R.J. Bachmann D.A. Kingsley J. Offi R. D. Quinn, G. M. Nelson and R.E. Ritzmann. Insect design for improved robot mobility. In R. Dillmann K. Berns, editor, *4th International Conference on Climbing and Walking Robots*, pages 69–76. CLAWAR, Professional Engineering Publishing, September 2001.

[12] S. Schulz, C. Pylatuik, and G. Bretthauer. Walking machine with compliance joints. In *4th International Conference on Climbing and Walking Robots (CLAWAR)*, pages 231–236, Karlsruhe, September 2001. Professional Engineering Publishing Limited.

[13] H.-J. Weidemann, F. Pfeiffer, and J. Eltze. The six-legged tum walking robot. In *Proceedings of the 1994 IEEE International Conference on Intelligent Robots and Systems*, pages 1026–1033, Munich, 1994.

[14] H. Witte, R. Hackert, K. Lilje, N. Schilling, D. Voges, G. Klauer, W. Ilg, J. Albiez, A. Seyfarth, D. Germann, M. Hiller, R. Dillmann, and M.S. Fischer. Transfer of biological priciples into the construction of quadruped walking machines. In *Second International Workshop On Robot Motion And Control*, Bukowy Dworek, Poland, 2001.

Intelligent Autonomous Systems 7
M. Gini et al. (Eds.)
IOS Press, 2002

Concept for Human Robot Co-operation Integrating Artificial Haptic Perception

Catherina Burghart, Sadi Yigit, Oliver Kerpa, Dirk Osswald, and Heinz Woern
Institute of Process Control & Robotics,
University of Karlsruhe,
D-76128 Karlsruhe, Germany.
{burghart, yigit, kerpa, osswald, woern}@ira.uka.de

Abstract. Human robot co-operation becomes more and more important when using robots in daily life and is a key note in the design of humanoid robots. The authors are members of the newly founded centre of excellence "Humanoid robots – co-operating and learning systems" at the University of Karlsruhe. Our group of researchers is responsible for the human robot co-operation, a haptic sensor system and the control of a robot hand. All three topics closely co-operate as the human robot co-operation (using a robot arm and a robot hand attached to it) is mainly based on the haptic sensor system within the first phase of the project. In this paper we present our first classifications for human robot co-operation and contacts between a robot and a human or an object, which both form the basis for our practical work. The control concept of our robotic system is depicted as well.

1. Introduction

At the University of Karlsruhe a new centre of excellence titled "Humanoid robots – co-operating and learning systems" has been founded which focuses on the perception and the modelling of the environment of a humanoid robot (including humans) and the co-operation between human and robot. Within this scope the authors of this paper are responsible for the design and transposition of the human robot co-operation, for the design and integration of haptic sensibility and the control of a five fingered robot hand. For this purpose we use a robot arm with seven degrees of freedom and a five fingered hand attached to it as first test bed. Later, results will be transposed to a humanoid robot designed by other groups within the centre of excellence.

Haptic sensors form the basis of the human robot co-operation within the first phase of the project. They are used as a means of a human user to directly and indirectly communicate his or her intentions to the robot. Additionally, the robot hand needs haptic sensors to control the grip of manipulated objects or to explore the texture of objects. The coupling of hand and arm implies the consideration of both arm and hand within different human robot co-operation schemes and a close co-operation between arm control and hand control.

In the following sections we first present our classification of human robot co-operation schemes. The third section features the haptic sensor system and the fourth section describes the control system and interaction of arm, hand, human- robot co-operation module and haptic sensor system.

2. Human Robot Co-operation

When co-operating with a human the behaviour of a robot differs due to the actual task. For example, a robot holding an object manipulated by a human must have another behaviour during this task and in case of an error than a robot holding and leading a human. On the other hand, such a distinction is not always necessary. For example, a robot needs no different way of behaviour when opening a cupboard or a refrigerator. There are different kinds of tasks that need a different behaviour of the robot, whereas other tasks require the same behaviour of a robot. The first step to implementing the different, task dependent ways in robot behaviour is a classification of human robot co-operation schemes.

Our classification has been created under the aspect of which situations require different behaviours of the robot. Then this classification has been used to find primitives of interpretation enabling the robot to identify each class by a specific pattern from its sensor data, internal robot data and robot environment. The classification is depicted in Figure 1.

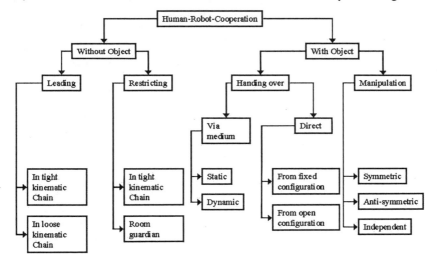

Figure 1: Classification of possibilities of Human-Robot-Cooperation.

In our classification we first made a distinction whether an additional object is involved in the co-operation or not. If no object is involved there are two classes called *leading* and *restricting*. In the class *leading* the robot is leading the human (e.g. robotic guide dog) or the human is leading the robot (e.g. teaching). This can happen while the robot and the human have a tight connection like a grip or while they have a loose connection like a hand lying on the arm of the co-operation-partner, which results in different kinds of reactions (especially in case of an error) and opens on two new different classes. In the class *restricting*, the robot restricts the movements of the human (e.g. separating the human from dangerous areas) or the human restricts the movements of the robot (e.g. stopping the robot as a additional concept of safety). This can either be done while the robot and the human are touching each other in a tight grip or - within the class *room guardian* - while one partner of the co-operation is prohibiting the other form entering an area by crossing his trajectory with parts of his own body in case the other would enter the area. In all classes there is no object directly involved in the co-operation, but this does not mean that there is no object involved at all i.e. a robot can carry an object while it is lead by a human to another place.

If there is an object directly involved in the co-operation there are two possible schemes of interaction between human and robot. Either they are just handing over the object or they both manipulate the object at the same time.

In case of *handing over* there is the possibility of a *direct handover*, which means form hand(s) to hand(s), or the possibility to transfer the object by *using a medium*. The direct handover requires different ways of behaviour if the object is presented in a fixed configuration that has to be opened during the handover (i.e. a closed hand), or if the object is presented in a way that the other partner of the co-operation can just take the object. These classes can be classified into smaller and more specific classes by focusing of the kind of grip(s) used to hold the object. The handing over of an object by using a medium can be achieved by placing it somewhere in the working space of the partner of the interaction or by throwing (or dropping) it into the partner's hand.

In case of the object manipulation, there are three different classes according to the movement and forces of the participants of the co-operation in aspect to each other. Those can be *symmetric, anti-symmetric* or *independent*. In the *symmetric* case, the movement of the robot and the movement of the human are the same to reach the aim of the co-operative task, like carrying, pushing or pulling something together. In the class *anti-symmetric object manipulation* the movement of robot and human and the forces generated by them are working against each other. Examples for anti-symmetric object manipulations are things like opening a screw plug or tearing off a piece of paper. If the robot and the human are doing different manipulations on the same object or work on different objects that have a connection to each other, this kind of interaction is found in the class *independent object manipulation* i.e. filling a glass held by the robot or taking something from a plate carried by the robot.

All of the classes have the potential to be split into further subclasses; we already have eleven different classes, most of them resulting in different ways of behaviour of the participants; the behaviour depends on the part of the task done by the robot and by the human, which all in all results in about twenty-two different interactions schemes.

The classification of co-operation schemes is basis of the specification of so called interpretation primitives. These primitives combine data and patterns of sensors, the robot, the environment and the human in order to specify and interpret the different steps of a specific co-operation between human and robot. Some of these data are gained by artificial haptic perception.

3. Artificial haptic perception: pre-classification of mechanical contacts

The model of the artificial haptic perception is the human sense of touch. It is often considered to consist only of the tactile sense of the skin which is only a part of it. The *bathyaesthesia* or *deep sensibility* is also an integral part of this human sense. It includes the senses of force, position and movement of the different parts of the human body [1]. These haptic information –tactile ones and forces- in combination with information about the context of the robot and its environment allow a classification of a detected contact with the environment (Fig. 2 and Fig. 3).

In the case of human robot co-operation the class of *control contacts* is relevant. A contact identified as a *control / positioning* contact can initialise a co-operation task but the robot needs further information which task is demanded. These information have to be given by other senses e.g. the auditory sense; limiting to haptic perception it would be imaginable to give commands directly using the tactile sense as an input interface which matches defined input patterns with defined commands.

In the following classifications a sensor system comprising surface and depth sensibility of a robot is considered. Surface sensibility is achieved by using tactile sensors in foil design applied to the robot arm and hand with different spatial resolution. The palm of the robot hand and the finger tips are supplied with sensors at relevant point to support hand control. Depth sensibility is realised by acceleration sensors and a force torque sensor in the wrist.

3.1 Classification of contacts

The main purpose is the classification of contacts according to the quality of *intention*. We are proposing three classes of contacts which cover all possibilities of interaction between a robot and its environment: *collisions, control contacts* and *task contacts*. These main classes can be split in generic sub-classes which can be partly divided again in specific contact patterns e.g. the sub-class *grip* can be divided in a taxonomy proposed by Cutkosky and Wright [2] with related contact patterns. In Fig. 2 the classes and sub-classes of contacts are shown and will be explained in the following.

Collisions can be described as contacts that are not intended at all, neither from the robot nor from its environment. The differentiation of this class in collisions of the robot with humans or objects is making sense if there are different reactions of the robot in such cases.

Control contacts are initiated by the robot's environment with the purpose to give information or commands to the robot. The sub-class *positioning* is quite near to the model of intuitive human-like interaction because it does not require further information but contacts and forces. Using the tactile sensing as an input interface for defined commands via contact patterns –contacts called *command inputs*- is not quite that kind of human-like interaction but there are imaginable situations in which it would make sense e.g. in loud environments or simply when the robot does not have other senses.

Task contacts are contacts that are related with the actual task of the robot and therefore they are expected and mainly initiated by the robot. In the case of *grips* these are specific contact patterns for each particular grip. *Exploration* contacts are initiated by the robot and expected in a certain area of its body e.g. a fingertip but a further differentiation by means of properties to be explored is necessary e.g. the shape of an object.

Support describes contacts which are expected as part of a support task e.g. the robot supporting a person with its arm.

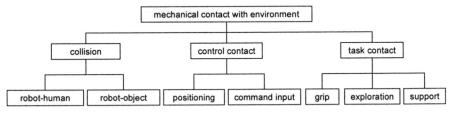

Figure 2: Contact classes and sub-classes

3.2 Parameters of the formal description

To perform a classification a formal description of these contact classes is needed. The parameters on which this description is based include information about the context of the robot and its environment and, naturally, characteristics of the sensor signals. As seen above in the description of the contact classes appears the quality of *intention* or *expectation* so we have chosen the following four parameters concerning the context:

Operation mode of the robot: the robot has a task to perform or not (it is *active* or *passive*). When the robot is *passive* it is in a stand-by-mode and is waiting for an input ("lurking mode"). When it is *active* it has a task to perform with a related contact pattern ("no contact" represents also a contact pattern e.g. for the task "go from point A to point B").

Expectation of the robot: the perceived contact is *expected* or *not expected*. When the robot is active a contact is *expected* if it is matching the related contact pattern; if not it is *unexpected*. When the robot is passive all contacts are *unexpected*. To perform this differentiation the actual contact pattern has to be given to the contact classification unit.

Status of the environment: whether there are autonomously moving objects (humans or other robots) in the workspace of the robot (*dynamical*) or not (*static*). If this information is not available, the default status of the environment has to be *dynamical* for safety reasons.

Intention of environment: a contact can be *intended* or *unintended* by the environment.

The combination of these four parameters with two possibilities each results in sixteen sets of parameters. Half of these sets do not make sense e.g. a *passive* robot does not expect any contacts, therefore any contact is *unexpected;* and to provoke any contact the environment has to be *dynamical*. In the following diagram (Fig.3) the reasonable combinations with their derived contact classes are shown.

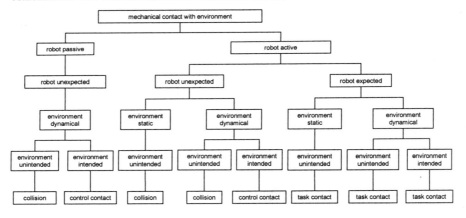

Figure 3: Parameter sets of contact classes

As it can be seen, the three contact classes can be distinguished on the basis of these four parameters. The sub-contact classification has to be performed on the basis of the characteristics of the sensor signals.

In the case of *task contacts* the related pattern must be given to the classification unit.

A possibility to differentiate *collisions* of the robot with a human or an object could be the acceleration measured in defined points on the robot. This integration of acceleration sensors in the haptic perception system will be investigated. In the field of robotic control this approach already has been tested [3].

In the class *control contacts* two sub-classes have to be differentiated. The contact patterns for direct input of commands are defined (e.g. in a kind of morse code) but somehow the input has to be initiated. Using only a part of the robot body as input interface e.g. the back of the hand in combination with an initiation pattern like "three taps" would be a possibility. To recognize a *positioning* contact the tactile pattern can be used. To lead a robot the user has to grip a part of the robot e.g. the arm. This grip results in a tactile pattern that has opposing areas which would be one possibility to recognize the initiation of a *positioning* task.

Furthermore the classes *collision* and *control contact* must be differentiated if environment information is not given. Measurement of accelerations, forces in the joints and the analysis of tactile patterns - whether they are probable for a collision or not - will be investigated.

4. Control System

The control system of the robot must be specially designed to be able to fulfill the desired tasks. For cooperative tasks the physical interaction between a human and the robot is a key issue. Because robots are usually much stronger and mechanically more robust than a human being, this physical interaction is potentially dangerous for the human. Therefore the control system has to guarantee the safety of the human at all costs.

Generally a control system has to perform the three steps 'perceive', 'process' and 'act' to fulfill tasks like object manipulations or movements. But not only a robot performs these steps, but also a human being. While the robot uses its sensors to 'perceive' the human uses his or her senses. The 'processing' is done by software on a microprocessor for the robot while the human uses the central nerve system. And finally the robot 'acts' using actuators like motors and the human uses his or her muscles. For the human-robot-cooperation with physical interaction the acting of the robot is perceived by the human and the acting of the human is perceived by the robot. Therefore the human closes the control loop from acting to perceiving for the robot and vice versa [4].

Figure 4 shows the proposed control system for the robot used for the human-robot-cooperation schematically. The approach is based on the 'classical' hierarchical control architecture [5], [6]. The figure also shows the human involved in the cooperation as described in the previous paragraph. At the highest level shown in the figure is the human-robot-cooperation module. This module decides what to do, i.e. how to 'act', according to:

- the classification of possibilities of human-robot-cooperation described in section 2
- the interpreted sensor data, like e.g. the classification of mechanical contacts described in section 3
- context information about the environment from the internal model.

The 'acting' commands from the human-robot-cooperation module are then forwarded to the high and low level control of the different actuators. The complete robot planned in this project will consist of a 'body', i.e. a mobile platform, equipped with two arms, two hands and a head. The respective sub-control systems for these sub-systems are not considered here, because their specifics are not relevant for the human-robot-cooperation (see e.g. [7]). For the first phase of the project a simpler mechanical system will be used, consisting of one fixed robot-arm equipped with a humanoid hand with up to 5 independent fingers.

The sub-control systems of body, arm and hand control the different actuators of the robot respectively. These mechanical actuators, like motors, pneumatic cylinders etc., physically interact with the human cooperation partner. This interaction can be more or less mechanically direct, as discussed in section 2.

The human perceives the acting of the robot with his or her senses, processes the perception and reacts by actuating his or her muscles. This reaction is in turn perceived by the robot with its sensors. To perform the physical interaction safely for the human all relevant physical sizes must be sensed. Therefore not only the usual position sensors are needed but many additional kinds of sensors, like e.g. tactile, force, torque or acceleration sensors.

The raw sensor data must be preprocessed and prepared for further evaluation. This sensor data is then used as feedback signal for the control systems of the body, arms or hands. Additionally the data can be stored in the internal context model of the environment. The preprocessed data is also used to classify the mechanical contacts as discussed in section 3. This step hence interprets the information contained in the sensor data in order to perceive 'what is going on' in the environment. This information can as well be stored in the internal context model of the environment, but is mainly used by the human-robot-cooperation module. This closes the overall control loop.

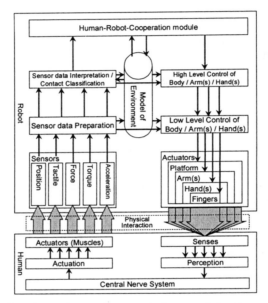

Figure 4: control system for cooperation between robot and human

5. Conclusion

Human robot co-operation involves the robot's perception and interpretation of human actions. For this purpose we have set up a classification of possible human robot co-operation schemes as a basis to design interpretation primitives for robot control; which considers co-operation with or without an object, leading a partner or restricting the movements of a co-operation partner, and the different manners of manipulating the same object at the same time.

Closely related to the classification of the human robot co-operation schemes is the classification of the robot's contacts with its environment, especially considering the intention of the environment and the robot's expectation of commands according to patterns defined by the interpretation primitives of the human robot co-operation unit. Here, haptic sensors are one of the basic sensor systems to pass the appropriate information and a pre-classification on to the human robot co-operation unit.

The present test bed for our research comprises a robot arm and a humanoid hand attached to it as well as different tactile sensors for surface sensibility and haptic sensors for depth sensibility. In our paper we have also presented the concept of the control system involving robot arm, robot hand, human, and the robot's artificial haptic sensor system.

Future work comprises the definition of interpretation primitives for human robot co-operation and the planning and transposition of co-operative tasks. Different sensors are tested and integrated and sensor patterns as well as contact patterns have to be specified. In a later phase of the project further sensors are to be included in the human robot co-operation. All newly designed methods will be tested with our testbed.

Acknowledgement

This research was performed at the Institute of Process Control and Robotics, headed by Prof. Dr.-Ing. H. Wörn, at the University of Karlsruhe. The work is being funded by the German Research Foundation, as it is part of the special research program "Humanoid Robots – Learning and Co-operating Systems".

References

1 Schmidt R.F., Thews G.: "Physiologie des Menschen", 27.Aufl., Springer, 1997.
2 Cutkosky M.R., Wright P.K.: "Modeling Manufacturing Grips and Correlations with the Design of Robotic Hands", Proc. IEEE Int. Conf. Robotics and Automation, pp. 1533-1539, 1986.
3 Stelter J.: "Verbesserung des Positionierverhaltens und der Bahntreue eines Industrieroboters durch den Einsatz von Beschleunigungssensoren", Shaker Verlag, 2001.
4 Geiser, G. (1990): Mensch- Maschine-Kommunikation. Oldenbourg Verlag, 1990.
5 B. Magnussen, Infrastruktur für Steuerungs- und Regelungssysteme von robotischen Miniatur- und Mikrogreifern, Fortschritt-Berichte VDI Reihe 8, Nr.567, Düsseldorf: VDI-Verlag, 1996.
6 T. Fischer and H. Woern, Structure of a robot system: Karlsruhe Dexterous Hand II, Mediterranean Conference on Control and Systems, 1998.
7 Dirk Osswald, Heinz Wörn, Mechanical System and Control System of a Dexterous Robot Hand; *Proceedings of the IEEE-RAS International Conference on Humanoid Robot, Tokyo, Japan, 2001.*

Intelligent Autonomous Systems 7
M. Gini et al. (Eds.)
IOS Press, 2002

Probabilistic, On-line Construction of Memories

Brendan Burns, Paul R. Cohen
University of Massachusetts
Amherst, Massachusetts, 01002.
{bburns,cohen}@cs.umass.edu

Abstract.
We describe a probabilistic technique for unsupervised construction of object models and demonstrate that an agent can construct models that distinguish objects. In contrast to previous model-based object recognition techniques, our method performs unsupervised clustering. The agent develops a set of models based solely upon its previous experience. The domain of object recognition is sequences of snapshots, rather than individual images, allowing an agent to recognize objects once it has learned how appearances change over time. Empirical results with simulated and real-world objects are presented.

1 Introduction

Our goal in exploring object memory and recognition is to have autonomous agents build symbolic representations of things and relations between things in the environment. Our approach is dynamical, that is, we build models of how the appearances of objects change with viewpoint. Rather than the instantaneous question "which object is this?" we are interested in recognizing similar objects in temporally distinct experiences. We are concerned less with whether the agent can identify an object as, say, a chair, and more with the agent's ability to identify two or more sequences as experiences with the same object. We also require that an agent build its memory with little or no outside supervision. Unsupervised methods allow the agent to develop representations which *it* sees as useful, and ensures that object models contain information that an agent needs, information that might not have been anticipated by programmers. These goals constrain what and how the agent learns about objects. It should build a memory of objects incrementally, adding new models as well as merging existing models, as information becomes available through experience. The memory-building process must be implemented in online algorithms, since the agent exploring its environment needs to take advantage of new information immediately.

2 Object Memories for Recognition

Two subtasks in the development of an object memory system are recognition of objects given models, and the development of new models. The former serves the latter by establishing how well a current experience matches a current model and, when the match is poor, suggesting adding a new model. This section describes our approach to object recognition.

Object recognition is performed not on sequences of images but on sequences of snapshots. For a particular image i, creating a snapshot s_n involves applying a generalizing function $g(i) = s_n$ which maps i to a general snapshot class. We call the range of this generalizing function $g()$ the *snapshot vocabulary* since it describes the set of snapshots which an agent may possibly view. Snapshot sequences are created by moving through the world and at each time step t, recording the output of the function $g(i_t)$, where i_t is the agent's view at time t. A sequence is written $S = \{s_1, \ldots, s_n\}$ where each s_n is in the range of $g()$

Recognition is implemented in a straightforward probabilistic framework: Given a sequence S and a set of models $\{M_1 \ldots M_n\}$, find M_i such that $argmax(P(M_i|S))$. That is, select the model with the highest probability of representing S.

2.1 Object Models

Object models encapsulate the probabilities of snapshots in sequences. An object model M_i includes two functions: First, $f_{M_i} : (S, S) \to \Re$ such that $f_{M_i}(s_k, s_n) = P(s_k \to s_n|M_i)$, that is, the probability that s_n follows s_k in a time series S, given that the object described in the model M_i generated S. This is the *transition probability* of two snapshots s_k and s_n. Second, $f'_{M_i} : S \to \Re$ where $f_{M_i}(s_k) = P(s_k|M_i)$, the probability that a snapshot s_k is observed in a sequence that was generated by the model M_i. f'_{M_i} is called the *appearance probability* of a snapshot s_k.

2.2 Recognizing a Sequence

Given a new sequence of snapshots S, recognition involves finding the most probable model, that is, the model M for which $P(M|S)$ is maximum. From Bayes' rule, $P(M|S) = P(M)P(S|M)/P(S)$. For the purpose of maximizing $P(M|S)$, it is not necessary to know the value $P(S)$, as it depends on S, not on M. Thus, $P(M|S) \propto P(M)P(S|M)$. We can easily calculate the latter conditional probability as follows:

$$
\begin{aligned}
P(S|M_i) &= P(s_1 \to s_2 \to \ldots \to s_n|M_i) \\
&= P(s_1|M_i)P(s_1 \to s_2|M_i) \ldots P(s_{n-1} \to s_n|M_i) \\
&= f'_{M_i}(s_1)f_{M_i}(s_1, s_2) \ldots f_{M_i}(s_{n-1}, s_n)
\end{aligned}
$$

To finish the calculation we only need to estimate the probability of the model. For the recognition task the probabilities of a model M_i, $P(M_i)$ is $\frac{1}{n}$ when there are n models.

Thus given a set of models and a sequence of snapshots S, we calculate $P(M_i|S)$ for each model M_i and select the one with the highest probability. S is recognized as the object corresponding to M_i.

3 Building Memories

Object memory is simply a set of models. This section discusses how the models are constructed, incrementally, with experience. First we describe how a sequence of snapshots becomes a model, then we discuss how models are modified through experience.

From a sequence of snapshots S_k, memory construction produces a model M_j that represents three kinds of probabilities observed in S_k: Transition probabilities, appearance probabilities, and probabilities of unobserved transitions.

Let $|S_k|$ denote the number of snapshots in the time series S_k, then for each transition between snapshots $s_m \to s_n$, the *transition probability* $P(s_m \to s_n|S_k)$ is estimated by

$Count(s_n, s_m, S_k)/(|S_k| - 1)$, where $Count(s_m, s_n, S_k)$ returns the number of times the transition $s_m \to s_n$ is observed in S_k.

The *appearance probability* of s_n in a sequence S_k is estimated by $Count(s_n, S_i)/|S_k|$ where $Count(s_n, S_i)$ returns the number of times the snapshot s_n appears in the sequence S_i.

In the absence of any observations (e.g. if $Count(s_m, s_n, S_i) = 0$ or $Count(s_n, S_i) = 0$) the corresponding probabilities are given as a small non-zero constant, otherwise an unobserved transition would generate an estimate of zero for the probability of the sequence that contains it.

Thus from any sequence of snapshots we can estimate f_{M_i} and f'_{M_i} (Sec. 2.1). We can construct a new model of the object that generated a sequence. In the next section we show how sequences are incorporated into existing models.

3.1 Augmenting Memories

If an agent recognizes a sequence of snapshots S_k as belonging to a model M_j then it makes sense that it should use this new information to update and improve M_j. Generalizing transition probability estimates to a *set* of sequences $S_1 \ldots S_n$ which match a model M_k, the probability of a transition $s_n \to s_m$ is estimated by:

$$P(s_n \to s_m | M_k) = \frac{\Sigma_{i=1}^n Count(s_n, s_m, S_i)}{\Sigma_{i=1}^n (|S_i| - 1)}$$

This *generalized transition probability* is the sum of the number of times a particular transition occurs in each sequence divided by the total number of transitions in the set of sequences.

The *generalized appearance probability* for a model M_k comprising $S_1 \ldots S_n$ is estimated by:

$$P(s_n | M_k) = \frac{\Sigma_{i=1}^n Count(s_n, S_i)}{\Sigma_{i=1}^n |S_i|}$$

This is the number of times a snapshot, s_n, is observed in a set of sequences, divided by the total number of snapshots observed in the set of sequences.

3.2 Adding Memories

Given a set of models $M_1 \ldots M_n$ and a sequence S, memory addition is the process which decides whether to add S to an existing model or create a new model. We select the alternative that results in the most probable set of models given all observed sequences [4].

We start by calculating $P(M_k | S_1 \ldots S_n)$, the probability of a model M_k given a set of series $S_1 \ldots S_n$ which M_k has recognized. From Bayes' rule this is equivalent to $P(M_k)P(S_1 \ldots S_n | M_k)/P(S_1 \ldots S_n)$. In this domain, all objects and orientations that generate experiences are equiprobable, i.e., $\forall i, j P(S_i) = P(S_j) = c$, and the sequences are independent, so $P(S_1 \ldots S_n) = P(S_1) \ldots P(S_n) = c^n$. Thus,

$$
\begin{aligned}
P(M_k|S_1 \dots S_n) &= P(M_k)P(S_1 \dots S_n|M_k)/c^n \\
&= P(M_k)P(S_1|M_k) \dots P(S_n|M_k)/c^n \\
&= P(M_k)\Pi_{i=1}^n P(S_i|M_k)/c \\
&\propto P(M_k)\Pi_{i=1}^n P(S_i|M_k)
\end{aligned}
$$

Now in order to calculate which set of models is more probable, we must generalize this probability to $M_1 \dots M_k$. Let $S_{j1} \dots S_{jn}$ be the set of sequences that are classified as belonging to the model M_j. Let n_j be the size of this set. Generalization is complicated by the fact that the denominator in each of the expressions of $P(M_k|S_1 \dots S_n)$ depends on the number of series recognized by model M_k. We aproximate the generalization by assuming that $\forall i, j c^{n_j} \approx c^{n_i} \approx c'$. Thus

$$
\begin{aligned}
P(M_1 \dots M_k|S_{11} \dots S_{kn}) &= \Pi_{j=1}^k P(M_j)\Pi_{i=1}^{n_j} P(M_j|S_{ji})/c \\
&\approx \frac{1}{c'}\Pi_{j=1}^k P(M_j)\Pi_{i=1}^{n_j} P(M_j|S_{ji}) \\
&\propto \Pi_{j=1}^k P(M_j)\Pi_{i=1}^n P(M_j|S_{ji})
\end{aligned}
$$

By using this formula to compare the two sets of models (one with S added as a new model and one with S merged) we choose the set which is most probable and the system moves on to the next experience.

4 Experiments

The first experiment tested the quality of object recognition using stored memories, while the second tested the agent's ability to develop memories autonomously.

Each experiment was run in both a three-dimensional simulated environment and a real robotic platform. The former involved a simulated camera circling a point in three-space, keeping the camera focused on that point. Experiments were run with the center of the camera located randomly within a half-meter of the object being observed.

In our first experimental condition the objects being observed were a cube, a pyramid and a wedge (fig. 1). To see how our technique scalesto larger numbers of objects we ran a second condition with sixteen objects built from cubes (fig 2). The task of object recognition in this domain was complicated by the fact that all objects are identical to two or more objects when viewed from some angles. For example the silhouette of the wedge appears identical to the pyramid when viewed from one angle, and when it is turned ninety degrees it has the same appearance as the cube. Shared views would be fatal to any object recognition techniques that do not represent the possibility that an object has multiple appearances from different perspectives. Because our approach models the dynamics of snapshots of objects, it can cope with shared views.

The robotic environment consisted of a Pioneer II robot in a play-pen with three-dimensional objects in it. The robot had a low-level controller that caused it to approach and circle an object while maintaining the focus of its camera upon the object.

The only difference between the testbeds was the origin of pixel maps of the (simulated or real) world. All visual analysis, image extraction, snapshot creation and object

Figure 1: The three objects in the first simulation experiment

Figure 2: Four of the sixteen objects in the second simulation experiment

recognition was performed by the same program. We believe that results for both platforms are comparable; the only important difference is reduced (not eliminated) noise in the simulated environment.

4.1   Experiment 1

The first experiment consisted of providing the agent with an object memory constructed from brief (60 seconds) experiences with each object and a single experience with an unknown object which the agent was required to classify as one of the objects in memory or unknown.

4.1.1   Results and Discussion

First we discuss results obtained with simulated objects. For the first set of simulated objects (cube, wedge, pyramid) the overall accuracy of the recognition algorithm is 97%. With the larger set of sixteen objects the overall accuracy is 80% In both cases these averages were over multiple instances of each object; four instances of each object in the first set and ten of each in the second.

The random focus in these experiences meant that at times large portions (up to 50%) of the objects were outside of the agent's view. This resulted in previously unseen snapshots which created transition probabilities not present in the object models. The algorithm is robust to this level of noise in its observations. Additionally when the algorithm made mistakes it made them by misclassifying objects that share common snapshots. For instance, the wedge was mistaken for the cube and the pyramid was mistaken for the wedge, but the cube was never mistaken for the pyramid.

The lower accuracy in the set of sixteen objects is probably due to the high degree of similarity between objects in the set. For each object there was another that differed from it by the placement of only a single cube (less than 25% of its total volume), and as a result, many objects were identical from many viewpoints. Also, objects in the second set are more complex than the cube, wedge, and pyramid in the first set; which may make the recognition task more difficult if the process of generalizing viewpoints into snapshots eliminates important details.

The overall accuracy in the robotic domain was 75%, although this result is based on only 20 trials because it is difficult to work with the robot platform. The snapshot sequence captured by the robot was similar to the randomly-centered sequence of simulated objects in that objects were often partially outside of the camera's frame. The robot is incapable of making a perfect circle of a point, so sequences of images of an object differed in scale and orientation across experiences.

4.2 Experiment 2

The second experiment was designed to test the agent's abilities to develop memories. The agent was exposed to a sequence of snapshots and given the choice to recognize the sequence as a known object and incorporate the sequence into a known object model, or build a new object model from the sequence. The agent started with no memories and was scored on both the accuracy of the models it learned and the number of models learned relative to the actual number of objects in the environment.

4.2.1 Results and Discussion

In the simple simulated environment, populated with only a cube, wedge and pyramid, and given twelve experiences with each, the system constructed four memories, one each for the cube and the wedge, and two for the pyramid. This redundancy did not greatly affect the accuracy of the memory, if recognizing an instance of the pyramid as either model is considered "correct" and recognizing a non-pyramid as one or the other pyramid model is considered "incorrect". Since the classification of pyramids was nearly evenly split between the two models, a model merging operation would probably eliminate the redundancy in models and not reduce accuracy.

With the second set of sixteen objects the system constructed 40 memories when exposed to 160 experiences, ten with each object. This means that 24 excess memories were created, though eleven of the sixteen objects were represented by only one model. The five objects that were duplicated were the same ones the system had the most difficulty recognizing in the pure recognition task described earlier.

In the real world data set, the memory construction created seven models for three objects (two models of a man, two of a giraffe and three of a dog). Again, a model merging component would be useful, as the models of a single object shared many snapshots.

5 Related Work

Statistical and probability-base methods have existed in the pattern recognition community for a long time ([1], [8]). Although our approach is similar to previous efforts in some ways, the differences are interesting: First, unlike much work in pattern recognition, our method is unsupervised (but see [6] which demonstrates autonomous development of three-dimensional memories of objects represented as line segment features.) Our method autonomously builds models without any prior information about the number or distribution of objects in the environment. Second, we model objects as first-order Markov chains of snapshots, that is, as probability distributions over how the appearances of objects change over time. Third, pattern recognition usually works on fixed-length vectors of values. In our technique, there is no required length of experience needed for recognition.

One of the first attempts to introduce probability into the process of three-dimensional object recognition is [3], in which probabilistic rankings were used to reduce the size of the model search space. [2] gives a good overview of model-based approaches to three-dimensional object recognition. Much earlier vision work involved monolithic object representations, but lately people have favored snapshot-based representations [5]; for instance, [9] developed models which are sets of snapshots and performed recognition by interpolation between views. Other work breaks up an object into its constituent parts for recognition ([7]) but uses two-dimensional images and extracted features which are consistent across multiple views.

The BCD algorithm [4] should also be mentioned since it introduces a general probabilistic method to autonomously cluster sequences of discrete data. These clusters are analogous to our memories and in fact BCD influenced the development of our theory. Unlike BCD, which reconsiders the total number of models at every step, our system considers only the choice between adding an experience as part of an existing model and constructing a new model. As we mentioned earlier, some form of incremental merging, possibly not after every experience, would no doubt improve performance.

6   Conclusions and Future Work

This work can be improved in several ways. Currently, snapshots represent the shape class of an object. Snapshots that represent aspects of activities such as rotational velocity and camera angle could provide more accurate representations of how objects' appearances change during activities. The uniform prior probabilities of models could also be replaced with something that better represents the real prior probabilities of objects. This would in effect bias model development, an interesting prospect. Finally, we intend to incorporate our model-development algorithm into an active exploration system that seeks to reduce its uncertainty about the object it is currently exploring.

Acknowledgments

This research is supported by DARPA under contract DASG60-99-C-0074. The U.S. Government is authorized to reproduce and distribute reprints for governmental purposes notwithstanding any copyright notation hereon. The views and conclusions contained herein are those of the authors and should not be interpreted as necessarily representing the official policies or endorsements either expressed or implied, of DARPA or the U.S. Government.

References

[1] P. Devijer and J. Kittler, editors. *Pattern Recognition Theory and Applications.* NATO/ASI, 1986.

[2] S. Edelman. Computational theories of object recognition, 1997.

[3] D. Lowe. Three-dimensional object recognition from single two-dimensional images. *Artificial Intelligence*, 31(3), March 1987.

[4] P. Cohen M. Ramoni, P. Sebastiani. Bayesian clustering by dynamics. *Machine Learning*, 2001.

[5] H. Bulthoff M. Tarr. Image based object recognition in man, monkey and machine. In H. Bulthoff M. Tarr, editor, *Object Recognition in Man, Monkey and Machine.* MIT Press, Cambridge, MA, 1998.

[6] R. Nelson. Three-dimensional recognition via twostage associative memory, 1995.

[7] R. Nelson and A. Selinger. A cubist approach to object recognition. In *Internation Conference on Computer Vision*, New Delhi, India, 1998. Narosa Publishing House.

[8] Jürgen Schürmann. *Pattern Classification: A unified view of statistical and neural approaches.* Wiley Interscience, 1996.

[9] S. Ullman. Three-dimensional object recognition based on a combination of views. In H. Bulthoff M. Tarr, editor, *Object Recognition in Man, Monkey and Machine.* MIT Press, Cambridge, MA, 1998.

Intelligent Autonomous Systems 7
M. Gini et al. (Eds.)
IOS Press, 2002

Exploiting Multi-robot Geometry for Efficient Randomized Motion Planning

Stefano Carpin   Enrico Pagello[†]
The University of Padova
Department of Electronics and Informatics
Intelligent Autonomous Systems Laboratory
via Gradenigo 6/A 35131 Padova - ITALY
{shamano,epv}@dei.unipd.it
[†]also with LADSEB - National Research Council
corso Stati Uniti 4 35127 Padova - ITALY

Abstract. Randomized motion planners proved to be a viable solution to problems characterized by an elevate number of degrees of freedom. In this paper we illustrate how it is possible to exploit the geometrical properties of autonomous multi-robot systems to improve the performance of such methods. The underlying idea is to sample the configuration space of the single robot and to build a roadmap for it. Then, a subset of the samples and the single robot roadmap are used to quickly build samples in the composite configuration space of the multi-robot as well as a roadmap over them. The proposed solution is formalized and simulation results illustrate the speed up gained while constructing the roadmap.

1   Introduction

In this paper we address the problem of planning the motion of a group of autonomous robots operating in an indoor environment. Every robotic agent is assumed to be a mobile platform which has to autonomously move without colliding with other robots and obstacles. This task is one of the fundamental ones when designing a multi-robot system, but, in spite of this, there is still a lack of well accepted and viable solutions. It is well known that this problem can be reduced to that of planning the motion of a single robot system obtained by the *composition* of the single robots which compose the team ([14]). It is also known that this approach suffers from dealing with high dimensional configuration spaces, and this turns out in high processing times. For this reason a number of *non-standard* techniques have been introduced, trading off performance with solution quality. Among these, randomized techniques proved to be highly effective, especially in the case of high dimensional configuration spaces. We used randomized motion planning in the case of multi-robot systems, with the idea of exploiting simple geometrical properties of such systems to reduce the time needed to build the data structures used by those algorithms. This framework can be applied in the case where all the robots of the team have the same shape, or in the more likely case that they can all be approximated with the same bounding sphere or cylinder.

The paper is organized as follows: section 2 gives the details about related works in the area of randomized motion planning, and 3 illustrates the formal details of our proposal. In 4 we describe how we implemented the simulation framework and the quantitative results we obtained. In 5 we discuss some of the extensions we are evaluating as well as their motivation, and in 6 we offer the conclusions.

2   Related work

Planning the motion of a robot is an inherent difficult task, as shown in [5]. For this reason a number of methods have been introduced to deal with robots operating in high dimensional configuration spaces. One of the possible approach is to trade off completeness for speed. An interesting approach stems from the definition of *probabilistic complete algorithms* ([17]), i.e. algorithms which find a solution with some degree of probability, and such that this value increases as we allot more time to the planning algorithm. The novel approach introduced in [12] is based on this idea and proved to be highly efficient. In this *probabilistic motion planning* framework a two stages procedure is applied. In the first stage a set of samples (vertices) in the configuration space is built. Then, if a free path between two vertices is found, an edge connecting them is generated. In the second stage the so produced graph (also called roadmap) can then be searched to compute paths between two points in the configuration space, provided that they can be joined to the roadmap. Thus the problem of path planning is reduced to graph search. This approach is probabilistic complete, i.e. by growing the number of random samples, and then the processing time, it will eventually find a solution if one exists. Afterwards a number of refinements has been proposed, in order to get both further gains in performance ([1],[2]) and higher probability of success even in case of difficult environments ([3],[4],[11],[16]). Randomized motion planning has also been applied in multi-robot systems ([18]) with success. One of the most attractive aspects of randomized motion planning comes from the fact that the method is not devoted to a particular class of robots, but is rather applicable to every type of robots, so it is easily usable in multi-robot systems as well as for robotics manipulators and so on.

3   Efficient construction of the roadmap

The goal of this paper is to develop a framework for planning the motion of a multi-robot system composed by a N mobile platforms operating in a shared environment. Our work is based on [18], which proved to be very efficient. We assume that each robot is holonomic and can be approximated with the same bounding cylinder and that they operate in a two dimensional environment, their workspace W is R^2. Of course a robot works in a three dimensional space, but since we ignore the third dimension and we analyze just its motion on a flat surface, we assume its workspace is two dimensional. Moreover, the configuration space C_s of each robot is R^2, and each robots shares the same space of free configurations C_{sFree}. The configuration space of the multi-robot C_m is then R^{2N}.

Planning a motion in the composite configuration space C_m using a randomized motion planner would involve to build a graph $G = (V, E)$, the probabilistic roadmap, and then a search on this graph (these two phases are also called *learning* and *query* phases). In a straightforward application of this method, the first step of the learning phase is a random sampling of C_m followed by a validation, i.e. a test to determine if the samples belong to the space C_{mFree} of free configurations of the multi-robot. The goal is to produce the a set of vertices V. Let K be the cardinality of the set V. The i-th sample is

$$((x_{i1}, y_{i1})(x_{i2}, y_{i2}) \cdots (x_{iN}, y_{iN})) \in C_{mFree} \subseteq R^{2N}$$

All over the paper we assume that a couple of values (x, y) represents the center of the base of the bounding cylinder of a robot. Thanks to the hypothesis that each single robot shares the same space of free configurations, we have

$$(x_{ij}, y_{ij}) \in C_{sFree} \quad 1 \le i \le K, 1 \le j \le N$$

This is a consequence of the fact that we are dealing with a multi-robot system, so that the geometrical properties of C_{mFree} are strictly related to those of C_{sFree}. This observation can be exploited to reduce the time needed to build the set V.
The idea is to build a preliminary set S of s samples coming from C_{sFree} and then to use those samples to build the set V of samples in C_{mFree}. Figure 1 illustrates the procedure.

```
 1: S ← ∅
 2: while |S| < s do
 3:     p ← random (R²)
 4:     if p ∈ C_sFree then
 5:         S ← ⋃ {p}
 6:     end if
 7: end while
 8: V ← ∅
 9: while |V| < K do
10:     Build c ∈ R^2N selecting N random samples in S
11:     if c ∈ C_mFree then
12:         V ← ⋃ {c}
13:     end if
14: end while
```

Figure 1: Fast generation of the set of vertices V from the set of samples S

The need to test if the composed configuration c belongs to C_{mFree} (line 11), is due to the fact that even if all the samples in S belong to C_{sFree}, their composition does not necessary do. Indeed if the Euclidean distance between the samples s_j and s_j is less then $2R$, where R is the radius of the bounding cylinder, every configuration which includes both s_j and s_j will not be a configuration in C_{mFree} (see figure 2).

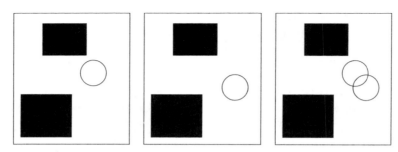

Figure 2: A two robots example of inappropriate configuration composition. Even if both the configurations of the single robots belong to C_{sFree}, their composition is not a configuration in C_{mFree}

An additional point worth of discussion is the size $s = |S|$ of the single robot configuration samples set. Of course a sort of trade off is necessary. A small number of samples is of course preferable, in order to speed up the generation of the set S, but this can turn out in a longer time spent to generate the set V, since a lot of composed configurations will be discarded because of inappropriate compositions. Next section will illustrate the results coming from different choices of this value. A fast generation of the set S is however not sufficient to get a significant speed up of the learning phase, because the time consuming stage is building the set of edges E. But thanks to the fact that V has been built upon S, we can preprocess S in order to efficiently build E. Given S, we first build the set ES, the set of edges over S. The set ES is defined as follows:

$$ES = \{(s_i, s_j) : s_i \in S, s_j \in S \; AND \; dist(s_i, s_j) < D \; AND \; \overline{s_i s_j} \in C_{sFree}\}$$

where $dist(s_i, s_j)$ denotes the Euclidean distance between the two samples and D is a threshold, so that we do not try to join two samples if they are too far. An edge connecting two vertexes is added if the associated samples can be connected by a straight line (this is quite a common choice in PRM based algorithms). Thus (S, ES) is the random roadmap for the single robot. Two samples in V can be joined by an edge only if the corresponding components have already been joined by an edge in ES. Then, given the configurations in C_{mFree}

$$v_i = (s_{i1}, \cdots, s_{iN}) \in S^N$$

and

$$v_j = (s_{j1}, \cdots, s_{jN}) \in S^N$$

they can be connected only if

$$(s_{ik}, s_{jk}) \in ES \qquad 1 \leq k \leq N$$

These conditions are however not sufficient, because it could be the case that two robots collide while the multi-robot is moving from v_i to v_j. For this reason we assume that the transition, i.e. the motion, between two configurations in V is performed in N steps. At step t the t-th robot moves from (x_{it}, y_{it}) to (x_{jt}, y_{jt}) and all the others are stationary (this strategy is sometimes called *lazy motion strategy*). With this assumption it is easy to check if (v_i, v_j) can be inserted in E, the set of edges over V. With reference to [18], the set E is the equivalent of the *flat super-graph*, while the motion strategy is the equivalent to the so called *discretised coordinated path*. Figure 3 gives the details of this procedure.

The main difference between our approach and the one presented in [18] is that while in [18] the choice is on the size of the single roadmap and then this is "projected" into the multi-robot roadmap, with our approach the focus is directly on the size of the composite roadmap, since the size of the the single robot roadmap is deducted from $|V|$. Thus it is not necessary to go through the efficient but complicated multi-level super-graph step.

4 Experimental results

The proposed idea has been implemented and tested to verify its effectiveness. The software has been developed in C++ and has been tested on different hardware platforms (SGI, Sun and PCs). All the simulations discussed in this section have been performed on a Linux based Athlon 900 MHz PC with 256 Mb of RAM.

In the simulated framework we supposed to operate with circular robots performing in

```
 1: ES ← ∅
 2: for each pair (s_i, s_j) ∈ S × S do
 3:    if dist(s_i, s_j) < THRESHOLD then
 4:       if segment s̄_i, s̄_j is in C_{sFree} then
 5:          ES ← ⋃(s_i, s_j)
 6:       end if
 7:    end if
 8: end for
 9: E ← ∅
10: for each pair (v_i, v_j) ∈ V × V do
11:    if (v_i, v_j) is in C_{mFree} then
12:       for k = 0 to N do
13:          if (v_{ik}, v_{jk}) ∈ ES then
14:             E ← ⋃(v_i, v_j)
15:          end if
16:       end for
17:    end if
18: end for
```

Figure 3: Fast generation of the set of edges E from the set S

an environment with polygonal obstacles. The most interesting result comes from the time spent to build the set E. Because we are dealing with a randomized algorithm we analyze the average of the numerical data we collected. Figure 4 compares the *classical randomized* algorithm with our proposal. Path planning is performed for three robots operating in the environment illustrated in figure 5. By classical randomized method we mean the algorithm that builds the roadmap $G = (V, E)$ operating directly in the composite configuration space R^{2N}. We compare the average time spent to build and edge, for different values of s, which is the size of the set of samples S. The y values are then the ratio between the time spent to build the set E and its size. This because different runs in the same environment will produce graphs with a different number of edges as a consequence of the random sampling, so it is not fair to just compare the time spent to build sets of different cardinality.

The obtained speed up is evident. After the construction of the roadmap the problem of searching a path between the initial and goal configurations is a *single-pair shortest path* problem over a graph with positive edges, and then the classical Dijkstra algorithm ([10]) is used. Of course, also the time spent to build the set V benefits from using the samples in S, but this is not a massive improvement, since the bottleneck of roadmap methods is the construction of the set of edges (the gap is of two degrees of magnitude). Figure 6 compares the global execution times for the two algorithms over a set of 100 trials. The size of set V is 3500 and the problem involves three robots (thus a search in R^6). With this size, both algorithms got a 100% success rate. One final issue concerns s, the size of the set S and the consequent success rate of the algorithm. The success rate of the algorithm is the number of positive outcomes it gets related to the number of trials. An heuristic hint is to choose s proportional to the square of the logarithm of $|V|$. Figures 7 and 8 illustrate the success rate of the algorithm stemming from this choice. Data are averaged over 100 trials for each different choice of the size of V.

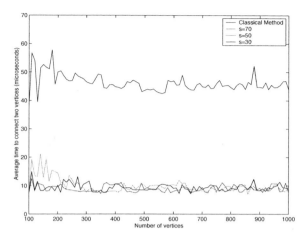

Figure 4: Time comparison between classical randomized motion planner and our algorithm for different values of s

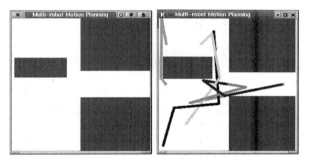

Figure 5: The Environment used for the tests discussed in this section and the computed solution for a 5 robots motion planning problem

5   Future work

We are currently investigating a number of issues concerning the proposed framework. In the past ([6]) we explored the possibility of getting a better performance utilizing a distributed computational approach. This idea comes from the observation that currently used multi-robot systems are distributed systems, since every robot is equipped with an on board powerful processing unit. Thanks to the availability of this set of computational units it would be nice to let them cooperate to solve time consuming tasks like motion planning. The distributed approach could be used both for graph building and searching. The query phase can be speeded up using a parallel version of the Dijkstra algorithm ([13]). We are also evaluating if it is possible to build the roadmap using a distributed approach. Another possibility is to let every computational node, i.e. every robot, solve the whole problem using the randomized approach, and then choose the best solution. This approach already proved to be effective ([7],[8],[9]), but does not take into any account the geometrical consequences that comes from dealing with a system which is some sort of *replication* of simpler identical (or similar) units, as we did in this framework.

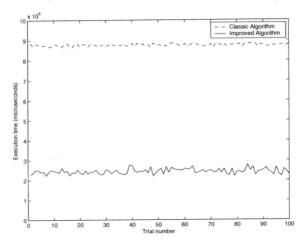

Figure 6: Time comparison between classical randomized motion planner and the proposed algorithm

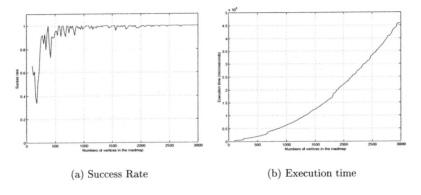

(a) Success Rate   (b) Execution time

Figure 7: Three robots problem

Another issue concerns how, and if, it is possible to use this framework in a dynamical scenario. Indeed, all over the paper we supposed to deal with a static environment. Recently (see [15] for example) some techniques have been introduced in the field of randomized motion planning. Because they are defined over the domain independent configuration space, they can be directly applied in our framework. However, once again we are evaluating if the geometrical peculiarities of a multi-robot system can be used to get further improvements.

6   Conclusions

In this paper we illustrated how it is possible to exploit some geometrical aspects of systems composed by a set of mobile platforms in order to get a more efficient implementation of randomized motion planners. We introduced a framework for the efficient generation of random roadmaps in the composite configuration space starting

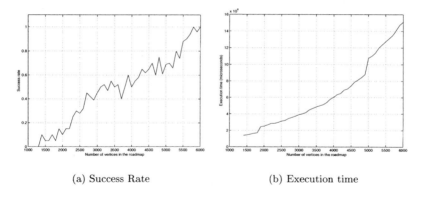

(a) Success Rate   (b) Execution time

Figure 8: Five robots problem

from a roadmap built in the configuration space of the single robot. The approach can be applied in the case of holonomic robots which can be all approximated with the same bounding sphere or cylinder. Simulations give evidence of the speed up obtained with this technique.

Acknowledgments

This research has been partially supported by the Ministry for Education and Research, by the Italian National Council of Research, and by the Italian Energy Agency (ENEA Parallel Computing Project).

References

[1] N.M. Amato, L.K. Dale. *Probabilistic Roadmap are Embarassingly Parallel.* Proc. of the IEEE ICRA, 688-694, Detroit (USA), 1999

[2] R. Bohlin, L.E. Kavraki. *Path Planning Using Lazy PRM.* Proc. of the IEEE ICRA, 521-528, S. Francisco (USA), 2000

[3] V. Boor. M.H. Overmars, A.F. van der Stappen. *The Gaussian Sampling Strategy for Probabilistic Roadmap Planners*, Proc. of the IEEE ICRA, 1018-1023, Detroit (USA), 1999

[4] M.S. Branicky, S.M. Lavalle, K. Olson, L. Yang. *Quasi-Randomized Path Planning.* Proc. of the IEEE ICRA, 1481-1487, Seoul (Korea), 2001

[5] J. Canny. *The Complexity of Robot Motion Planning*, MIT Press, 1988

[6] S. Carpin, E. Pagello. *A Distributed Algorithm for Multi-robot Motion Planning.* Proc. of the 4th European Conference on Advanced Mobile Robots, 207-214, Lund (Sveden), September 2001

[7] S. Caselli, M. Reggiani. *ERPP: An Experience-based Randomized Path Planner.* Proc. of the IEEE ICRA, 1002-1008. San Francisco(USA), 2000

[8] D. Challou, D. Boley, M.Gini, V. Kumar. *A Parallel Formulation of Informed Randomized Search for Robot Motion Planning Problems.* Proc. of the IEEE ICRA, 709-714, Nagoya (Japan), 1995

[9] D. Challou, D. Boley, M.Gini, V. Kumar, C. Olson. *Parallel Search Algorithms for Robot Motion Planning*, in K. Gupta, A.P. del Pobil (Eds.), *Practical Motion Planning in Robotics*, John Wiley & Sons, 1998

[10] T.H. Cormen, C.E. Leiserson, R.L. Rivest. *Introduction to algorithms*, MIT Press, 1990

[11] D. Hsu, L.E. Kavraki, J.C. Latombe, R. Motwani, S. Sorkin. *On Finding Narrow Passages with Probabilistic Roadmap Planners*, in P.K. Agarwal, L.E.Kavraki, M.T. Mason (Eds.), *Robotics : The Algorithmic Perspective : The Third Workshop on the Algorithmic Foundations of Robotics*, A.K. Peters, 1998

[12] L.E. Kavraki, P. Švestka, J.C Latombe, M.H. Overmars. *Probabilistic Roadmaps for Path Planning in High-Dimensional Configuration Spaces*, in IEEE Transactions on Robotics and Automation, 12(4), 566-580, 1996

[13] V. Kumar, A. Grama, A. Gupta, G. Karypis. *Introduction to Parallel Computing.* The Benjamin/Cummings Publishing Company, 1994

[14] J.C Latombe. *Robot Motion Planning*, Kluwer, 1991

[15] P. Leven, S. Hutchinson. *Towards Real-Time Path Planning in Changing Environments*, in Bruce Donald, Kevin Lynch, Daniela Rus (Eds.), *Algorithmic and Computational Robotics: New Directions*, A.K. Peters, 2001.

[16] G. Song, S. Miller, N.M. Amato. *Customizing PRM Roadmaps at Query Time.* Proc. of the IEEE ICRA, 1500-1505, Seoul (Korea), 2001

[17] P. Švestka, M.H. Overmars. *Probabilistic Path Planning*, in J.P. Laumond (Ed.), *Robot Motion Planning and Control*, Lectures Notes in Control and Information Sciences 229, Springer, 1998

[18] P. Švestka, M.H. Overmars. *Coordinated path planning for multiple robots.* Robotics and Autonomous Systems, 23(3)125-152, 1998

Intelligent Autonomous Systems 7
M. Gini et al. (Eds.)
IOS Press, 2002

VISUAL NAVIGATION OUTDOORS:
THE ARGOS PROJECT

ENRIC CELAYA and CARME TORRAS
Institut de Robòtica i Informàtica Industrial (CSIC-UPC)
Llorens i Artigas 4-6, 08028-Barcelona
celaya@iri.upc.es, torras@iri.upc.es

Abstract. Visual navigation in unstructured, previously unknown, environments is investigated in the project ARGOS. This paper describes the goals of the project, the guidelines we follow to face each aspect of it, and the motivations that led us to work with a walking robot. The current state of the project in the areas of legged locomotion, landmark-based navigation and vision is succintly described.

1 Introduction

Visual navigation, that is, reaching a target using visual information, is a capability that most animals exhibit in different degrees, and should be considered as a basic one for any autonomous mobile robot. The goal of the ARGOS project[1] is to develop a system that, after placing the robot in an unstructured, previously unknown environment, allows the user to select a target in the scene as captured by the cameras of the robot, and start a navigation process to reach the target with no further human intervention.

Posed in this way, the problem faced is so general that it cannot be expected to be solved in all cases. Our purpose is to push the set of solvable situations from the simplest (e.g., target always visible and no obstacles in the way), towards the more complex ones (difficult terrain forcing long detours where the target gets occluded).

In this paper we describe the approach we follow to advance in our understanding of visual navigation, and the current state of our work in the aspects of legged locomotion, landmark-based navigation, and vision. While our work in each one of these aspects has been reported in specific papers, our purpose here is to give a global view of the project as a whole.

2 Robot autonomy

It is agreed that autonomy is a desirable property for a robot, and specially for a mobile one. It is clear that autonomy implies the ability to work without close control or supervision by a human, but in our opinion, there are other aspects that must also be considered. A first aspect is related to the diversity of environmental situations the robot can face. Only if the situations to which the robot is able to respond in a sensible way are varied enough, or even unpredictable, we say that the robot is autonomous.

A second aspect of robot autonomy has to do with the complexity of the robot itself. On the one hand, the robot sensory system has to be complex enough to distinguish

---

[1] Carried out jointly with the Institut d'Investigació en Intel·ligència Artificial (CSIC), and financed by the Spanish research agency (CICYT) under project DPI2000-1352-C02-01.

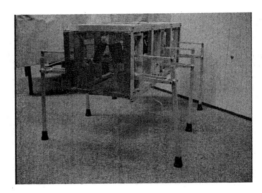

Figure 1: The six-legged robot ARGOS.

between the different environmental situations that require different actuations. On the other hand, the robot's repertoire of actions must be varied enough so that its response can be tailored to each situation.

We are interested in autonomous robot navigation in unstructured outdoor environments. The choice of this kind of environments grants the diversity of situations needed to look for autonomy. Concerning the robot, we prefer to use legged locomotion since it provides a much richer set of motor actions than wheeled locomotion, and thus, the opportunities to deal with autonomy are increased.

To deal with the problem of controlling a complex robot in a complex environment, we have decomposed the task of visual navigation in 3 main subtasks: Piloting, landmark-based navigation, and vision. Each of these subtasks is further decomposed in more specialized processes, as described in the following sections.

3   Legged robot locomotion: The pilot

The task assigned to the pilot is to move the robot in the direction specified by the navigation level while negotiating obstacles and terrain difficulties found in the way. In the pilot, no map information is used, but only reactive behaviors responding to the information provided by local sensors.

We have built a six-legged robot (Fig. 1) within this project. In order to have full mobility, each leg has three degrees of freedom, which gives a total amount of 18 for the robot. Following a behavior-based approach, we have decomposed the pilot task in 4 main subtasks [1]:

3.1   Posture control

The purpose of posture control is to keep the body in the most appropriate position with respect to the current feet locations. Keeping an optimal posture improves robot stability and mobility at the same time. The action of posture control consists in the execution of simultaneous movements of all legs in such a way that feet positions remain fixed, so that the net effect is a displacement of the body with respect to them.

3.2   Terrain adaptation

The task of terrain adaptation consists, first, in keeping contact with ground at any time with as many feet as possible and, second, in adapting the body position to avoid collisions with obstacles and unstable situations. Whenever a foot lacks ground contact, it is made to move down vertically until the contact is recovered. Note that even though this layer does only provide a mechanism to move legs down, the average leg elevation is maintained constant by virtue of the posture control described above.

When the robot detects an obstacle, its body is displaced in the opposite direction to avoid the contact. Similarly, if the robot stays in an inclined position, the body is moved up-hill in order to increase its stability.

3.3 Gait generation

This is the layer specifically devoted to walking. Two simple rules are used to determine which legs can perform a step: the *stability rule*, that forbids to rise a leg whenever this would result in losing ground contact with two neighboring legs at the same time, and the *efficiency rule*, that gives priority to legs nearer to reach their workspace limit.

3.4 Direction control

The task of this layer is to drive the robot in the direction provided by the navigation level. It is assumed that the driving commands specify the local trajectory as an arc of circumference with given center and radius. The action of this layer consists in defining the target position (usually called the AEP, or anterior extreme position) for each stepping leg, according to the local trajectory.

When an obstacle forbidding the advance of the robot in the desired direction is found, the trajectory is modified to avoid the obstacle.

4 Landmark-based navigation

The whole navigation system is implemented as a multiagent system whose activity is regulated through a bidding mechanism, in which each agent bids to gain access to the limited resources of the robot, according to the priority attributed to its task in any given situation [5].

The navigation task is decomposed in the following three ones: target reaching, map management, and path planning.

4.1 Target reaching

This layer drives the robot directly towards the target when it is perceived by the vision system and there are no big obstacles in the path to it.

4.2 Map management

We have adopted the use of a topological map, instead of a metric or a grid one. The approach we follow is that proposed in [4], that is based on the relative positions of landmarks in order to estimate the location of the target using a method called the *beta-coefficient system*.

4.3 Path planning

When the target is not visible by the robot, its position is estimated from that of the currently visible landmarks, and the robot is directed towards it. If a large obstacle is found ahead, the path is modified by directing the robot towards a provisional new target selected using the topological map.

5   Vision

The vision system we have designed to comply with the specific demands of visual navigation for legged robots is arranged around three main modules: landmark extraction and recognition, egomotion computation, and terrain characterization.

5.1   *Landmark extraction and recognition*

Detecting *salient* and *stable* landmarks is required to build the topological maps used by the navigation module.

Saliency outdoors should be addressed in a radically different way than in indoor environments, where geometric features (corners, straight lines, etc.) are the usual option. In natural settings, color and texture seem more appropriate characteristics. Inspired by a model of visual attention, we have devised a pyramidal system that computes multiscale saliencies based on color opponencies and texture orientations.

Attaining color *stability* outdoors is a big challenge, since the acquired visual information is strongly dependent on lighting geometry (direction and intensity of light source) and illuminant color (spectral power distribution), which change with time within the day, seasons, and atmospheric conditions. Noting that what was needed was not constancy of colors themselves, but constancy of color opponencies, we have devised a novel procedure called *multiscale color ratios*. For instance, the red-green color ratio is $R_c G_s / R_s G_c$, where R and G are red and green center or surround regions, depending on the subscript. The yellow-blue and white-black opponencies are dealt with likewise. The interesting point about these ratios is that they are intrinsically invariant to intensity and color normalizations and, as a consequence, they are immune to the inclusion/exclusion of objects in the scene [7].

The other type of features, namely texture orientations, are computed using Gabor filters. Each feature (color opponency or texture orientation) has an associated Gaussian pyramid that permits computing saliencies at different resolutions. By comparing centers and surrounds at high-resolution levels, e.g. between levels 2 and 5, visual saliencies of relatively small targets are found, while at lower resolution levels, e.g. levels 4 and 7, relatively large salient regions are found. Thus, with this technique, it is possible to detect salient objects within a wide size range, for example, from small stones to big trees.

The salient regions in the image are subsequently analysed to obtain visual signatures (based essentially on textures and contours), capable of identifying them as an existing or a new landmark. The description of the entire pyramidal system, as well as some results can be found in [6].

5.2   *Egomotion computation*

Since the odometry of legged robots outdoors is quite unreliable, a requirement for the vision module is to provide position estimation in the short run. This is complementary to the long-term localization based on landmarks.

Estimating camera motion from optic flow is computationally costly. Thus, we have explored the cheaper option of deriving the 3D egomotion from the deformation of an active contour fitted to a landmark (e.g., the target), by relating the affine deformation in 2D to the translation and rotation in 3D. Note that the translation along the optical axis is recovered up to a scale factor. Results of this module are reported in [2] and the particular case of a zooming camera appears in these proceedings [3].

5.3   Terrain characterization

Classifying terrains according to their degree of traversability is a very challenging task. In particular, vegetation poses many problems. Although a depth map may be useful, it cannot distinguish between different types of obstacles, some of which may be traversable. So far, we have just worked on the obtention of a depth map, and future efforts will be devoted to complementing it with additional information.

The results of our egomotion computation module permit computing the epipolar lines, which can be used to guide the search for point matches. This leads to the obtention of qualitative depth maps, such as those presented in [2].

6   Conclusions

We are addressing the problem of autonomous visual navigation in outdoor, unstructured environments. Our approach to deal with the challenging complexity of this task is to decompose the task itself in order to solve separately different aspects of it, instead of simplifying the environment on which this task is performed. Legged locomotion provides the required complexity to respond to the diversity of environmental situations that the robot must face.

The main contribution of the project in legged locomotion is an adaptable walking algorithm that allows the robot to follow arbitrary trajectories in very rough terrain. Concerning landmark-based navigation, a multiagent system is used to manage the map and locate the target. In the vision area, a novel technique to detect potential outdoor landmarks has been developed, whose stability in front of lighting variations and slight viewpoint changes has been demonstrated. Moreover, an egomotion estimation procedure based on active contours has proven to be reliable enough, with much less computation than the usual methods based on optic flow.

References

[1] Celaya, E. and Porta, J.M., "A Control Structure for the Locomotion of a Legged Robot on Difficult Terrain", *IEEE Robotics and Automation Magazine*, Vol. 5, No. 2, June 1998, pp. 43-51.

[2] Martínez, E. and Torras, C., "Qualitative vision for the guidance of legged robots in unstructured environments", *Pattern Recognition*, 34(8), 2001, pp. 1585-1599.

[3] Martínez, E. and Torras, C., "3D motion recovery while zooming using active contours", Proc. *7th Intl. Conf. on Intelligent Autonomous Systems (IAS-7)*, Marina del Rey, USA, March 2002.

[4] Prescott, T.J., "Spatial representation for navigation in animats", *Int. Adaptive Behavior*, 4(2), 1996, pp. 85-125.

[5] Sierra, C., López de Mántaras, R., Busquets, D. "Multiagent Bidding Mechanisms for Robot Qualitative Navigation", *Lecture Notes in Artificial Intelligence*. Springer, March 2001.

[6] Todt, E. and Torras, C., "Detection of natural landmarks through multiscale opponent features", *15th International Conference on Pattern Recognition*, Barcelona, Spain, 2000, vol. 3, pp. 988-991.

[7] Todt, E. and Torras, C., "Color constancy for landmark detection in outdoor environments", Proc. of the *4th European workshop on advanced mobile robots* (EUROBOT'01), Lund, Sweden, Sept. 2001.

Intelligent Autonomous Systems 7
M. Gini et al. (Eds.)
IOS Press, 2002

Automatic color landmark detection and retrieval for robot navigation

L.A Cornelissen[1], F.C.A. Groen
Informatics Institute, University of Amsterdam
The Netherlands,
lode.cornelissen@cmg.nl, groen@science.uva.nl

Abstract. Landmarks play an important role in the navigation of mobile systems like office robots. The robot position can be found from these landmarks. In this paper a method is presented for automatic detection and retrieval of color landmarks. Measures for the uniqueness of a region are investigated in relation to color invariants and landmark retrieval. Results show the potential of the presented method for selection and retrieval of color landmarks. Best results are obtained with the maximum intersection measure for landmark detection and with the rgb and $c_1c_2c_3$ color spaces for landmark retrieval.

1 Introduction

In mobile robot navigation the question "where am I?" is not unimportant. A system looses its position from odometry after a while due to slip, collisions etc. Vision can be an important source of information to locate the robot again.

Localization with vision can be based e.g. on appearance modeling (Kröse [1]) or on landmarks. In this paper we will focus on automatic localization based on landmarks, as landmarks are less sensitive to dynamic objects in a scene than appearance modeling. Recently Schiele et al [2] introduced a technique which combines both approaches and is also robust the major occlusions.

The idea is that a robot explores a number of rooms or is guided through the rooms and automatically finds and stores landmarks. When the system sees the landmarks later again, it can retrieve its position. Swain and Ballard [3] introduced a technique called color indexing to retrieve landmarks. Enneser and Medioni [4] improved this technique by using local histograms. It proved to be a powerful technique to find known colored objects in scene images. Given a color space (e.g., red, green and blue), a color histogram is obtained by quantizing the image colors and counting the number of times each discrete color occurs in the image. Histograms are invariant to translation and rotation about the viewing axis, and change only slowly under change of angle of view, change in scale, and occlusion [3]. The sensitivity to change of illumination can be reduced by the use of color invariants or by color image normalization.

Local color histograms L(k,l) are calculated at successive image positions (k,l) and compared to the model histogram M of the landmark we are looking for. The position of the best

---

[1] Currently with CMG, The Netherlands,

matching histogram L(k,l) gives the location (k,l) of the landmark in the image. The local histograms are compared based on the weighted histogram intersection. Let M_i be the number of counts in bin i of the model histogram and L_i be the number of counts in bin i of the local histogram at position (k,l). The weighted histogram intersection is given by:

$$I = \sum_i w_i \min(L_i, M_i)$$

The weights w_i express the specificity of colors for the landmark. The most interesting colors are those present in a landmark and sparse in the whole image. When H is the histogram of the whole image and M is the model histogram, a good choice for w_i is:

$$w_i = \frac{M_i}{H_i}$$

2 Color Models and color invariants

Important for color landmarks is the variation of the colors of a landmark as function of the observation angle, surface orientation and illumination. Swain and Ballard [3] showed that histograms are invariant to translation and rotation about the viewing axis, and change only slowly under change of angle of view, change in scale, and occlusion. Histograms can be made invariant to illumination by using color invariants.

In general the total observed reflection is the sum of a diffuse and a specular reflection component. In case of diffuse reflection light will be reflected with equal intensity. It does not matter from which position the surface is observed. The intensity only depends upon the angle between the illumination direction $\vec{s}$ and the surface normal $\vec{n}$ of the object. The (Lambertian) diffuse reflection D is defined as:

$$D_C = \alpha I_C k_C (\vec{n} \cdot \vec{s})$$

in which C stands for the color component, I_C is the intensity, k_C is the diffuse reflection coefficient.

In case of specular reflection most of the light is reflected in one direction $\vec{r}$, which is the mirrored illumination direction around the surface normal $\vec{n}$ in such a way that $\vec{r}$, $\vec{n}$, and $\vec{s}$ are planar. The light is also reflected in a region around $\vec{r}$. The specular reflection S_C according to Phong is given by:

$$S_C = I_C k_S (\cos\theta)^q$$

in which k_S is the specular reflection coefficient, θ is the angle between the observation direction and $\vec{r}$, q is a constant depending upon the shine of the material.

A detailed discussion of color invariance and image normalization can be found in Funt et al. [5] and Geuzebroek at al. [6]. In this paper we will limit ourselves to a number a simple invariant color representations [7] to explore the trade-off with non-invariant color representations in discriminative power.

2.1 Color space of Swain and Ballard

Swain and Ballard [3] propose 3 parameters describing the color space, in which the parameter *bw* is the intensity. By leaving out this parameter in the construction of the color histogram, these histograms become less sensitive to the intensity, but the intensity is still present as scale factor in *rg* and *by*.

$$rg(R,G,B) = R - G$$
$$bw(R,G,B) = R + G + B$$
$$by(R,G,B) = 2B - R - G$$

2.2 Normalized RGB

A simple way to make the color parameters invariant to the intensity is to divide the R,G,B values by the intensity. The representation becomes invariant to the orientation of surfaces, illumination direction and illumination intensity for diffuse reflection. Two parameters are sufficient to describe the color space, as the sum of r, g and b is constant.

2.3 Hue and Saturation

Also the well known HIS model is invariant to intensity in its representation of the color plane by hue H and saturation S. Hue only depends on k_c, which makes it invariant for both diffuse and specular reflection. The saturation is invariant for diffuse reflection only.

$$H(R,G,B) = \arctan\left(\frac{\sqrt{3}(G-B)}{(R-G)+(R-B)}\right)$$
$$S(R,G,B) = 1 - \frac{\min\{R,G,B\}}{R+G+B}$$

2.4 $c_1 c_2 c_3$

The parameters $c_1 c_2 c_3$ are given by:

$$c_1(R,G,B) = \arctan\left(\frac{R}{\max\{G,B\}}\right)$$
$$c_2(R,G,B) = \arctan\left(\frac{G}{\max\{R,B\}}\right)$$
$$c_3(R,G,B) = \arctan\left(\frac{B}{\max\{R,G\}}\right)$$

These parameters are intensity invariant in case of diffuse reflection.

2.5 l_1, l_2, l_3

Each expression defining colors in the same plane as the diffuse reflection **D** and the specular reflection **S** is invariant to the reflection model. The l_1, l_2, l_3 features fulfill this requirement [7] and are intensity invariant for both types of reflection.

$$l_1(R,G,B) = \frac{(R-G)^2}{(R-G)^2 + (R-B)^2 + (G-B)^2}$$

$$l_2(R,G,B) = \frac{(R-B)^2}{(R-G)^2+(R-B)^2+(G-B)^2}$$

$$l_3(R,G,B) = \frac{(G-B)^2}{(R-G)^2+(R-B)^2+(G-B)^2}$$

3   Landmark detection and retrieval

The color parameters described in the previous section cannot be used directly to construct color histograms. Some parameters give unreliable results for pixels with a low intensity. Pixels with a value lower than 5% of the intensity range are disregarded in the calculation of the histograms [7]. As a result the number of entries in the histograms differs and normalized histograms are used.

We will introduce now four measures, which express the uniqueness of a local area in an image based on color histograms.

3.1 Pixel uniqueness

This simple measure calculates the uniqueness of the pixel colors for a region R of the size of the expected landmarks. The uniqueness of a pixel color is based on the normalized color histogram of the whole image H(c). The pixel uniqueness of a region R is defined as

$$P_R = \sum_{k \in R} \frac{1}{H(c(k))}$$

in which c(k) is the quantized color of pixel k

3.2 Summed intersection

Instead of the uniqueness of the individual colors in a local neighborhood, the uniqueness of the distribution of the colors in a local region could be used. Such a measure can be obtained by dividing the whole image into regions and by calculating the intersection of the local color histogram of a region R with that of all other regions. The sum of these intersections is taken as measure S_R.

$$S_R = \sum_{j \neq R} \frac{\sum_{i=1}^{n} \min\{M_R(i), M_j(i)\}}{\sum_{i=1}^{n} M_j(i)}$$

3.3 Threshold intersection

When colors from a quite unique region are also present in other large areas in the image, taking the previous sum may give a large contribution. As a result these areas are not recognized as unique. To prevent this we could count the number of regions within the image, which have comparable histograms. When the histogram intersection is below a certain threshold ε we see them as comparable. This results in the threshold intersection measure T_R

$$T_R = \sum_{j \neq R} U\left(\sum_{i=1}^{n} \min\{M_R(i), M_j(i)\}, \varepsilon \right)$$

in which $U(a,\varepsilon) = 1$ if $a > \varepsilon$ and 0 otherwise.

3.4 Maximum intersection

Another possibility is not to threshold the intersection, but to take the maximum of the intersections with all other regions. When a region is unique this maximum should be low.

$$K_R = \max_{j \neq R} \frac{\sum_{i=1}^{n} \min\{M_R(i), M_j(i)\}}{\sum_{i=1}^{n} M_j(i)}$$

4   Localization based on landmarks

We will assume that the mobile robot is driving through an office environment over the flat surface of a floor and that the camera is mounted parallel to that floor. We are interested in the position of the robot on the ground floor. When the robot moves it perceives the landmarks at successive locations of its trajectory. Given the motion of the robot (Δx, Δy) and its rotation (θ) at the two successive positions where an image is taken, the 3D position of the observed landmarks can be calculated with a standard stereovision technique. The projected positions of the landmarks on the ground floor are stored in the coordinate system of the robot. The translation and rotation between the successive coordinate systems is where we are looking for: the motion of the mobile robot with respect to the landmarks.

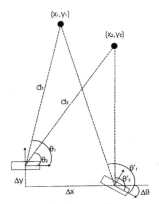

Figure 1: Landmarks seen from two robot positions

Assume that the projected positions on the ground floor of n landmarks $\{r_i, s_i\}$ are calculated with respect to some room coordinate system during exploration. Given the newly observed position of the landmarks when the robot enters the room again, the position and orientation of the robot can be calculated from the displacement $\{r_0, s_0\}$ and rotation φ of the current coordinate system with respect to the exploration coordinate system. The parameters can be found by minimizing the mean squared error E between both coordinate systems

$$E = \sum_i \{x_i - f(r_i, s_i)\}^2 + \{y_i - g(r_i, s_i)\}^2$$

$$f(r_i, s_i) = \cos\varphi(r_i - r_0) + \sin\varphi(s_i - s_0)$$

$$g(r_i, s_i) = -\sin\varphi(r_i - r_0) + \cos\varphi(s_i - s_0)$$

5   Results

5.1   Influence of color space and bin size

To investigate the influence of the different color spaces and the bin sizes 3 landmarks were retrieved from a number of video image sequences. Results are given in table 1. Max landmark peak value and min landmark peak value give the maximum and the minimum peak values of the 3 landmarks. Max background gives the maximum value of a non-landmark peak over all image sequences. When a background peak is higher than the minimum landmark peak in an image, errors can be introduced. The amount of overlap between the landmark range and the background range is given in the last column. This overlap should be preferable zero.

Color Space	bins/parameter	max landmark peak value	min landmark peak value	max background peak value	Overlap in %
RGB	8	134	28	42	13
RGB	16	190	17	37	11
Swain rg,by,bw	16	180	24	48	15
Swain rg,by	16	144	35	66	28
Swain rg,by	32	161	39	56	14
Swain rg,by	64	172	34	48	10
rgb	16	154	28	62	27
rgb	32	166	42	39	0
rgb	64	173	43	33	0
$c_1 c_2 c_3$	16	151	51	51	0
$c_1 c_2 c_3$	32	170	42	36	0
$c_1 c_2 c_3$	64	169	38	37	0
l_1, l_2, l_3	16	106	15	78	70
l_1, l_2, l_3	32	113	25	70	51
l_1, l_2, l_3	64	114	25	72	53
hs	8	116	20	34	14
hs	16	160	25	27	1
hs	32	172	25	30	3

Table 1. Retrieval results for the different color parameters and quantization values. Three landmarks are present in the images.

The rgb and $c_1 c_2 c_3$ parameters give the best results. These parameters are intensity invariant for diffuse reflection. The l_1, l_2, l_3 parameters are invariant for both diffuse and specular reflection but this invariance also decreases the discrimination considerably. For the rgb and

$c_1c_2c_3$ parameters 32 bins/parameter are better than 16 bins, but 64 bins does not improve anymore. In summary best retrieval results are obtained with rgb and $c_1c_2c_3$ parameters for which 32 bins per parameter are required.

5.2 Landmark detection

To compare the four different measures for landmark detection (section 3), two criteria had to be met.
1) Detection ability: the method should be able to find the characteristic areas in the image
2) Background suppression: the background should get a low score, so that characteristic areas could be well distinguished.

The method was tested on video image sequences from an office scene. As color parameters rgb, $c_1c_2c_3$ and l_1,l_2,l_3 were used. The number of bin for a parameter was 32: the optimal value found for the retrieval. In table 2 the results are summarized.

Criterion	Pixel uniqueness	Summed intersection	Threshold intersection	Maximum intersection
1: detection ability	-	+	+	+
2: background suppression	++	+/-	+/-	++

Table 2. Detection results of the 4 uniqueness measures

The pixel based method scores well in the background suppression, but is not able to locate always-usable landmarks. The other methods are capable to find usable landmarks, but only the maximum intersection gives good background suppression, and is clearly preferable.

The difference between the 3 color spaces can be observed in the amount of background suppression, for which the l_1,l_2,l_3 method scores best. For the preferable choice: the maximum intersection, however, the influence of the color space is small.

(a)   (b)
Figure 2 Color image (a) and the detected landmark (b)

5.3 Localization

The average error in the distance to the landmarks was 0.11 m, where the distance between the robot and the landmarks was varied between 1.7m and 5.9 m. The calculated robot position for a typical scene is given in figure 3. The calculated positions are shown relative to the central position 5 by triangles and the real positions by dots.

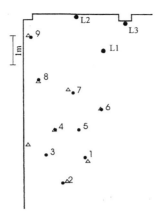

Figure 3. Calculated and real robot positions from landmarks relative to position 5

6   Conclusions

In this paper we have shown that automatic detection of landmarks is possible and can be used to retrieve robot position and orientations. To automatically detect landmarks the best results are obtained with the *maximum intersection* measure. The measure gives good landmarks and the best background suppression. The best retrieval results are obtained with the *rgb* and $c_1c_2c_3$ color parameters with *32* bins/parameter.

References

[1] B.J.A. Kröse, N. Vlassis, R. Bunschoten, and Y. Motomura. A probabilistic model for appearance-based robot localization. *Image and Vision Computing*, 19(6):381-391, 2001.

[2] B. Schiele, J.L. Crowley, Recognition without correspondence using multidimensional receptive field histograms, *Int. Journal of Computer Vision*, 36: (1) 31-50, 2000.

[3] M.J. Swain, D.H. Ballard, Color indexing, *Int. Journal of Computer Vision*, 7(1):11-32,1991.

[4] F. Ennesser & G. Medioni, Finding Waldo, or focus of attention using local color information. *PAMI*, 17, (8):805-809, 1995

[5] B. Funt, G. Finlayson, Color constant indexing. *IEEE Trans. Pattern Anal. Machine Intell.*, 17(5):522-529, 1995

[6] J. M. Geusebroek, R. van den Boomgaard, A. W. M. Smeulders, and H. Geerts. Color invariance. *IEEE Trans. Pattern Anal. Machine Intell.*, 23(12):1338-1350, 2001.

[7] T. Gevers, A.W.M. Smeulders, Color based object recognition, *Pattern Recognition* 32:453-464,1999

Intelligent Autonomous Systems 7
M. Gini et al. (Eds.)
IOS Press, 2002

Biomimetic Actuators –
The Need for and use of 'Soft' Actuators

S. Davis, N.G. Tsagarakis, and Darwin G.Caldwell

Dept. of Electronic Eng. University of Salford, Manchester M5 4WT, UK.
D.G.Caldwell@salford.ac.uk

Abstract

Traditional robot design has been concerned primarily with the development of structures and mechanisms that have high accuracy and speed but at the expense of high mass and power requirements and limited human interaction. Recent advances in computational power however have allowed lightweight and highly flexible structures, similar to those found in biological creations, to be used in robot design. This has led to the development of bio-mimetics were the trend is to try to emulate the 'soft' compliant structure of muscle, bone, tendons and skin and combine this with the power, robustness, accuracy, and endurance of mechanical drives.

This paper will study the use of pneumatic Muscle Actuators (pMAs) as a soft actuators that can macroscopically replicate much of the action of natural muscle. The actuators will be tested in antagonistic pairs and used to show an application of a robot primate (dimensionally comparable with a female gorilla) with a mass of less than 25kg constructed using light flexible. materials. The paper will draw conclusions based on this work and suggest a path for future development

1. INTRODUCTION

Organic systems have long fascinated scientists and engineers with the attributes of the natural system forming goals that are often set as a target for emulation. This is particularly true in robotics where the close similarity of anatomy and function to humans has further enhanced this goal of replicating natural actions. But attempts at functional (and on occasional operational) duplication of humans and animals require developments in mechanical/kinematic design, sensory systems, Artificial Intelligence, computing, control, communications and actuation/power systems [1-7], and this complexity makes their construction a substantial challenge.

Considering the physical structure of a creature its is clear that traditional robot and mechanism design has been concerned primarily with the development of systems that are rigid metal structures with inflexible drives that are highly predictable given the historic limitations in mathematics and control theory. However, it is clear from organic systems that light flexible structures and designs might offer many advantages if they could be used effectively. It is increasingly believed that the development of certain core technologies for the next generation of robots will aim to make extensive use of these biologically inspired paradigms which will result in an evolution from rigid structures to softer more flexible and redundant machines. An area that is increasingly being considered in this regard is the actuation system with suggested goals relating to the development of soft, self repairing actuators that can interact with people in a more user friendly and indeed safer manner.

Within this overall biomimetic robotic challenge although there are many areas which will need to be addressed particularly attention in this paper will focus on:

i) the provision of a soft compliant drive system.

ii) the development of a lightweight mechanically anthropomorphic design based around a gorilla skeletal layout.

This paper will initially study the actuation requirements for a biomimetic replication showing how pMAs (pneumatic Muscle Actuators) can emulate much of the action of natural muscle. The mechanism for integrating and controlling the actions of the joints using the pMAs will be studied. The design of the mechanical structure, making extensive use of composite materials, will then be introduced showing the construction of a low mass entity with dimensions comparable with a adult female gorilla. The paper will finally draw conclusions based on this work and suggest a path for future development.

2. Biomimetic Actuation

Organic muscle is unique in its ability to provide power for motion on land, in the sea and in the air, in climatic conditions varying from desert to polar icecaps and in creatures ranging in size from the whales to microbes. Further, operation as antagonistic pairs provides the ability to modulate stiffness, and position, which is vital for safe interaction activities, gives more natural motion and control, and enables energy conservation through the spring elements of the actuators. Unfortunately on the negative side natural muscle is not an engineering technology, and is prone to fatigue and damage - characteristics that are not suited to machine operation.

The goal of this research and the general bio-mimetic trend in robotics is therefore to try to emulate the 'soft' compliant structure of muscle, bone, tendons and skin and combine this with the power, robustness, accuracy, and endurance of mechanical drives.

A particularly interesting actuation system from the perspective of bio-mimetic systems are pneumatic Muscle Actuators (pMA) derived from McKibben muscles developed in the 1950's [8]. In attempting to use these actuators to duplicate natural muscle it is important to compare the attributes of both muscles, table 1 [13].

Parameter	Biological Muscle	PMA Muscle
Displacement	35%	35%
Force/cm^2	20-40N	100-500N
Power/weight	40-250W/kg	500-2kW/Kg
Efficiency	45-70%	32-50%
Rate of Contraction	25-2000%/s	35-700%/s
Bandwidth		Muscle size dependent. Up to 5Hz
Control	Good	Fair – Good. 1% in both force and displacement
Operation in water	Yes	Yes
Temperature range	0-40°C	-30 - +80°C
Robustness	Excellent	Fair – Good
Self repair-regeneration	Yes	No
Antagonistic Operation	**Yes**	**Yes**
Compliance/Impedance Control	**Yes**	**Yes**
Energy Source	Chemical	Pneumatic
Environmental Safe	Produces CO_2	No biproducts
Scalable from	μm – m	cm –m
Linear Operation	Yes	Yes

Table 1. Comparison of pMA and Natural Muscle

3. Actuation System

The pMA format is a two-layered cylinder in which there is an inner containment liner (often in the form of an elastomeric material), an outer flexible double helix layer of braided material such as nylon, Kevlar, polyester etc and endcaps that seal the open ends of the muscle. The detailed construction, operation, and mathematical analysis of these actuators can be found in [9-12].

This basic structure of the muscles gives the actuator a number of desirable characteristics:

i). This muscle can be made in a range of lengths and diameters with increases in sizes producing increased contractile force.

ii). Actuators have exceptionally high power and force to weight/volume ratios.

iii). The actual achievable displacement (contraction) is dependent on the construction and loading but is typically 30% of the dilated length - this is comparable with the contraction achievable with natural muscle [1].

iv). Being pneumatic in nature the muscles are highly flexible, soft in contact and have excellent safety potential. This gives a soft actuator option which is again comparable with natural muscle.

v). Controllers developed for the muscle systems have shown them to be controllable to an accuracy of better than 1% of displacement. Bandwidths for antagonistic pairs of muscles of up to 5Hz can be achieved. Force control using antagonistic pairs of muscle (compared with muscle action) is also possible [1].

vi). When compared directly with human muscle the contractile force for a given cross-sectional area of actuator can be over 300N/ cm^2 for the PMA compared to 20-40N/cm^2 for natural muscle [8,11].

vii). The actuators can operate safely in aquatic or other liquid environments and are safe in explosive/gaseous states.

viii). Lateral and rotational tolerances mean that accurate alignment is not a problem and rapid low tech. construction is viable with this actuator.

3.1 Actuator Attachments And Joint Torque Control

Within animals and by analogy in a biomimetically inspired robot, all joint motion is achieved by producing appropriate antagonistic torques. In animals this action is transmitted through tendons to the joint while power to the robotic joint is distributed through cables and pulleys driven by the pMAs. This antagonistic scheme requires two actuators for each joint working in opposition to control the position of the joint and thus, effectively provides constraints in the rotation movements. An equivalent pulley model is shown in figure 1.

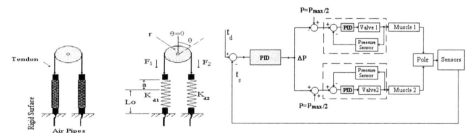

Figure 1. Torque transmission scheme & model.   Figure 2. Joint Torque Control scheme

Each antagonistic scheme includes a high linearity potentiometer for position sensing and a strain gauge torque sensor.

To represent the above antagonistic scheme a model of the pneumatic muscle is required [14]. For this system pneumatic actuators are modelled as pure springs with variable stiffness (compliance) K_{d1}, K_{d2}. Thus, the forces developed by the actuators for an angle θ are given by

$$F_1 = K_{d1} \cdot (a + r \cdot \theta) \qquad (1)$$
$$F_2 = K_{d2} \cdot (a - r \cdot \theta) \qquad (2)$$

where r is the radius of the pulley, θ the rotational angle and a is the dilation of each actuator from the actuator length L_0 when it is in its neutral position ie $\theta = 0°$, figure 1. To achieve the maximum controllable range of motion, a has been set equal to half of the maximum displacement.

In determining the stiffness, it is considered that the stiffness of each actuator consists of two components one a constant to represent the elasticity of the rubber K_e, and one a variable function of the air pressure [11]. This gives the stiffness coefficient as:

$$K_{di} = K_p \cdot P_i + K_e \quad i = 1,2 \tag{3}$$

K_p is the stiffness per unit pressure $K_p = \dfrac{dK}{dP}$ and approximates a constant within the range of the muscle motion as defined in [11]. At any time the torque developed at a joint is given by

$$T = (F_2 - F_1) \cdot r \tag{4}$$

It can be seen from (1), (2), and (4) that torque can be developed even with zero pressure in the two actuators due to the elasticity of the rubber. The closed loop joint torque transmission scheme shown in figure 2 was adopted. Where τ_d is set to be equal to the optimal state feedback. The torque control loop uses the torque error to calculate the amount of pressure change in the two muscles of the antagonistic pair. The command pressure for the muscles at each cycle are given by

$$P_1 = \frac{P_{max}}{2} - \Delta P \quad (5) \quad P_2 = \frac{P_{max}}{2} + \Delta P \quad (6)$$

Where ΔP is computed using a PID control law

$$\Delta P = K_{p1} \cdot e + \frac{1}{T_i} \int e + T_d \cdot \dot{e} \tag{7}$$

and

$$e = \tau_d - \tau_s \tag{8}$$

is the joint torque error, P_{max} is the maximum pressure within the pneumatic muscle actuators. Therefore the torque developed by the muscles becomes:

$$T = 2 \cdot r \cdot a \cdot K_p \cdot \Delta P - 2 \cdot r^2 \cdot (K_p \cdot P_{max} + K_e) \cdot \theta$$

The joint torque data is provided from the integrated strain gauge joint torque sensor.

Control of the actuator requires regulation of the air flow for which any valving system can be used. For operations described in this paper the muscles are energised by 8 port Matrix (758 series) valves operating in a pulse width modulation structure. The Matrix valves have a pulsing frequency of up to 200Hz, with 120Hz selected. Development of an adaptive controller and details of the design can be found in [10]. Although the valves are designed as flow control valves, by incorporating a pressure sensor into the valve inlet, closed loop pressure control is achieved. By varying the duty cycle of the pulse width modulator, regulation of the air flow can be achieved and thus, the amount of muscle contraction can be controlled. The above closed loop joint control scheme compensates not only the actuator shortcomings such as actuator elasticity but also for joint friction etc and enhances the quality of the response of the antagonistic scheme to torque commands.

4.   Biomimetic Design Principles For A Primate Robot

As a test bed for the biomimetic actuation principles outlined above and to further extend the concepts of biomimetic design to the skeletal "bone" structure, a robotic gorilla was constructed with the following goals:

i). Replication of the mechanical structure of a primate to provide a platform on which aspects of bipedal and quadrapedal locomotion could be compared paying particular reference to the dextrous capacity of the system and the ability to manipulate objects and complete handling tasks.

ii). The primate design should produce a mechanical structure composed of links, joints, drives and sensors that is light, flexible, strong, energy efficient and robust. It should have highly complex functionality without the need for high structural and component complexity and precise mechanical tolerances.

It should be noted, as with the actuators, that the aim of the work is replication of functionality of the biological structures, attributes and behaviours and not direct duplication on a cellular or even molecular level.

4.1. Legs

The primate has two legs and two arms which are used in conjunction to produce a quadrupedal gait, each leg consists of 7 degrees of freedom although only three of these are duplicated on the robot.

4.1.1 Ankle
The robot has a one degree of freedom ankle which provides foot up/down (plantar/dorsi-flexion) motion (±30°). The lateral (inversion/eversion) motion is not included to reduce mechanical complexity. The pMAs used to actuate the ankle have a maximum contractile force (at 500kPa) of 2500N allowing the robot to balance when in a bipedal stance.

4.1.2 Knee
The knee has 1 dof with the muscles used to actuate it located in the upper thigh with the antagonistic pair operating across a pulley located at the point of knee rotation by means of a timing belt, this provides motion in the range 0-120° .

4.1.3 Hip
The hip of the gorilla is the most complex joint having three degrees of freedom, however the robot has only one degree of freedom that of flexion-extension. The muscles to actuate the hip joint are located within the centre of a pelvis. Motion is transferred to the leg along a shaft on to which the upper leg a mounted. The muscles used to drive the hip have a maximum contractile force of 2500N and torque is applied to the shaft by a timing belt and pulley. The diameter of the pulley is 30mm which provides the desired range of leg motion (10-120°) and 75Nm of torque which is enough to support the mass of the robot. A gorilla is not capable of straightening its leg and therefore the robot's leg never reaches its singular configuration.

4.2 The Upper Body

The main feature of the upper body is a single rigid spinal column onto which a ribcage is mounted. The ribcage consists of three ribs interconnected by a central sternum forming a protective enclosure which houses the robots control hardware. The bottom of the spine is mounted on a 2 dof universal type joint which in turn attaches to the pelvis thus providing flexibility in both the frontal and sagittal planes (±30°). Motion in the frontal plane is produced by muscles mounted to the sternum and the rear of the spine and the muscles which provide sagittal motion run along both sides of the pelvis.

4.3 The Arms and Hands

Like humans the gorilla has 7 dof arms with dextrous hands providing a total of more than 20 degrees of freedom in each arm.

4.3.1 The Shoulder
The robot has two degrees of freedom in the shoulder which provide flexion/extension (±180°) and abduction-adduction (0-90°). The muscles to produce the flexion/extension motion are attached in a parallel antagonistic pair located along the side of the ribcage. To de-couple the

two motions the muscles to provide abduction-adduction motions are located in the upper arm, this is not anatomically accurate but the functionality is duplicated without the need for complex tendon routing. Again motion is produced using timing belts and pulleys with the maximum torque in each dof being 75Nm.

4.3.2 The Elbow
The elbow is a 1 dof joint powered by a muscle pair located in the upper arm which provides motion in the range 60-180°.

4.3.3 The Hand
The gorilla uses its hand for two tasks walking and object manipulation. When walking the hand is closed in a fist and the first metatarsals of each finger produce a flat paw on which to walk. The robot has an opposal thumb and three fingers, however as the robots primary function is that of walking, to reduce computation the three fingers are linked and so the hand has only one controllable dof. The hand is able to perform a simple grip and has been designed in a way so as to allow the robot to climb a ladder or hang from "monkey bars".

4.4 The Head

The robot is fitted with a head mounted on a 2 degree of freedom pan and tilt neck. The head will provide a mounting point for navigation sensors, stereovision cameras or auditory sensing which may be added to the robot in the future.

4.5 Material Selection

The robot is produced from composite materials with the major limbs being made from lightweight glass reinforced plastics (GRP) and the load bearing sections and interconnecting brackets being formed in aluminium. This resulted in a robot being produced with a mass of only 24kg including control hardware, actuators and valves. The completed robot is shown in figure 3.

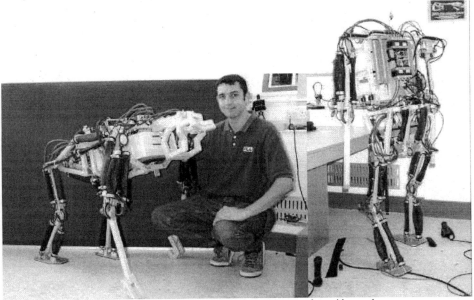

Figure 3 – The final primate skeletal structure complete with muscles.

5. Control Hardware

To control the position and stiffness of each joint the pressure in each antagonistic muscle pair must be controlled as described in section 3. In this instance 3-3 valves are selected to control air flow. Each port is a normally closed three positions valve meaning a muscle can be actively filled, actively vented or in a position where no filling or venting occurs. Provided there are no air leaks in the system this means that the robot can stay in any stable stationary position without using any energy (pneumatic or electric) indefinitely. Pressure sensors are integrated into the end cap of each muscle as it provides a more accurate reading than is obtained if the sensors are located remotely from the muscles. This is due to the Bernoulli effect and pressure loses along a pipe. Integrated sensors also eliminate the need for additional air lines and connectors. In addition to pressure sensors each joint is fitted with a rotary potentiometer to measure its angle.

The main onboard controller for the robot takes the form of a half size ISA card based industrial PC with a Pentium 233 processor with 32MB RAM. The processor card plugs into a backplane which has two additional cards attach to it, which provide 16 single ended and 32 bipolar analogue inputs with 12 bit resolution and 72 lines of digital I/O. The robot also carries a hard drive although this will ultimately be replaced by flash "disk on a chip" memory. The onboard PC runs DOS and all control routines are programmed in C. Figure 4 shows a block diagram of the robot's hardware.

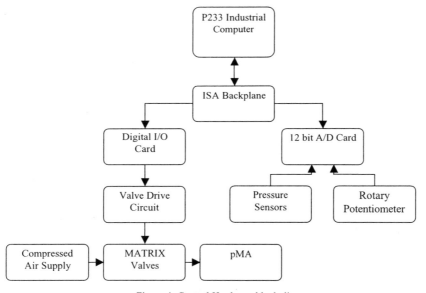

Figure 4- Control Hardware block diagram.

The main justification for using a PC based controller rather than microcontrollers is it provides a multipurpose platform which is well suited to future modification. This means the robot can be used as a test platform in other robotic disciplines such as navigation etc.

The robot is powered from an external power source however ultimately this will be replaced by on board power.

6. The Gait To Be Implemented

Initial experimentation has implemented a wave gait on the robot as it is the most stable of gaits. However with the number of degrees of freedom the robot possesses alternative gaits are also been tested including a three legged gaits which allow one arm to be used to carry objects. The distribution of mass of the robot primate is similar to the biological entity with the larger proportion of the robot's mass being in the upper body, this means the body needs to shift the centre of mass into the support triangle before lifting a leg. The gorilla is able to walk on its hind legs for short distances and the robot has been designed in a manner that it too will be able to stand on two legs, however bipedal walking is not an aim.

7. Conclusion/Further Work

This work has shown how a complex anthropomorphic structure – a primate robot can be constructed using 'soft' biomimetic systems involving compliant actuators and lightweight structures. In respect of the actuators this paper has shown a mechanism for compliance regulation in the control of the joints and how this can be used to regulate the nature of contact and interactions. The total structure has show that biomimetic design presents a striking contrast to conventional designs in terms of robot mass and therefore the safety, power requirements, and potential efficiency.

Future work will further investigate the bio-mimetic functionality of the actuators and how this can be applied to operation of the humanoids/primates and the provision of a 'soft' system that will fulfil the safety and performance requirements needed in machine human interactions. Work will also investigate the use and provision of on-board power generation and storage.

Acknowledgements
The authors wish to acknowledge and thank EPSRC for their support for this work.

8. References
[1] H. Inoue, " Whither Robotics: Key Issues, Approaches and Applications", IROS'96, pp. 9-14, Osaka, Japan, Nov. 1996.
[2] D.E.Whitney, " Historical Perspective and state of the Art in Robot Force Control", Int. J. Robotics Research, Vol. 6, No.1, pp3-14, 1989.
[3] R. Brooks, "Behaviour Based Humanoid Robotics", IROS'96, pp. 1-8, Osaka, Japan, Nov. 1996.
[4] F. Kanehiro, I. Mizuuchi, K.Kayasako, Y. Kakiuchi, M. Inaba, and H. Inoue, 'Development of a Remote Brained Humanoid for Research on Whole Body Action", IEEE Int. Conf. On Robotics and Automation, Leuven, Belgium, pp.1302-07, May 1998.
[5]. T. Marita, K. Shibuya, and S.Sugano, "Design and Control of a Mobile Manipulation System for Human Symbioic Humanoid: Hadaly 2", IEEE Int. Conf. On Robotics and Automation, Leuven, Belgium, pp.1315-20, May 1998.
[6] K.Hirai, M.Hirose, Y,Haikawa, and and T.Takenaka, "The Development of Honda Humanoid Robot", IEEE Int. Conf. On Robotics and Automation, Leuven, Belgium, pp.1321-26, May 1998.
[7] D G.Caldwell, N. Tsagarakis, W.S.Yin and G.A.Medrano-Cerda, "Soft" Actuators - Bio-mimetic Systems for a Bipedal Robot, CLAWAR 98, pp 279-84, Brussels, 26-28 Nov. 1998.
[8]. H.F.Schulte, "The Characteristics of the McKibben Artificial Muscle", In the Application of External Power in Prosthetics and Orthotics, National Academy of Science, NRC, Appendix H, pp94-115, 1961.
[9] D.G. Caldwell, "Natural and Artificial Muscle Elements as Robot Actuators", Mechatronics, Vol. 3, No. 3, pp. 269-283, 1993.
[10] D.G. Caldwell, G.A. Medrano-Cerda, and M.J. Goodwin, "Control of Pneumatic Muscle Actuators", IEEE Control Systems Journal, Vol.15, no.1, pp.40-48,Feb. 1995.
[11] P.Chou and B. Hannaford, "Measurement and Modeling of McKibben Pneumatic Artificial Muscles", IEEE TRANSACTIONS On Robotics and Automation Vol 12, No 1,February 1996.
[12]. N. Tsagarakis and D. G. Caldwell "Improved Modelling and Assessment of pneumatic Muscle Actuators", ICRA 2000, San Francisco, USA May 2000.
[13] H.Gray et al, "The Concise Gray's Anatomy (Wordsworth Collection)", 1997.

Intelligent Autonomous Systems 7
M. Gini et al. (Eds.)
IOS Press, 2002

Automated Laser-based Profiler for Sewer Inspection

Olga Duran, Kaspar Althoefer, Lakmal D. Seneviratne
Department of Mechanical Engineering, King's College, Strand, London WC2R 2LS, UK
Email: olga.duran@kcl.ac.uk

Abstract

Standard sewer inspection techniques are mainly based on CCTV systems. The performance of such systems relies on an off-line and manual recognition of defective areas to be carried out by an engineer. Advances in the automation task become a priority in order to avoid the cost incurred in such a manual process. An automated inspection method to obtain a complete picture of the sewer condition is proposed in this work. The focus of the research is the automated identification and classification of discontinuities in the internal surface of the pipe. The method consists of several processing stages including image segmentation into characteristic geometric features and potential defect regions. Automatic recognition and classification of pipe defects are carried out by means of an artificial neural network (NN). Experiments in a realistic environment have been conducted and results are presented.

1. Introduction

The sewer network is one of the biggest infrastructures of many industrialised countries, reaching 5000 km of sewers per million of persons. For example, the total sewer network of the United Kingdom is estimated to be 250,000 km long [11]. Some sewers are now 100 years old, and at least half the total number of sewers is over 50 years old, and around 20% of them are damaged in some way. According to a study by an UK-based company, there are approximately 5,000 collapses and 200,000 blockages per year in the United Kingdom [10]. Considering that only 0.1 % of the sewer system is replaced every year, these figures are expected to increase by 3% every year. Early detection of pipe defects may prevent severe failures that could involve environmental catastrophes and high economic costs.

Due to the small diameter of sewers, humans can not access and directly inspect a large part of pipes in the municipal sewer system. Remotely controlled inspection devices based on mobile robots or wheeled platforms equipped with sensors are used instead. The standard approach is to have a closed circuit television (CCTV) camera fitted on a mobile platform that travels through the pipe recording images onto a videotape. The recorded images are assessed off-line by an engineer. One of the major drawbacks of this technique is the exorbitant amount of information generated, that can reach about 3 hours of video for the inspection of 1 km of line [8]. Moreover, the time required by the engineer to assess the pipe condition depends on the number of defects. Owing to the variability and the time consumption incurred by the human-based assessment process, automation of this part of the inspection task becomes an important issue.

Some research has been carried out in the field of automated sewer inspection. For instance, the Civil and Environmental Engineering group, at Concordia University (Canada), is developing a model for automating the process of identifying surface defects from digitised video images of underground water and sewer pipes, using image analysis, pattern recognition and neural networks [7]. Another example, is PIRAT (Pipe Inspection Real-Time Assessment Technique), an experimental system developed by CSIRO (Commonwealth Scientific and Industrial Research Organization) and Melbourne Water, Australia. The PIRAT platform measures and analyses the internal geometry of the sewer. Automatic recognition, rating and classification of pipe defects are carried out by means of artificial intelligence software tools. Images are segmented into potential defect regions and characteristic geometric features. Region classification is performed by a feed-forward, off-line trained neural network classifier. A rule-based system interprets, rates and reports the classified regions as pipe defects [2].

Another drawback of current CCTV-based inspection systems is that only non-flooded pipes can be inspected. Considering that in typical conditions, the lowest water level in a sewer is around 5%, other sensors are needed to inspect the flooded part of the pipe. While single techniques have been used for other types of pipe inspection [12], inspecting the two media (liquid and gas) in sewers is more complex than, for example, gas or water pipelines inspection [9]. The investigation reported in this paper is part of a research programme that aims to develop intelligent autonomous agents that can travel inside sewer pipes and inspect them for defect detection. A main research challenge in developing autonomous systems is to create a robust sensor system capable of monitoring the pipe above and below the water line. The objective of this programme is to study and develop multi-sensor systems and to create intelligent sensor fusion and sensor data processing algorithms. It is envisaged that a system based on a laser-profiler and CCD camera measures the surface geometry of the drained part of a sewer, while a sonar scanner measures the flooded part [1]. This paper presents research progress on the inspection and automated condition assessment of the non-flooded part of the pipe, using a structured light source and a ring-pattern projector. The described method makes use of an intelligent classification stage that provides clear identification of defective pipe segments. The research work presented in this paper mainly aims at the identification of holes. Work to classify other faults such as radial cracks, longitudinal cracks and obstacles is still in the experimental stage.

The different processing stages such as data acquisition, image processing and classification are described in the following sections. In Section 2, the transducer that is used to acquire the images is described. The segmentation of the image into regions of interest will be shown in Section 3. The classification step is shown in Section 4. Finally, conclusions are given in Section 5.

2. CCTV-based laser profiler

CCTV images are often difficult to interpret. Spatial reference is needed if faults are to be located, and illumination is not always sufficient to allow fault identification. This is specially the case with small cracks and holes in the pipe surface (fig. 1).

In this work, a laser-based profiler, which can easily be incorporated into existing camera-based systems, is used to cope with these problems. The transducer consists of an optical pattern generator that projects rings of laser light onto the wall, illuminating a pipe section. The optical ring generator is made of an assembly of a laser generator and diffractive diffuser optics. The experimental rig used consists of a 1 mW-semiconductor laser diode of 635 nm wavelength and a diffractive diffuser generating a ring pattern. The projected circular patterns of light are read by a calibrated CCD camera, and stored in a computer via a frame grabber (fig. 2). Tests were conducted on PVC pipe sections with inner diameters of 260 and 300 mm. Holes were pierced into the walls of the pipe segments in order to simulate defects.

Local discontinuities, such as cracks or holes can be detected by analysing the light intensity of the projected rings. At points where discontinuities occur, the laser light is scattered resulting in changes of intensity levels in the acquired camera image (fig. 2). By analysing these intensity levels defects can be detected. Furthermore, as the geometric characteristics of the cone of light are fixed for a given diffuser optic, the location of the surface under investigation relative to the platform is known. Consequently discontinuities can be identified and located.

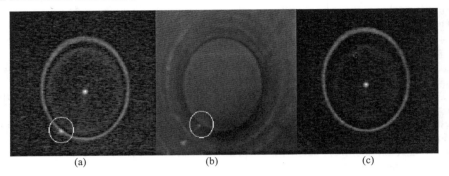

(a)   (b)   (c)

Figure 1: Images of a pipe with a 2 mm hole. The location of the hole is highlighted with a small circle.
a) Image acquired using the laser-based transducer, where the crack can be easily identified.
b) Image acquired with a conventional CCD camera and lighting system. Note that the white spot next to the circle is created by a light reflection; it is not a defect, but could be identified incorrectly as one.
c) Image acquired using the laser-based transducer, corresponding to the location of the non-defective point caused by light reflection as referred to in b). This image shows that the proposed method correctly presents this point as non-defective.

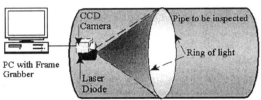

Figure 2: Experimental set-up, where a camera and a laser projector are placed next to each other. The camera images are read into a PC by means of a frame-grabber.

3. Image Segmentation

A sequence of image processing stages is used to identify the location and specific shape of the ring profile and to separate it from the image background. Once a ring profile is properly extracted, a feature extraction algorithm scans the profile. This process is repeated for each of the acquired profiles, instead of searching the entire and potentially very large pipe wall image for defects. This approach has been shown to considerably reduce the amount of data to be processed, and consequently the computation time.

The projection of a circle into the camera image is an ellipse. Also, any misalignment of the projector device with respect to the pipe centre will cause the ring projection to degenerate into an ellipse. Fitting the acquired pixels to an ellipse equation becomes a need. Clustering methods such as Hough transform were tried, but besides some advantages like high robustness to occlusions and lack of pre-segmentation requirements, they are computationally complex and result in multiple solutions. In this work, a highly efficient specific ellipse-fitting algorithm is used. It is preceded by a signal-conditioning and filtering stage, which enhances the quality of the images and facilitates the identification of the ellipse [4].

3.1 Signal conditioning

Prior to applying the conic-fitting algorithm, image quality needs to be enhanced. Low-pass filtering steps are applied, followed by an edge detector [4]. One essential operation to cope with lighting variations is histogram adaptation. In our case, the intensity levels of all the image pixels are adjusted to span the entire available intensity range (0,255). This is an adaptive image processing step where the pixels intensities are mapped in such a way that all input intensities that fall below the average intensity value (background) are mapped to zero intensity, and all intensities higher than the average are "expanded" to span the entire image intensity range. The effect is that the resulting image is considerably brightened, and the dynamic range of the dark portions of the original image is widened, making it much easier to identify the circle of light. A median filter is applied then, so that noise is attenuated and edges are preserved. Finally, a canny edge detector has been found most efficient to pre-segment the image by finding the edges of the ring of light [4].

3.2 Ellipse extraction – conic fitting

A method has been used in this work to efficiently fit the image to a conic. It is based on the ellipse fitting method proposed by Fitzgibbon [3], improved by a method suggested by Halýr [5]. The approach is a non-iterative algorithm based on a least squares minimisation of the algebraic distance between the data points and the ellipse. A quadratic constraint guarantees an ellipse-specific solution even for scattered or noisy data, or for a partial occlusion condition, which can occur when the sewer is partially flooded. Also to improve the behaviour of the algorithm in the presence of noise, a weighted function is applied to the points before feeding them to the above algorithm, The method is based on a the representation of a conic by an implicit second order polynomial:

$$F(\bar{a}, \bar{x}) = \bar{a} \cdot \bar{x} = a \cdot x^2 + b \cdot x \cdot y + c \cdot y^2 + d \cdot x + e \cdot y + f = 0,$$

where $\bar{a} = [a \quad b \quad c \quad d \quad e \quad f]^T$ and $\bar{x} = [x^2 \quad x.y \quad y^2 \quad x \quad y \quad 1]^T$ and $F(\bar{a}, \bar{x})$ is the algebraic distance of a point (x, y) to the conic $F(\bar{a}, \bar{x}) = 0$. The fitting consists of minimising the squares of the distances, $\sum_{i=1}^{N} F(\bar{a}, \bar{x}_i)^2$, subject to the equality constraint, $4ac - b^2 = 1$, incoporating both scaling and ellipse-specific conditions.

3.3. Feature extraction

At that stage, the extraction of the intensity information along the segmented ellipse is completed. In order to achieve this, a partial histogram of the image is created using local averages of intensity along the ring profile. Subsequently, faults can be identified and located analysing successive histograms.

The feature extraction algorithm computes local averages of intensity along a defined number of segments of the ellipse. The local average intensity is computed using the for-

mula: $\dfrac{\sum \mu \cdot x}{\sum \mu}$, where x is the grey level of a certain image point, and μ the

frequency of that grey level. Since the image of the ring of light is wider than one pixel, the average is computed over a sliding window covering areas along individual ellipse segments. Figure 3 shows the results of the feature extraction process displayed as a partial image histogram. In the current experiment, a pipe of 260 mm of diameter is used. A 2mm hole has been pierced at 122 degrees from the horizontal axis. The partial histograms have been computed with steps of one degree along the ellipse segments. Successive histograms are computed from the pipe surface images that are taken while the platform travels along the pipe, to create an image of the pipe wall. Steps of 5 mm in the longitudinal direction were used in this experiment.

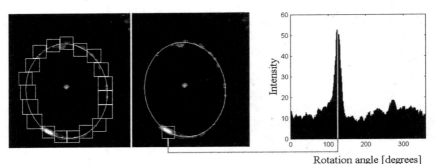

Figure 3: Feature extraction; a sliding window is applied along the ellipse. The peak indicates the location of a potential discontinuity.

4. Classification using a neural network

The aim of the neural network is to identify defective and non-defective pipe sections and to classify different defects in the surface of the pipe. This work focuses on the identification and classification of holes. In further work, the neural network is intended to classify between other discontinuities such as longitudinal cracks, radial cracks, holes and obstacles.

Sets of pre-processed data representing 360-degree profiles are fed into the network. The main aim of the pre-processing stage is to remove measurement noise from the training data, that could otherwise enter the NN algorithm and interfere in the learning process and probably complicate the classification task.

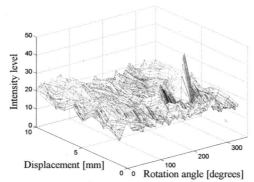

Figure. 4: Open-up representation of the inner surface of the pipe. The peak intensity shows the location of a potential hole/defect.

4.1. Pre-processing

Before feeding the neural network, the raw data is pre-processed to filter out non-relevant data. In our case the goal is the identification of holes in the surface of the pipe. The main interest here is to locate the points where the variation in intensity levels is high, both in the radial and in the displacement direction, regardless of the lighting conditions (fig.4).

The first step consists of emphasising sharp intensity variations (associated to possible holes). The computation in this stage is based on obtaining the partial derivatives $\partial f / \partial d$

and $\partial f / \partial \alpha$, with d being the displacement along the pipe in the longitudinal direction, and α being the rotation angle along a pipe profile. Both partial derivatives at the point where the hole is located present sharp peaks of amplitude (fig. 5). This operation effectively attenuates slow changes in intensity, that are primarily due to simple differences in illumination across the section under study. As a lateral effect, low-amplitude noisy data transitions will be emphasised as well. A low-pass moving average filter is applied to remove this high-frequency noise. Figure 5 shows sample pipe profiles after filtering. It can be noticed that defective sections present sharp peaks of intensity in both partial derivatives. Therefore, the product of them is computed and used in any further stage.

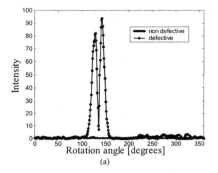 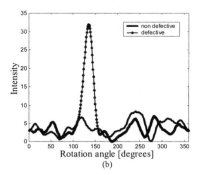

(a)   (b)

Figure 5: Pipe profiles representations after filtering. It can be noticed that defective sections present a sharp peak of intensity in both partial derivatives.
  (a) Partial derivative respect to the rotation angle of non-defective and defective sections
  (b) Partial derivative respect to the displacement of non-defective and defective sections

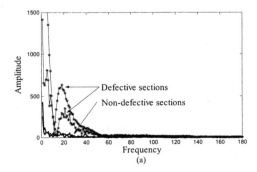 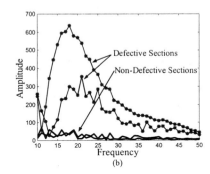

(a)   (b)

Figure 6: FFT of defective and non-defective sections:
  (a) Defective sections with faults at different spatial locations along the profile have similar fft signal.
  (b) Amplification of 6a, range on interest that is fed to the neural network.

Although the resulting data could probably be used to feed the NN (fig. 5), a further step is taken in order to reduce the amount of data supplied to the network. Figure 6 shows defective and non-defective pipe profiles together with their Fourier transforms. The peaks that can be observed in the mid-frequency range of the Fourier domain correspond to intensity variations due to physical defects. The interesting feature is the fact that a neural network can use this FFT data in order to consider defects, regardless of their position in the spatial domain along a certain profile. In other words, by using Fourier data, the NN does not need to be trained with a stream of sections that contain defects at all possible points in the pipe section (fig. 6).

4.2. Network structure

The network used in this work is a multi-layer perceptron (MLP). It consists of three layers composed of neurons. These neurons are arranged in such a way that each of them has a weighted connection coming from every neuron in the previous layer. Each neuron performs a summation of all its inputs and passes the value through a non-linear function before sending it to the output. Choosing the parameters for these operations is done through a training process. The training algorithm used in this work is based on backpropagation, which is widely used for classification problems. The training method is gradient descent with momentum and variable learning rate. The training rule is:

$$\Delta w_{ij}(n+1) = \eta (\delta_{pj} O_{pi}) + \alpha \Delta w_{ji}(n) ,$$

where n is the index for the presentation number, α is the learning rate , and Δ is a constant that determines the effect of the past weight changes [1,6].

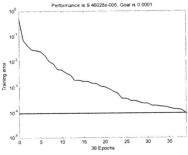

Figure 7: Training error.

Section	Target	Output	Error
Defective	0	-0.0433	0.0433
sections	0	0.0650	0.0650
Non-defective	1	1.0362	0. 0362
sections	1	1.0496	0. 0496

Table 1: Network response to unseen data

4.3. Experimental results

Experiments with different configurations of NN architectures and different number of hidden layer nodes were carried out. The best performance was achieved with a multi-layer perceptron with three layers: One first layer of 50 inputs, the second of 100 hidden nodes and finally one output node as the third layer. The input to the training algorithm was a stream of pre-processed measurement data, as shown in section 4.1, containing both faulty and non-faulty pipe sections. The training behaviour shows that this algorithm is able to classify the training data bringing the error down to 0.0001 after 39 epochs (fig. 7).

The ability of the data to cope with unseen data is also tested. Table 1 summarises the output results of the network for four pipe profiles not seen during the training. The target column specifies the value '1' or '0' corresponding to non-defective and defective sections respectively. The output column represents the output node. The error is the absolute difference between the target and the output values. These results show that the network is able to identify irregularities in the data, related to defective sections with holes, even if the data was not seen by the network beforehand.

5. Conclusions

A new laser-based inspection system for the automatic assessment of sewer conditions has been presented. The method is based on the projection of a laser-generated pattern onto the pipe walls, allowing the location of cracks in the inner surface of pipes by analysing the light intensity of the projected rings. Classification is achieved by means of an MPL NN. The trained network is capable of distinguish between defective and non-defective pipe sections. Tests have been conducted in plastic pipes with different inner diameters. Holes as small as 2 mm have been detected. Future work will focus on the classification of other discontinuities such as longitudinal and radial cracks.

This laser profiler as well as the suggested defect detection method can be easily incorporated into existing CCTV inspection systems, greatly improving their behaviour, especially under harsh lighting conditions. After these encouraging results, the next step will be to test the system in real pipes, in order to confirm its robustness and to improve the experimental set-up and the algorithms to cope with real conditions. Algorithms to retrieve the position of the platform with respect to the pipe are under research, since the platform is not expected to be always in the centre of the pipe. Besides that, the profile images could change, due to possible particles in the gaseous part of the pipe and roots penetrating through the walls of the pipe. To cope with such problems the algorithms and the experimental set-up could need further development to guarantee a clear representation of discontinuities.

The long-term objective of this research is to assess partially flooded pipes using a multi-sensor system. The transducer described in this paper inspects the non-flooded surface, while an ultrasonic sensor will inspect the flooded parts [1]. In order to get a complete image representing the condition of the entire surface of the pipe, sensor fusion can be carried out at the pixel, signal or feature levels. Moreover, the information acquired by the ultrasonic sensor will be used in order to improve the performance of the camera-based method described here, especially in the presence of noise or under conditions where the camera images are partially occluded.

References

[1] Bin Hussen, M. P., Althoefer, K. A., Seneviratne, L. D., Automated Inspection of Sewers Using Ultrasonic Sensors, *Proc. EUREL European Advanced Robotics Systems Masterclass and Conf.*, Vol.1, Salford, UK, 2000.
[2] Campbell, G. Rogers, K. Gilbert, J. Pirat – a system for quantitative sewer assessment, International No Dig'95, Dresden, Germany, 1995.
[3] Fitzgibbon, A. Pilu, M. Fisher R. Direct Least Square Fitting of Ellipses. *Pattern analysis and machine intelligence*, Vol. 21 Issue 5, pp 476-480, 1999.
[4] Gonzalez R.C. Digital Image Processing. Addison-Wesley, MA, 1987.
[5] Halir, R. Flusser J.: Numerically stable direct least squares fitting of ellipses. *The Sixth International Conference in Central Europe on Computer Graphics and Visualization*, Plzeò, pp. 125-132, 1998.
[6] Haykin, S. Neural Networks, A Comprehensive Foundation. Macmillan College Publishing Company, NY, 1994.
[7] Moselhi, O. Shehab-Eldeen, T. Automated detection of surface defects in water and sewer pipes. *Automation in Construction 8*, pp 581-588, 1999.
[8] Pace, NG. Ultrasonic surveying of fully charged sewage pipes, *Electronics and Communications Engineering Journal*, pp 87-92, 1994.
[9] Romero, A. Applications and benefits of using camera technology to internally inspect polyethylene main service piping", *American Gas Association Operations Conference*, Clevehand, Ohio, USA, May 1999.
[10] Roth, H, Schilling, K. Navigation and Control for Pipe Inspection and Repair Robots: *Proc of IFAC World Congress*, 1999.
[11] WEF Manuals & Reports FD-6, ASCE Manuals & reports on engineering, *Existing sewer evaluation and rehabilitation*, No. 62, 1994.
[12] Willke, T. Five technologies expected to change the pipe line industry, *Pipe Line & Gas Industry*, January 1998.

Intelligent Autonomous Systems 7
M. Gini et al. (Eds.)
IOS Press, 2002

Application of Vision in Simultaneous Localization & Mapping

Trevor Fitzgibbons and Eduardo Nebot*
Australian Centre for Field Robotics (* also with CRC CMTE)
Rose Street Building, J04
The University of Sydney, 2006

Abstract

This paper presents the strategies and algorithms for implementing video data into the Simultaneous Localization and Mapping (SLAM) problem, with the emphasis for outdoor applications. The video information is obtained from standard cameras. Natural feature detection algorithms are implemented to obtain relative bearing information. This information is then used in a Simultaneous Localization and map building framework to bound the dead-reckoning errors. Experimental results in an outdoor environment are presented.

1 Introduction

Localization through *a priori* map has been a solved problem for sometime, as has mapping from observations at known positions [1],[2],[3]. More difficult is the combination of localization and mapping, which is known as the Simultaneous Localization & Mapping problem, commonly referred to as SLAM. This infers no *a priori* information is known, and all localization is done as the map is built [4].

The extended Kalman Filter (EKF) can be used to solve a SLAM problem [4],[5],[6], as long as models can be provided for the vehicle's motion and sensors. Increased complexity comes from taking this application into outdoor environments [7], due to the difficulty of extracting and mapping natural landmarks.

The use of vision has been applied to localization and mapping. Extracting structure and motion from video [8] is a currently pursued field, which parallels the efforts of SLAM. The distinction between the two is that SLAM aims to carry its operation in a sequential manner, where 'structure and motion' is performed in batch mode.

The use of visual information for localization has been approached by [9],[10], who used a Monte Carlo filter to localize their position, and both [11],[12], using stereo-vision to aid in applying SLAM.

One way to use video information is by extracting bearing to natural features selected as targets. As such initialization can only be performed with at least two observations of the same landmark. This raises problems on data association and landmark validation. Furthermore, since all pixels are occupied by some feature in the environment, association between a known landmark contained in the map and its appearance in the image can be difficult. These problems are addressed with techniques that enable association using information derived from classical imaging methods and statistical validation.

The paper is structured as follows: Section 2 will provide information on the modelling of cameras and images. Section 3 introduces the SLAM problem and how the extended Kalman filter is applied. Section 4 discusses the selection process employed in obtaining well-conditioned features. Section 5 examines the problems with initializing for bearing-only SLAM. Section 6 looks at data association between landmarks and video images. Section 7 has the presentation of experimental results using the algorithms presented in this paper as used in an outdoor environment. Finally Section 8 presents a conclusion and future paths of this research.

2   Fundamentals of Cameras

The properties of the camera must be first understood for modelling it as a sensor and developing data association techniques. The advantages for using a camera are that provides 3-D information on the environment and delivers a large amount of information in each return.

The data delivered is a 2-dimensional image, formed from ray casting from the object to the camera's focal point and onto a CCD array. Each pixel value is a measure of the light intensity that is returned from the environment. This is made up of the amount of illumination that is incident to the scene and the amount of light reflected from the object itself. These two components are known as the illumination (α) & reflectance components (r), with the light intensity (p) being the product of the two [14]. As such the image model can be described as a function of the illumination and reflectance components;

$$p(u,v) = \alpha(u,v)r(u,v). \qquad (1)$$

The pixel values are then used to identify and associate landmarks as they are viewed.

The camera model works upon the principle that the image is a result of the projection of a point P onto the 'image' plane, typically the capturing CCD array. The perspective origin, O, acts as the origin of the reference frame, XYZ. The image plane lies parallel to the XY-plane at a distance known as the focal length, f, along the Z-axis. The point at which the Z-axis intersects with the image plane is known as the principle point, $[Cu, Cv]$. The position of the object on the image plane, p, is the projection of the pencil from P to the perspective origin, O.

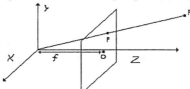

Figure 1:   Camera Reference Frame.

The camera model can then be expressed as

$$\begin{bmatrix} u \\ v \\ 1 \end{bmatrix} = K \begin{bmatrix} X/Z \\ Y/Z \\ 1 \end{bmatrix} \quad where \quad K = \begin{bmatrix} f_u & 0 & C_u \\ 0 & f_v & C_v \\ 0 & 0 & 1 \end{bmatrix} \qquad (2)$$

$$u = f_u \frac{X}{Z} + C_u \quad v = f_v \frac{Y}{Z} + C_v \qquad (3)$$

Where {X,Y,Z} are relative to camera reference frame, $\{f_u, f_v, C_u, C_v\}$ are the intrinsic parameters of the camera, and $\{u,v\}$ are the resulting coordinates of the image.

To develop a sensor model for localisation purposes, a point {X,Y,Z} needs to be converted into the global coordinates.

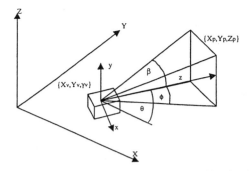

Figure 2:   Camera Model fixed to Vehicle, related to the Global Reference Frame.

For this we assume that the camera is fixed forward on the vehicle, so that the camera's Z-axis points in the direction of ϕ. The vehicle will move in the XY plane (2D), where the landmarks will be 3D. The coordinates can be converted to the bearing it makes to the landmarks, such that;

$$\theta_i = \tan'\left(\frac{u_i - C_u}{f_u}\right) = \tan'\left(\frac{X_i}{Z_i}\right) \qquad \beta_i = \tan'\left(\frac{v_i - C_v}{f_v}\right) = \tan'\left(\frac{Y_i}{Z_i}\right) \tag{4}$$

The same bearings relative to the vehicle are described as

$$\theta_i = \phi_L - \tan'\left(\frac{y_i - y_L}{x_i - x_L}\right) \qquad \beta_i = \tan'\left(\frac{z_i}{\sqrt{(x_i - x_L)^2 + (y_i - y_L)^2} \cos\theta_i}\right) \tag{5}$$

3   Simultaneous Localization and Mapping

The SLAM algorithm [6] addresses the problem of a vehicle with known kinematic, starting at an unknown position and moving through an unknown environment populated with some type of features. The algorithm uses dead reckoning and relative observation to detect features, to estimate the position of the vehicle and to build and maintain a navigation map. With appropriate planing the vehicle will be able to build a relative map of the environment and localize itself. If the initial position is known with respect to a global reference frame or if absolute position information is received during the navigation task then the map can be registered to the global frame. If not the vehicle can still navigate in the local map performing a given task, explore and incorporate new areas to the map.

The system will detect new features at the beginning of the mission and when the vehicle explores new areas. Once these features become reliable and stable they are incorporated into the map becoming part of the state vector.

The state vector is now given by

$$X = \begin{bmatrix} X_L \\ X_I \end{bmatrix}$$

$$X_L = (x_L, y_L, \phi_L)^T \in R^3 \tag{6}$$

$$X_I = (x_1, y_1, ..., x_N, y_N)^T \in R^{2N},$$

where $(x,y,\phi)_L$ and $(x,y)_i$ are the states of the vehicle and features incorporated into the map respectively. Since this environment is consider to be static the dynamic model that includes the new states becomes:

$$X_L(k+1) = f\left(X_L(k)\right) + \gamma$$
$$X_I(k+1) = X_I(k) \tag{7}$$

An efficient method to solve this problem are presented in [7].

4 Feature Selection

The type of feature that is to be used in the map, should be detectable and easily distinguishable within the image. Working with an outdoor environment, the landmarks available usually consist of trees and buildings, which are reasonable to use as they are stationery and not closely grouped.

The corner detection technique used in this work is described in [13]. This involves calculating the 2x2 gradient matrix, G;

$$G = \begin{bmatrix} \Sigma g_x^2 & \Sigma g_x g_y \\ \Sigma g_x g_y & \Sigma g_y^2 \end{bmatrix}, \tag{8}$$

and then taking its eigenvalues. Large eigenvalues will indicate that the point is above the image noise, and if both are within a relative scale, then the point is well-conditioned.

For outdoor navigation, the selection of good features is hampered by non-uniform illumination. As such any comparison of pixel values or their gradients will not only be comparing the structure of the scene but also the illumination, shown in (1) & (9)

$$g_x = r \frac{\partial \alpha}{\partial x} + \alpha \frac{\partial r}{\partial x} \qquad g_y = r \frac{\partial \alpha}{\partial y} + \alpha \frac{\partial r}{\partial y}. \tag{9}$$

The eigenvalues derived from the G matrix cannot be compared to gain sight into the structure of the scene unless the illumination of the scene is known. This has shown that an area with little structural distinctiveness can outweigh those with high structure but shaded.

To address this problem in practical applications, it is assumed that the illumination is constant in the small region surrounding any given pixel. If we let the illumination be a constant value I, for this region, such that

$$\alpha(u,v) = I \tag{10}$$

then the gradient functions are proportional to the gradient of the scene structure (11)

$$g_x = I \frac{\partial r}{\partial x} \qquad g_y = I \frac{\partial r}{\partial y} \tag{11}$$

Direct ratio comparisons can then be done to get a relationship of the scene structure for that region. This was applied as a criteria in eliminating features that were selected due to large values of α, $d\alpha/dx$ & $d\alpha/dy$. The ratio comparison used requires the ratio of the average of a pixel's neighbours over the pixel in question. This provides an intuitive reasoning that if the pixel in question was a corner then its value would be notably higher than its neighbours.

$$C_{u,v} = 1 - \frac{average\left(q_{neighbours}\right)}{q_{u,v}} \tag{12}$$
$$= 1 - \frac{\Sigma q_{i,j}}{8q_{u,v}}.$$

where $q_{i,j}$ is the pixel value.

This ratio however does not carry any information on the strength of the corner, and thus should not be used to compare to other features in the image.

5   SLAM Initialization

Although the selection process will return a well-conditioned feature, it must be verified as being a stable landmark, since a single observation provides no information whether the feature is stationary. Furthermore, since the camera provides only bearing angles to the feature in question, initial estimations of its position will require at least two observations.

In the case where two observations are made of the same landmarks, then the position can be calculated as

$$[x_p, y_p, z_p] = f(x_1, y_1, \gamma_1, u_1, v_1, x_2, y_2, \gamma_2, u_2)$$

let :

$$\theta_i = \gamma_i - \tan^{-1}\left(\frac{u_i - C_u}{f_u}\right) \quad \alpha_i = \gamma_i - \tan^{-1}\left(\frac{y_p - y_i}{x_p - x_i}\right) \tag{13}$$

$$x_p = \frac{-y_1 + y_2 + x_1 \tan\theta_1 - x_2 \tan\theta_2}{\tan\theta_1 - \tan\theta_2}$$

$$y_p = (x_p - x_1)\tan\theta_1 + y_1$$

$$z_p = \left(\frac{v_1 - C_v}{f_v}\right)\left(\sqrt{(x_p - x_1)^2 + (y_p - y_1)^2} \cos\alpha_1\right),$$

where the first observation is $Z_1 = [u_1\ v_1]$ at pose, $X_1 = [x_1\ y_1\ \gamma_1]$, and the second, $Z_2 = [u_2\ v_2]$ at $X_2 = [x_2\ y_2\ \gamma_2]$.

To initialize these landmarks and to verify the choice in landmark, a 3-point verification method is employed.

1. Upon each camera observation, viable features are extracted and stored. The process of searching for matches to known landmarks is run, and any feature that lay within the search windows are eliminated from the list. This is done to prevent selection of landmarks that could be mistaken in later image searches, and to prevent initializing a second instance of an existing landmark in the map.

2. The remaining observations with their poses are added to a list of previous observations. From this list, a landmark position estimate is made for each possible pairing with its associated covariance. Possible pairing of observations are those that:

    a) the bearings converge to a point within both the pose's field of view (i.e. in front of the vehicle)

    b) the poses are far enough to pair well-conditioned observations for the calculation

    c) are from a different feature set (i.e. different time instance).

    d) Their appearance is similar, as found with the application of the correlation test.

3. Then for each pair, a third observation is required to verify that it is a valid landmark. When comparing to other observations, the estimated observation is calculated from the estimated landmark position and the pose of the camera for the compared observations, along with its covariance. Taking the innovation from the actual and estimated observations, a chi-squared test can be performed, where χ^2 of 7.38 is used for a confidence of 95%.

Validated results are added to the map, and covariance used in the re-calculation of the state covariance matrix.

6   Data Association between Landmarks and Images

The data association between a mapped landmark and a detected observation is usually verified using the statistical information in the EKF. The approach taken was to use a Chi-squared test as the bounding function for the coordinates of a search window. An equation by which the Chi-squared value is calculated by, is

$$\chi^2 \geq v^T S^{-1} v, \qquad (14)$$

where v is the innovation, $v = \left[(u_{obs} - u_{est}) \quad (v_{obs} - v_{est}) \right]^T$. Knowing that the observation has 2-degrees-of-freedom, the Chi-squared level for 95% confidence is 7.38.

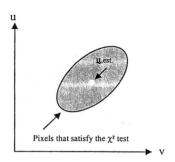

Figure 3:   Innovation Covariance projected onto Image Plane.

The bounding function from the test can be expanded to represent an ellipsoid region in the image defined by

$$7.28 \geq \frac{1}{\det(S)} \left(S_{22} \Delta u^2 - (S_{12} + S_{21}) \Delta u \Delta v + S_{11} \Delta v^2 \right). \quad (15)$$

$$\Delta u = u_i - u_{est} \quad \Delta v = v_i - v_{est}$$

As long as the pixel coordinates lie within this ellipse, it satisfies the chi-squared test and is thus a valid candidate for a match. Using this, the ellipse becomes the bounds for a search window, with each pixel that lies within tested for a match. To determine if a given pixel matches to the landmark, a correlation test is performed on the pixel, comparing it and its neighbours to the last 3x3 image of the landmark.

If the pixel correlation is high enough to satisfy a match, the feature is taken as a possible observation of the landmarks, and is added to a list of individually validated observations. From this list, the set with the highest number of jointly-compatible matchings is obtained by employing the Joint Compatibility Branch & Bound technique [15]. This filters out mismatches made in the initial matchings.

7   Experimental Results

The experimental tests were done on a utility vehicle, modified to carry a sensor suite. The images were captured on an ELMO TSP481 CCD camera, through a Matrox Meteor II video capture board. The dead reckoning data was obtained from a wheel encoder, situated on the rear left wheel, and the steering through an LVDT attached the steering rack. The true path was obtained through a G24 Ashtech GPS set in differential mode with an accompanying base station nearby. Logging was done by a 400MHz Celeron computer.

Two data sets were used in the testing of this algorithm: the first on a hockey field within university grounds and the second from a farm. These were chosen for their variety of landscape and potential landmark features.

Using the criterion described in the previous section for feature selection, a variety of results were obtained. On the hockey field, these features were predominantly building vertices and the corners of windows. Figure 4a shows one such frame from this run. On the farm, trees were more commonly selected, particularly their canopies and branches, as shown in figure 4b.

(a)   (b)

Figure 4 – a) Corners selected in Hockey Field, C >= 0.65. b) Corners selected in Outdoor Environment, C >= 0.65.

Without the ratio comparisons, the majority of selected features were taken along the edges of regions, which did not provide unique point features for the algorithm.

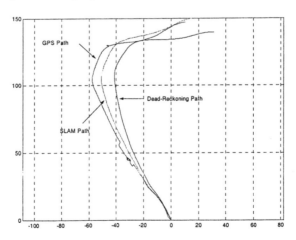

Figure 5:   Vehicle Path and Feature Map.

Figure 5 shows the localization results from the farm test data. Also shown is the DGPS path, accurate up to 2m, as well as the Dead-Reckoning path for comparison. After 200m of travel, the SLAM path diverges by 6m, whereas the dead-reckoning ends up being 22m out. Some of this can be attributed to the use of a 2D vehicle model, where the terrain has notable contours.

The Joint compatibility test showed that of the 1342 individual matches made, only 318 were deemed suitable for updating the system. This was partially due to the rejection of whole sets of observations if there was any ambiguity to their validity, as well as bad initial data association.

8 Conclusions

This work provides a template for the use of cameras in SLAM as bearing only sensors. The concepts covered were: modelling the camera for bearing observations, selection of point features for outdoor environments, initialization of landmarks from bearing observations, and the application of statistical and image processing techniques for data association.

References

[1] Elfes A., "Occupancy Grids: A Probabilistic framework for Robot Perception and Navigation", PhD. Thesis, Department of Electrical Engineering, Carnegie Mellon University, 1989.

[2] Stentz A., Ollis M., Scheding S., Herman H., Fromme C., Pedersen J., Hegardorn T., McCall R., Bares J., Moore R., "Position Measurement for Automated Mining Machinery", *Proc. of the Int. Conferrence on Field and Service Robotics*, August 1999, pp 299-304.

[3] Durrant-Whyte Hugh F., "An Autonomous Guided Vehicle for Cargo Handling Applications". *Int. Journal of Robotics Research*, 15(5): 407-441, 1996.

[4] Leonard J., Durrant-Whyte H., "Simultaneous Map Building and Localization for an Autonoumous Mobile Robot", *Proc. of IEEE Int. Workshop on Intelligent Robots and Systems*, pp 1442-1447, Osaka, Japan, 1991.

[5] Williams SB., Newman P., Dissanayake MWMG, Rosenblatt J., and Durrant-Whyte H., "A Decoupled, Distributed AUV Control Architecture", *31st International Symposium on Robotics 14-17 May 2000, Montreal PQ, Canada*.

[6] Guivant J., Nebot E., Baiker S., "Localization and Map Building Using Laser Range Sensors in Outdoor Applications", *Journal of Robotic Systems*, Volume 17, Issue 10, 2000, pp 565-583.

[7] Guivant J., Nebot E.M., "Optimization of the Simultaneous Localization and Map Building Algorithm for Real Time Implementation", *Proc. of IEEE Transaction of Robotic and Automation*, vol 17, no 3, June 2001 pp 242-257.

[8] Dellaert F., Seitz S., Thorpe C., Thrun S., "Structure from Motion without Correspondence", *Proc. of IEEE Computer Soc. Conf. on Computer Vision and Pattern Recognition (CVPR'00)*, June, 2000

[9] Dellaert F., Burgard W., Fox D., Thrun S., "Using the Condensation Algorithm for Robust, Vision-based Mobile Robot Localization", *Proc. of the IEEE Int. Conf. on Computer Vision and Pattern Recognition*, Fort Collins, CO, 1999.

[10] Fox D., Burgard W., Dellaert F., Thrun S., "Monte Carlo Localization: Efficient Position Estimation for Mobile Robots", *Proc. of the Sixteenth National Conf. on Artificial Intelligence (AAAI'99).*, July, 1999.

[11] Davidson A., Murray D., "Mobile Robot Localization Using Active Vision", *European Conf. on Computer Vision (ECCV)* 1998

[12] Lacroix S., Jung I., Mallet A., "Digital Elevation Map Building from Low Altitude Stereo Imagery", 9th *Symposium on Intelligent Robotic Systems (SIRS)*, Toubouse, July 2001

[13] C. Tomasi & T. Kanade., "Shape and Motion from Image Streams: A Factorization Method - Part 3 : Detection and Tracking of Point Features", Tech. report CMU-CS-91-132, Computer Science Department, Carnegie Mellon University, April, 1991.

[14] R. C. Gonzalez & P. Wintz, *Digital Image Processing: 2ed.*, Addison-Wesley Publishing Company, 1987.

[15] J. Neira & J.D. Tardos, "Data Association in Stochastic Mapping: The fallacy of the Nearest Neighbour", *W4: Mobile Robot Navigation and Mapping workshop at IEEE International Conference on Robotics and Automation*, 2000.

Intelligent Autonomous Systems 7
M. Gini et al. (Eds.)
IOS Press, 2002

Robots in Formation Using Local Information

Jakob Fredslund
Computer Science Dept.
Aarhus University
Aabogade 34
8200 Aarhus N, Denmark
chili@daimi.au.dk

Maja J Mataric
Interaction Lab, Computer Science Dept.
University of Southern California
University Park
Los Angeles, CA 90089, USA
mataric@cs.usc.edu

Abstract. We study the problem of achieving global behavior in a group of robots using only local sensing and interaction, in the context of formations. The goal is to have N mobile robots establish and maintain some predetermined geometric shape. We report results from 40 experiments with physical robots, showing the viability of our approach. The key idea is to keep a single *friend* at a desired angle by panning the camera and centering the friend in the image. We present a general analytical measure for evaluating formations and apply it to the position data obtained from two data gathering lasers tracking the robots during the experiments.

1 Introduction

This paper describes our continuing work [6] on the problem of achieving global behavior in a group of robots using only local sensing, with *formations* as an instance of that general problem. By *local* we mean that the robots do not know the position of other robots, except what they can sense themselves locally; in fact, in our algorithm each robot uses only the relative position of one other robot. The goal is to have N mobile robots establish some predetermined geometric shape, then maintain or re-form that shape, or change to another shape, while negotiating obstacles and experiencing occasional fallouts of group members. We have devised a simple, general, robust, decentralized, behavior-based algorithm that solves the problem for N robots each equipped with sonar, laser, camera, and a radio link for communicating with other robots. We also developed a general set of *global* quantitative criteria for evaluating the formations. In [6] we validated the algorithm largely through simulation; here we present an improved algorithm and results from applying this algorithm *and* the quantitative evaluation criteria to a group of four physical mobile robots.

2 Related Work

A variety of approaches have been proposed to create global behavior in a group of mobile robots. In [10], a robot soccer-playing team is described that has a minimalist behavior-based control system with only a few basic behaviors. From their interaction, two different group formations emerge, enabling seemingly 'willed' offensive and defensive team play. With formations, however, a more rigid and reliable structure is needed from the group of robots. Each robot has to somehow determine its spot relative to the position of its peers. In [2], three ways of doing this are identified: *neighbor-referenced*, where the robot decides its position relative to one predetermined neighbor, *unit-center-referenced*, where the robot references itself to the centroid of all robots, and *leader-referenced*, where the robot uses the position of a predetermined leader. Each robot determines other robots' positions by dead reckoning, GPS, or by direct perception, and its own coordinates in the global coordinate system are broadcast to all robots. Experiments were done with both simulated and real robots, but this high reliance upon a centralized world view and the need to transmit coordinates between robots might have a negative impact on performance, as the paper states.

In the *Leader-Following Control* mode, one of the modes in a general high-level framework for programming multi-robot systems [1], each robot references itself to one neighboring robot, using only locally available information, maintaining a certain angle, ψ, and distance, l, to it. Thus, the needed information is position and orientation of one robot close by and within line of sight. An experiment with physical robots (though only two) is reported, where the follower robot keeps a pre-set heading and distance to the lead robot. The follower uses a camera and color-blob detection to identify the lead robot and its heading and distance. In [3], all robots have a predetermined set of 'attachment sites' spread uniformly around the body, and the formation emerges as the group 'snaps' into shape with robots being 'pulled' towards the nearest attachment site. Depending on the angular offset of the attachment sites, different formations are possible. The approach is validated in simulation; however, since there is no one 'right' spot for each robot due to the symmetrical nature of the attachment sites, several configurations with the same attachment sites are possible, while only a specific one may be desired. Other researchers have studied formations in simulation using more theoretical approaches enabling formal performance analysis, e.g., [4, 5].

There is thus a spectrum of strategies, ranging from simple, purely local ones out of which global formations emerge, to more involved ones relying to varying extent on global knowledge, typically a global coordinate system or knowledge of other robots' positions and headings. The former category is characterized by minimalism and robustness but a lack of any guarantees that the desired formation will actually emerge; the latter category by reliability and efficiency but also a need for global knowledge and computation. In [9], Parker defines what 'global knowledge' could mean: knowledge of 1) global goals, and/or 2) of the actions and intentions of other robots. Within this framework, a robot knowing what formation and with how many robots it is supposed to participate in would be Type 1 global knowledge, whereas its knowing the globally required formation heading or whether another robot is about to evade an obstacle would be global knowledge of Type 2. She illustrates how the addition of global knowledge can improve system performance through formation simulations with four robots.

Of the two works that resemble our own the most, [2] does not demonstrate neighbor-referenced formations with real robots using only local information, and [1] showed leader-following with two robots at a fixed angle. We demonstrate our formations with four real robots that dynamically change the angles they are keeping to their neighbor to switch between formations and adapt the formation if the group size changes.

3 Algorithm

We have in our approach sought simplicity yet reliability through local sensing and minimal communication. Generality is also a primary goal; traditionally, the four formations studied are diamond, column, line (abreast), and wedge, but our algorithm works for almost any geometric shape. Our key idea is this: every robot positions itself relative to one designated neighbor robot, its *friend*, using some appropriate *friend-sensor*. To keep the algorithm for maintaining this position simple and general, the robot *pans* its friend-sensor some number of degrees pertaining to the current formation; thus, maintaining a place in the formation is simply keeping the friend in the center of the sensor's field of view – for all formations. Each robot has a unique ID number that it broadcasts reguarly as a heart-beat message; other robots can detect this ID. From the heart-beats, each robot knows how many robots are participating in the formation (N), and their IDs. One robot is the *conductor*, deciding the heading, and thus not following any friend (the term *conductor* is analogous to *leader* in the literature). All other robots follow a *friend*, and so all robots serve as "local leaders", and all are also followers (except the conductor). The conductor broadcasts a byte designating which formation to do, f, along with its own ID. This is an example of Type 1 global knowledge, as defined in [9]. The conductor does *not* broadcast its heading. Thus, the robots are organized in a *chain of friendships*, which is always kept sorted by ID. Since the conductor defines the formation heading, it should have a clear line of sight, and since it is the fix-point of the formation, it should be as much in the center as possible, so as to minimize congestion during formation switching. Therefore, for the *centered formations* (all except column), the robot with the middle ID (of those currently alive) is the conductor. For the column, the robot with the lowest ID is conductor, leading the formation. Hence, depending on N and f, any robot might serve the conductor duty.

This approach offers several nice implications. First, once the conductor starts moving, the only way for other robots to keep a stable position relative to their friend is by finding the friend's heading. In this way, the conductor 'drags' the whole formation into place just by going its own way. No global heading needs to be agreed upon, it is solved by self-organization. Since any robot can be the conductor, what seems a centralized element really is not. If the conductor fails, or if N otherwise changes, another robot can take over the role. Second, since the algorithm is basically 'keep your friend in the center', a switch between centered formations is easily done by gradually panning the friend-sensor into the appropriate angle (*camangle*); the change in position results automatically.[1] Third, there is no global coordinate system and hence no communication of coordinates. The behavior-based controller consists of three concurrent behaviors and a module holding state data (Figure 1). Each is described in turn.

The state module *WhatDoIKnow*: Here resides all state information: the robot's own ID, the total number of robots N, the table of IDs, *lessThanMe* (see below), the current formation f, and *camangle*. The behaviors manipulate and make use of this information as follows.

The *channelNListener* **behavior:** This behavior receives the heart-beat messages from the other robots and maintains N and the table of IDs in *WhatDoIKnow*. This information is used to calculate *lessThanMe*, the number of live robots with IDs lower than this robot's own ID (needed since IDs need not be consecutive). If N changes, the robot might be promoted to be the conductor. This happens if its *lessThanMe* value becomes equal to $\lfloor N/2 \rfloor$, in which case it is the middle robot.[2] Conversely, it could also be demoted if its *lessThanMe* is no longer $\lfloor N/2 \rfloor$. In this case it looks up a friend in the table of live IDs.

The *channelCListener* **behavior:** This behavior receives formation messages from the conductor and updates the formation variable, f, if necessary. If f changes, the robot might have to pan its camera to a new angle. The correct angle, *camangle*, is calculated from a simple geometric relationship of *lessThanMe* and N. For any N, the robots' respective *camangle*s will result in a formation that is either uniform or incomplete; i.e., attempting a diamond with 5 robots will result in an incomplete 3x3 diamond, not

---

[1]Switching to/from the column are special cases: they involve switching the conductor between the leftmost and middle robots, so the robots between the two find a new friend on the opposite side.

[2]In case of the column, the front robot should be the conductor, and so the robot whose *lessThanMe* is 0 will get the promotion.

an overcrowded 2x2.

Also, *channelCListener* handles messages from other robots that have encountered an obstacle and are evading it. If a robot detects an obstacle in its path (see below), it will swerve to evade that obstacle and send out a warning, a *swerve* message, with its ID and a value indicating the turn angle and direction of the turn. Other robots, not necessarily sensing the obstacle, will react to this message by making a swerve of solidarity of equal turn angle and direction, if the sender is swerving their way (if they themselves sense an obstacle, their own avoidance behavior takes precedence. As the leader is in front, it will have a clear line of sight and thus should always resolve any deadlocks by just going its way, even-

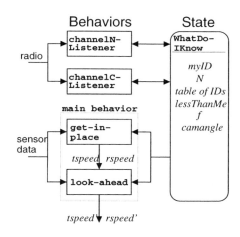

Figure 1: The controller.

tually dragging other robots with it). This principle is documented in the O_1 experiment reported in Table 1 below. Thus, robots share with each other knowledge about upcoming obstacles, categorizing our system as Type 2, following [9].

The main behavior: A robot R moves by setting two parameters, translational and rotational speed (*tspeed* and *rspeed*). In the main behavior, R cycles through a control loop that reads the sensors, sends out its heart-beat message, sets *tspeed* and *rspeed* to their default values (20 mm/sec and 0 deg/sec, respectively), and then passes them to the *get-in-place* sub-behavior.

The *get-in-place* sub-behavior: If R is the conductor, it pans its camera straight and modifies neither *tspeed* nor *rspeed* — unless it has just circumnavigated an obstacle or made a swerve of solidarity. In this case it will gradually modify *rspeed* so as to return to the heading it had before the interruption. This heading is stored in *WhatDoIKnow*. If R is not the conductor, it will first locate its friend; R identifies its friend's ID by N, f, and the table of live IDs. If other robots are swerving its way to evade obstacles, R makes a swerve of solidarity. Otherwise, R will first get close to its friend: it does so by panning its camera straight ahead (yielding a straight path towards the friend) and speeding up. Once appropriately close, R will start panning the camera towards *camangle*, the right angle with respect to the current formation, and make small corrections to *rspeed* and *tspeed* so as to center its friend in the image. These suggested values of *rspeed* and *tspeed* are then given to the *look − ahead* sub-behavior.

The *look-ahead* sub-behavior: A central element of *look-ahead* is the *aheadbuffer*. From *tspeed* and *rspeed*, a bounding box for the resulting movement is calculated and a buffer is added: the width of the robots, `robot-size`, on the sides, and *aheadbuffer* in the front. Then, it is checked if any obstacles are found within this bounding box. If so, a correction is made to *tspeed* and *rspeed* proportional to the proximity of the sensed obstacle. Thus, *aheadbuffer* induces immediate collision avoidance. In addition, if it is set to a high value, it allows R to look far ahead for obstacles, resulting in an elegant, smooth avoidance behavior. However, R cannot keep *aheadbuffer* high if it is not yet in its place in the formation, as it may have to get close to other robots. Consequently, *aheadbuffer* is set *low* (to `robot-size`) if R has not yet been in place in the formation, if R has been out of place for a long time, or if R has other robots in front of it when moving in the formation (derived from N, *lessThanMe* and f). Otherwise, *aheadbuffer* is set to a higher value proportional to `robot-size` and N; a large formation needs more space to negotiate obstacles than a small one. A higher *aheadbuffer* gives more time to react, and more time yields more space. If R is the conductor, *aheadbuffer* is always set high. As a safety measure, another sensor (in our case sonar) can be used for lowest-level collision avoidance. Finally, the *look-ahead* behavior sends the revised $(tspeed, rspeed)$ command to the wheels, and R makes the corresponding movement.

4 Experimental Evaluation

Having reported extensive simulation results in [6], here we focus on data obtained from trials with four physical robots. We used ActivMedia Inc. Pioneer2 DX robots with the SICK

Figure 2: Switching from diamond to line.

LMS200 laser, sonars, and the Sony PTZ camera, running *Player* [7], a server and protocol that connects robots, sensors, and control programs across the network. The camera (with color-blob detection software) functioned as the friend-sensor; each robot wore a customized helmet with two fluorescent color stripes, identifying its ID to other robots. Figure 2 shows overhead images of a switch from diamond to line.

Since the robots share no global coordinate system, and the field of view of the overhead camera was too small, we had to find an observer's way of obtaining position data from our experiments. We used a tracking system, written by Andrew Howard (see robotics.usc.edu/player): two lasers monitored the four robots, detecting special reflective beacons mounted on top of the helmets. To be tracked, a robot had to be visible to at least one of the data gathering lasers, but occlusions occasionally occured, especially when switching between formations. As our lab has six lasers in total, only two were available for data gathering.

We propose the following *formation evaluation criteria* as a means of judging quantitatively the notion of being in formation:

<u>Definition1</u> Given the positions of N mobile robots, an inter-robot distance $d_{desired}$, a desired heading h, and a connected geometric shape $\mathcal{G}$ completely characterizable by a finite set of line segments and the angles between them, the robots are considered to be in formation $\mathcal{G}$ iff:

(1) *uniform dispersion*: $\exists d$, such that $\forall$ pairs of immediate neighbors (R_{i_1}, R_{i_2}) with distance $dist(R_{i_1}, R_{i_2})$, $|d - dist(R_{i_1}, R_{i_2})| < \epsilon_{d_1}$, and $|d - d_{desired}| < \epsilon_{d_1}$,

(2) *shape*: $\exists$ a 'stretch function' f with $f(\mathcal{G}) = \tilde{\mathcal{G}}$, such that $\forall$ angles $\theta \in \mathcal{G}$, $|f(\theta) - \theta| < \epsilon_a$, and such that $\forall$ robots R_i, with distance $dist(R_i, \tilde{\mathcal{G}})$ to $\tilde{\mathcal{G}}$, $dist(R_i, \tilde{\mathcal{G}}) < \epsilon_{d_2}$,

(3) *orientation*: $|f(h) - h| < \epsilon_a$; for small $\epsilon_{d_1}, \epsilon_{d_2}, \epsilon_a > 0$.

Criterion 1 states that the same distance should be kept between all neighboring robots. Criterion 2 states that it should be possible to lay out the desired shape over the position data and perhaps adjust the angles a little, so that all robots are close to this tweaked shape: no angle in the original shape must be stretched more than ϵ_a to make the data points fit. Criterion 3 states that the stretching from criterion 2 must not skew the heading too much. Note that by using the term "immediate neighbor", Definition 1 does not demand completeness of formations; this means that 6 robots can actually form an incomplete diamond. Note also that the measure is global in the sense that N robots are *not* considered to be in line even if the angular offset between neighboring robots is small, if in fact overall they form, say, an arc. I.e., it is not enough that all robots *locally* are keeping more or less the right angle to their friends; criterion 2 ensures a *global* quality of the group.

Our assumption is that robots start out in the right order with respect to the chain of friendships but not necessarily with their respective friends in the visual field. This is reasonable, given that the problem of aggregating robots into such a formation has already been empirically demonstrated [8]. Furthermore, following ideas from [11], it is possible through only local interaction to have N robots form a chain. Once the chain is established, the distribution of monotonic IDs could follow, and the matching of unique color helmets with IDs could be communicated. Our experiments were designed to document stability, robustness, obstacle avoidance, and switching between formations.

To show stability, the robots were first placed close to the desired formation. Using Definition 1, we then recorded the % of time they were in formation after establishing it *for the first time* ($ft2$). To justify this, we also did experiments showing that the robots could actually get into any formation from any initial configuration, as long as they started out in the right order.

Table 1: Experiments with 4 robots, five trials each. Numbers are % of time in formation after $ft2$, using Definition 1. O_1 and O_2 are obstacle avoidance experiments with 2 robots.

Diamond	Wedge	Line	Column	O_1	O_2
74.96	100.00	94.05	96.11	100.00	81.77
99.68	99.87	46.80	100.00	97.89	81.97
88.60	99.56	70.43	98.87	99.82	89.39
99.80	50.32*	34.71	100.00	97.80	21.00
100.00	49.76	64.88	100.00	93.92	26.73†

Table 1 displays a summary of our large body of collected tracking data (in total, we performed more than 40 real robot experiments). Definition 1 was used as evaluation measure with desired inter-robot distance $d_{dist} = 80$ cm, $\epsilon_{d_1} = (0.20 * d_{dist})$, $\epsilon_{d_2} = (0.08 * d_{dist})$, and $\epsilon_a = (0.08 * 2\pi)$. In other words, the dispersion of robots was set fairly loosely, allowing actual inter-robot distances to differ with up to 20% of d_{dist} (see Section 5 for a discussion of real-world influence on precision). Criterion 2 was applied more strictly: all robots had to be at most 8% of d_{dist} from the line segments they belonged to[3]. Angles were allowed to deviate 28.8 degrees. We used a simple line-fitting algorithm to fit data points to straight lines.

The stability of the four basic formations is shown in the four leftmost columns of Table 1. Column and Diamond are indeed very stable. Wedge seems to be either very stable or rather unstable, and the line can hardly be dubbed anything but unstable. These facts stem from two related problems.

Wedge and line share the feature that some robots have to pan their cameras $\pm$ 90 degrees to look for their friend. This orientation proved to be the most difficult to maintain; it is very hard for a robot R to realize if it is ahead of or behind its friend F, if their headings are slightly different. If R thinks it is ahead, it will speed up, possibly

---

[3]For the line, we set $\epsilon_{d_2} = (0.30 * d_{dist})$. Otherwise, at least one robot would often find itself too far away from the best-fit line.

resulting in an unstable, oscillatory course. An example of this (the run marked with a * in Table 1) is shown in Figure 3(b). The other problem is that the laser range is only ±90 degrees: if R is right next to F (when its camera is panned ±90 degrees), F will be at the boundary of R's laser range. If F drops beyond that boundary, R can no longer judge the distance to F accurately, and so again an oscillatory course can result.

In the O_1 experiment of Table 1, two robots formed an incomplete diamond, i.e., the follower, R, kept a 45-degree angle to the conductor, C. They had to negotiate a wall that was only in R's path. By the combined workings of the long *aheadbuffer* and the *swerve messages* sent out by R, C made swerves of solidarity in time to let R keep its position. Hence the robots maintained the formation while still evading the wall, as seen by the high %'es of experiment O_1 in Table 1. Due to lack of space, it was not possible to show global formation obstacle avoidance with more than 2 robots.

In O_2, the wall was in both of their paths, but this time the wall had a passage in the middle, not wide enough to allow the robots to pass it while in formation. As seen by the three still high %'es in Table 1, only for a short time did R lose its position before regaining it. The run marked with † is shown in figure 3(c); here R did not follow C through the passage in the wall, but instead went around it to the left, completely losing sight of C. Generally in this situation, R will attempt to get back to the heading it had when it was last in position, and so here it made a right curve tightly following the wall. Eventually it spotted C in the distance and managed to catch up and get into position for last 100 time steps. The 21%-run also demonstrated a recovery where R re-established the formation before the last 100 time steps.

In Figure 3(a), a switch from diamond to wedge is shown: when given the command to switch, the back robot of the diamond pans its camera to get its friend on its right rather than its left, sliding behind the others and ending up in place (compare with Figure 2). The robots tended to occlude each other during switching, so it was very hard to get

Figure 3: (a) Switching from diamond to wedge. (b) An unstable wedge. (c) Two robots split around an obstacle and then re-join.

tracking data, but by human inspection (not by the authors), the robots were reliably capable of switching between any centered formations, and between line and column. By the heart-beat messages, each robot knows who else is alive. That introduces robustness into our system, so that if a robot fails, the formation will adapt, and if necessary, a new robot will become the conductor. Since this works extremely reliably, both due to reliable (minimal) radio communication and reliable color helmet detection, we do not report robustness experiments here (see [6] for simulation data).

5   Conclusions

Our robots use only local sensing (i.e., they each know only the (relative) position of *one* other robot), and through simple communication they know the global goal: to do formation f with N robots. Also, robots inform each other of upcoming obstacles. Hence, our system is of Type 2, according to [9]. Our key concept is to follow a designated 'friend' robot at the appropriate angle and distance (similarly to [1, 2]), by using a panning camera, and thus simply keeping the friend centered in the image. This also enables easy switching between formations. Unique IDs and a protocol for minimalist radio communication provide robustness to drop-outs and help negotiate obstacles. A conductor that leads the way solves the problem of determining the friend's heading; by the nature of the algorithm, the only stable configuration is when all robots eventually have the same heading as the conductor.

Our method proved highly succesful for certain formations (diamond, column, and, to some extent, wedge), but our expectations of problems with the line were also confirmed. We have since, however, improved the algorithm and almost eliminated the oscillatory behavior described above. We validated the algorithm through 40+ experiments with physical robots[4]. Many real-world issues affect the efficiency of the algorithm. E.g., since the physical shape of the robots is not completely symmetric, when a robot perceived a distance of n millimeters to its friend, its center was usually not at distance n from its friend's center. Further, the extent of this error varied with the angle between the robots. Finally, however well our colored helmets worked, they were not perfect: depending on ambient light and perception angle, the center of their color stripes was not always aligned with the camera, thus introducing a small error in the angular positioning of a follower robot to its friend.

We believe our results show that having a working multi-robot system in simulation does not necessarily prove that the algorithm will work with real robots; many and varied experiments with several robots should be performed.

References

[1] R. Alur, A. Das, J. Esposito, R. Fierro, G. Grudic, Y. Hur, V. Kumar, I. Lee, J. Ostrowski, G. Pappas, B. Southall, J. Spletzer, C. J. Taylor, *A Framework and Architecture for Multirobot Coordination*, International Symposium on Experimental Robotics (ISER 2000), Hawaii, December 10-13, 2000

[2] Tucker Balch, Ronald C. Arkin, *Behavior-based Formation Control for Multi-robot Teams*, IEEE Transactions on Robotics and Automation, Vol. 14, No. 6, pp. 926–939, Dec. 1998.

[3] Tucker Balch, Maria Hybinette, *Social Potentials for Scalable Multirobot Formations*, IEEE Intl. Conf. on Robotics and Automation, Vol. 1, pp. 73–80, San Francisco, 2000.

[4] Qin Chen, J. Y. S. Luh, *Coordination and Control of a Group of Small Mobile Robots*, IEEE Intl. Conf. on Robotics and Automation, pp. 2315–2320, San Diego, 1994.

[5] Jaydev P. Desai, Vijay Kumar, James P. Ostrowski, *Control of Changes in Formation for a Team of Mobile Robots*, IEEE Intl. Conf. on Robotics and Automation, pp. 1556–1561, Detroit, May 1999.

[6] Jakob Fredslund, Maja J Matarić, *Robot Formations Using Only Local Sensing and Control*, IEEE Intl. Symp. on Computational Intelligence in Robotics and Automation, pp. 308–313. Banff, Alberta, Canada, July 29 - August 1, 2001.

[7] Brian P. Gerkey, Richard T. Vaughan, Kasper Støy, Andrew Howard, Gaurav S. Sukhatme, Maja J Matarić, *Most Valuable Player: A Robot Device Server for Distributed Control*, To appear: Proc. IEEE/RSJ Intl. Conf. on Intelligent Robots and Systems, Wailea, Hawaii, October 2001.

[8] Maja J Matarić, *Designing and Understanding Adaptive Group Behavior*, Adaptive Behavior, vol. 4, 1:50–81, MIT December 1995.

[9] Lynne E. Parker, *Designing Control Laws for Cooperative Agent Teams*, IEEE Intl. Conf. on Robotics and Automation, pp. 582–587, Atlanta, May, 1993.

[10] Barry Brian Werger, *Cooperation Without Deliberation: A Minimal Behavior-based Approach to Multi-robot Teams*, Artificial Intelligence 110 (1999), pp. 293–320.

[11] Barry Brian Werger, Maja J Matarić, *Robotic 'Food' Chains: Externalization of State and Program for Minimal-Agent Foraging*, From Animals to Animats 4: Proc. Fourth Intl. Conf. on Simulation of Adaptive Behavior, eds. Maes, Matarić, Meyer, Pollack, and Wilson, pp. 625–634, Cape Cod, MIT September 1996.

---

[4]See http://robotics.usc.edu/~agents/projects/formations.html for video footage.

Intelligent Autonomous Systems 7
M. Gini et al. (Eds.)
IOS Press, 2002

Emergence of Bipedal Locomotion Using a Polymorphic CPG Circuit

Akinobu Fujii
Dept. of Computational Science and Engineering, Nagoya University
Nagoya 464-8603, Japan
akinobu@cmplx.cse.nagoya-u.ac.jp
Akio Ishiguro
Dept. of Computational Science and Engineering, Nagoya University
Nagoya 464-8603, Japan
ishiguro@cse.nagoya-u.ac.jp
Peter Eggenberger
Artificial Intelligence Laboratory
University of Zurich
CH-8057 Zurich, Switzerland
eggen@ifi.unizh.ch

Abstract. In order to create agile locomotion for legged robots, so far various methods have been proposed with the concept of neural circuits, or so-called Central Pattern Generator (CPG). In contrast to these approaches in this article a polymorphic neural circuit is used instead, allowing the dynamic change of its properties according to the current situation in real time. To this end, the concept of neuromodulation is introduced. To verify the feasibility of this approach, this concept is applied to the control of a 3-D biped robot which is intrinsically unstable. Some simulation results and significant points derived from the obtained results are provided in this article.

1 Introduction

Legged robots show significant advantages over wheeled robots since they can traverse uneven and unstructured environments. However, due to the complicated interaction dynamics between the robots and their environments, it is in general extremely difficult to design controllers for legged robots.

On the other hand, natural agents can show adaptive and agile locomotion even under unstructured environments. Investigations in neurophysiology suggest that such remarkable locomotion (e.g. walking, swimming, flying) is generated by specific neural circuits, or so called *Central Pattern Generator* (CPG). Based on these findings, so far various methods have been proposed for legged-robots with artificial CPG controllers consisting of a set of *neural oscillators* [8, 6, 14].

In contrast to these approaches in which *monolithic* CPG neural circuits are used to control locomotion, we use a *polymorphic* neural circuit instead, allowing dynamic change of its properties according to the current situation in real time. This is because what we are intrigued with is not to simply generate stable locomotion patterns but to investigate how sensory information modulates locomotion patterns according to the current situation. To this aim, we introduce the concept of *neuromodulation* with a *diffusion-reaction mechanism* of chemical substances called *neuromodulators*. As there is no current existing theory about how such dynamic neural networks can be created, the evolutionary approach is the method of choice to explore the interaction among the neuromodulators, receptors, synapses and neurons.

Here as the initial step of the investigation, this article attempts to create neural controllers with a neuromodulation mechanism for 3-D biped robot in sagittal terrain. Simulations were carried out in order to verify the feasability of our proposed method. This article also provides some significant points derived from the obtained results. It is expected that these points will be very crucial particularly in the case of evolving controllers for biped robots which are intrinsically unstable.

2   Related Works

Concerning the evolutionary creation of controllers for legged–robots, so far various methods have been proposed. Beer et al. evolved dynamically recurrent neural networks for a hexapod robot[1]. Jacobi introduced the concept of *minimal simulation* to not only speed up the evolutionary process but also to bridge the gap between simulated and real environments. He applied this concept to evolve a controller for an octopod robot[7]. Gruau et al. proposed the *Cellular Encoding scheme* based on Genetic Programming, and implemented it to evolve a hexapod controller[5]. Kodjabachian proposed a *geometry oriented encoding scheme* called SGOCE and evolved a neurocontroller for a legged robot[9]. Ijspeert evolved a controller to mimic a salamanders' locomotion and the resultant controller could smoothly switch between swimming and walking[6].

There are mainly two points to be noted from the above–mentioned works. First, most of these methods are based on *monolithic* neural networks, that is the properties of the controllers such as synaptic weights are fixed once acceptable controllers for the given task are evolved. Second, as mentioned above, so far various methods have been proposed for quadruped, hexapod and octopod robots, whilst very few have investigated the evolutionary creation of controllers for biped robots in spite of their remarkable mobility. This is presumably due to their high instability.

3   Dynamically–Rearranging Neural Networks

3.1   Lessons from the biological findings

Interestingly, neuroscientific results suggest that biological networks not only adjust the synaptic weights but also the neural structure by blocking or activating synapses or neurons by the use of signaling molecules, or so–called neuromodulators. These findings stem from investigations made with lobsters' stomatogastric nervous systems in which certain active neurons diffuse neuromodulators which then rearrange the networks[12]. Note that the effect of a neuromodulator depends not only on these substances, but also on the specific receptors, which are expressed differently in different cells. The effect on a cell depends therefore on the interaction between the neuromodulator and the receptor and not just the neuromodulator alone.

3.2   Basic concept of Dynamically-Rearranging Neural Networks

The basic concept of our proposed dynamically–rearranging neural networks (hereafter: DRNN) is schematically depicted in Fig.1. As in the figure, unlike conventional neural networks, it is assumed that each neuron can potentially diffuse its specific (i.e. genetically–determined) types of NMs according to its activity. Moreover, each synapse has receptors for the diffused NMs. We also assume that each synapse independently interprets the received NMs, and changes its properties (e.g. synaptic weight). The way these changes are exerted on the synapses is also genetically–determined. By selecting cyclical interaction between the diffusion and reaction of NMs, we expect to be able to evolve adaptive neural networks, which show not only a seamless transfer from simulations to the real world but also robustness against environmental perturbations. It should be stressed that mechanisms for *situation evaluation, behavior generation* and *behavior regulation* can all be embedded within a monolithic neural network. Due to this remarkable feature, we expect that whole genetic information can be reduced rather than straightforward approaches.

In summary, in contrast to conventional Evolutionary Robotics approaches that mainly evolve synaptic weights and neuron's bias of neuro–controllers, in this approach the following mechanisms are evolved instead:

– Diffusion of NMs (i.e. when, and which types of NMs are diffused from each neuron?)

– Reaction to NMs (i.e. how do the receptors on each synapse interpret the received NMs, and modify the corresponding synaptic property?)

To determine the above parameters, a Genetic Algorithm (GA) is used. For detailed explanation of the DRNN, see [3, 4].

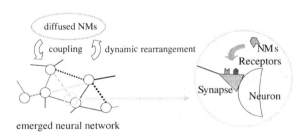

Figure 1: Basic concept of the DRNN.

4   Proposed Method

4.1   Task

Our final goal is to create adaptive controllers for biped robots that can appropriately cope with various situations, and to show the feasibility of the neuromodulation mechanism as well. Here as the initial step of the investigation, the following task is taken as a practical example: walking in sagittal terrain in a 3-D simulated environment.

4.2 Biped robot model

The 3–D biped robot model used in this study is illustrated in Fig.2. As in the figure this robot is composed of seven rigid links (i.e. torso, body, thigh, shank, foot) and seven joints. These joints are all independently driven by pairs of antagonistic actuators, *flexor* and *extensor* in order to take not only static torque but also the stiffness of the joints (for energy efficiency) into account. The robot has an angle sensor at hip joints which informs how the joint concerned rotates. In addition, an incline sensor which detects how the robot inclines exists in sagittal terrain.

The following simulations are implemented with the use of a simulation package provided from *MathEngine*[1] , which calculates the contacts between the rigid body and the ground and the body's dynamics as well within an acceptable time. The body parameters used in the following simulations are listed in Table 1.

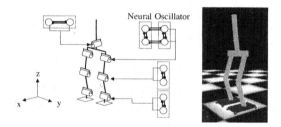

Figure 2: 3–D biped robot model used in the simulation.

Table 1: Body parameters of the biped robot.

part	size[m]	mass[kg]
torso	0.56	2.0
body	0.1×0.36($d \times w$)	1.0
thigh	0.44	1.0
shank	0.47	1.0
foot	0.31×0.21($d \times w$)	1.0

4.3 DRNN controller

Fig.3 schematically represents the structure of the DRNN controller for the biped robot. An oscillator–based neural controller for this purpose inspired by Taga's model[14] is used in this study. This controller consists of a set of neural oscillators each of which controls its specific joint.

4.3.1 Neural oscillator

In this study, a neural oscillator consisting of two neurons proposed by Matsuoka[10] is implemented. The dynamics of each neuron in this model is governed by the following differential equations:

$$\tau_r \frac{du_i}{dt} = -u_i - \sum_{j=1}^{n} w_{ij} y_j + s_i - b f_i \qquad (1)$$

---

[1]http://www.mathengine.com

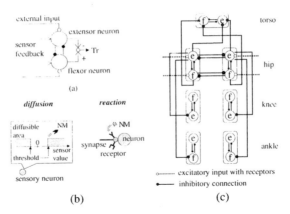

Figure 3: Controller for the biped robot: (a) neural oscillator; (b) neuromodulation mechanism; and (c) CPG model for the biped robot.

$$\tau_a \frac{df_i}{dt} = -f_i + y_i \qquad (2)$$

$$y_i(u) = max\{0, u\} \qquad (3)$$

where, u_i is the membrane potential of neuron i. f is a variable representing the degree of the self–inhibition effect of the neuron concerned. w_{ij} represents the synaptic weight of a connection from neuron j to neuron i. y is the output of the neuron. τ_r and τ_a denote the time constant. s is an external input (e.g. sensory feedback signal).

4.3.2   Neuromodulation Mechanism

In this study, for simplicity, only an incline sensory neuron is assumed to be able to diffuse its specific (i.e. genetically–determined) type of NM as long as the neuronal activity is within the diffusible area which is also genetically–determined (see Fig.3). On the other hand, as to the reaction mechanism receptors for the diffused NMs are assumed to be merely on the synapses of the external inputs in the hip oscillators. Due to this situation, the external inputs are dynamically modulated according to the concentration of the diffused NM. Furthermore, the number of NM types is set at four.

4.4   Incremental Evolution

In order to efficiently create adaptive controllers, an *incremental evolution scheme* is adopted. Here, in the first stage of the evolutionary process, each individual are tested on flat terrain, and the synaptic weights of the CPG circuit are the targets to be evolved. These synaptic weights evolved in this stage are used as default weights. Once acceptable a set of the synaptic weights is obtained, we move onto the second stage of the evolutionary process. Here the target to be evolved is a neuromodulation mechanism which modulates the previously–obtained default synaptic weights. In this stage, the incline sensor which can diffuse NMs is added to the neural controller in order to modulate the structure of CPG controller according to the current situation.

In order to create adaptive controllers, each individual is tested under a set of different environments, uphill and downhill. The evaluation criterion used for each evolutionary process is the resultant distance traveled during the evaluation period (30 seconds). During the evaluation period the trial will be forcibly terminated in case the robot falls down.

5 Results and Discussion

5.1 Walking on flat terrain

Fig.4 shows the locomotion of the best evolved agent obtained in the first stage. As the figure indicates, the robot can successfully walk on the flat terrain. However we observed that this evolved controller can not cope with unexperienced situations, uphill and downhill. This implies that the *pure* CPG neural circuit can not provide high adaptability.

Figure 4: Obtained locomotion through the 1st stage of evolution.

5.2 Walking on a slope

In order to investigate the adaptability of the evolved neuromodulation mechanism, we put the best evolved robot obtained in the second stage in the environments shown in Fig.5. Remember the environments experienced in the second stage are pure uphill and downhill slopes. Thus the robot has never experienced these environments through its evolutionary process. Interestingly, we observed that the robot changes the walking speed and step width as well to cope with these environments.

In order to visualize the adaptation mechanism, the transition of the synaptic weights and the torques applied to the hip joints during this walking are shown in Fig.6. As in the figure, the generated torques and the synaptic weights are dynamically modified according to the gradient of the terrain.

Figure 5: Test environments never-experienced in the evolutionary process.

5.3 Discussion

This section provides some significant points to be noted that are derived from the results mentioned above.

The first point is concerning the sensor information for CPG controllers. In the conventional CPG approaches, rhythmic sensory information (e.g. angle sensor) is indispensable in order to create oscillatory behavior. In other words, non–rhythmic sensory information is not appropriate for CPG controllers. However the diffusion–reaction mechanism of NMs can translate non–rhythmic sensory information to appropriate information for the CPG controllers (i.e. modulation of synaptic weights). This is a remarkable aspect of the concept of the neuromodulation mechanism for CPG models.

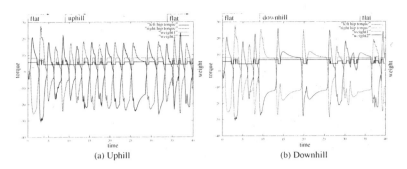

(a) Uphill   (b) Downhill

Figure 6: Transition of the torques outputs applied hip joints and synaptic weights.

The second point to be noted is concerning the embodiment issue. Obviously one way for creating natural gaits is to ameliorate the fitness function by taking not only the resultant distance the robot traveled but also the consumed energy during the locomotion into account. Another way, which is crucially important, is closely related to the embodiment issue, or so-called *passivity*. Torsten effectively showed that bipedal robots with passive joints (in his simulation, there exists no actuation at the knee joints) provide high *evolvability* compared to the embodiments without passivity[13]. Although his main stress is to increase evolvability, passivity is expected to contribute to improve energy consumption as well. Therefore, evolving controllers with the use of the embodiment which can potentially realize *passive dynamic walking*[11] would be a promising way toward creating natural gaits with low energy consumption as well as high evolvability. Recently, Steven et al. developed a wonderful 3-D passive dynamic walker[2]. Based on this consideration, we are currently constructing a simulated 3-D passive dynamic walker and trying to evolve DRNN controllers with this embodiment.

6   Conclusion and Future works

In this paper, we have investigated an evolutionary creation of an adaptive neuro-controller for a biped robot. To this end the concept of the neuromodulation widely observed in biological nervous systems has been introduced to legged robot controllers.

Currently we are also trying to create more natural gaits based on the embodiment that we potentially achieve passive dynamic walking. Furthermore, in order to persuasively show adaptability, we are going to investigate whether the same evolved DRNN controller can be seamlessly applied to different embodiments (e.g. changing body parameters such as mass, length of the leg etc.).

Acknowledgements

The authors would like to thank Auke Ijspeert at the University of Southern California, Josh Bongard and Chandana Paul at the AI Lab of the University of Zurich for their many helpful suggestions for the simulator MathEngine. This research was supported in part by a grant from the Japanese Ministry of Education, Culture, Sports, Science and Technology (No.C:12650142) and Research Fellowship of the Japan Society for the Promotion of Science for Young Scientists.

References

[1] R. D. Beer and J. C. Gallagher. Evolving dynamical neural networks for adaptive behavior. *Adaptive Behavior, vol.1, no.1, 91–122*, 1992.

[2] S. H. Collins, M. Wisse, and A. Ruina. A 3-d passive–dynamic walking robot with two legs and knees. *The International Journal of Robotics Research*, 2001.

[3] P. Eggenberger, A. Ishiguro, S. Tokura, T. Kondo, T. Kawashima, and T. Aoki. Toward seamless transfer from simulated to real worlds: A dynamically–rearranging neural network approach. *Proc. of the Eighth European Workshop on Learning Robot (EWLR-8), 4–13*, 1999.

[4] A. Fujii, A. Ishiguro, T. Aoki, and Petrer Eggenberger. Evolving bipedal locomotion with a dynamically-rearranging neural network. *Proc. 6th European Conference on Artificial Life(ECAL2001),509-518*, 2001.

[5] F. Gruau and K. Quatramaran. Cellular encoding for interactive evolutionary robotics. *Proc.of the Fourth European Conference on Artificial Life, ECAL97, MIT Press, 368–377*, 1997.

[6] A. J. Ijspeert. A neuromechanical investigaiton of salamander locomotion. *Proc. of International Symposium on Adaptive Motion of Animals and Machines,AMAM2000, CD–ROM*, 2000.

[7] N. Jakobi. Running across the reality gap:octppod locomotion evolved in a minimal simulation. *Proc. of Evorob98, Springer Verlag, 39–58*, 1998.

[8] H. Kimura and Y. Fukuoka. Biologically inspired dynamic walking of a quadruped robot on irregular terrain – adaptation at spinal cord and brain stem –. *Proc. of International Symposium on Adaptive Motion of Animals and Machines, AMAM2000, CD–ROM*, 2000.

[9] J. Kodjabachian and J.-A. Meyer. Evolution and development of neural networks controlling locomotion, gradient-following, and obstacle-avoidance in artificial insects. *IEEE Transactions on Neural Networks, 796–812*, 1998.

[10] K. Matsuoka. Mechanisms of frequency and pattern control in the neural rhythm generators. *Biological Cybernetics, 56, 345–353*, 1987.

[11] T. McGeer. Passive dynamic walking. *The International Journal of Robotics Research,9(2),62–82*, 1990.

[12] P. Meyrand, J. Simmers, and M. Moulins. Construction of a pattern–generating circuit with neurons of different networks. *NATURE, 351-2MAY, 60–63*, 1991.

[13] T. Reil and C. Massey. Facilitating controller evolution in morpho-functional-machines-a bipedal case study. *Proc. of International Workshop on Morpho-functional Machines,165–181*, 2001.

[14] G. Taga, Y. Yamaguchi, and H. Shimizu. Self-organized control of bipedal locomotion by neural oscillators in unpredictable environment. *Biological Cybernetics, 65, 147–159*, 1991.

Intelligent Autonomous Systems 7
M. Gini et al. (Eds.)
IOS Press, 2002

Control of the Pneumatic Muscle Driven Walking Machine Airbug

C. Hillenbrand, J. Albiez, T. Kerscher, K. Berns, R. Dillmann
Forschungszentrum Informatik an der Universität Karlsruhe (FZI)
Interactive Diagnosis and Service Systems
Haid-und-Neu-Str. 10-14, 76131 Karlsruhe, Germany
berns@fzi.de

January 14, 2002

Abstract

Pneumatic muscles as actuator offer several advantages due to their performance weight relation or to the passive compliance of this type of actuators. Passive compliance is an important component for the control of locomotion in rough terrain to absorb the power stroke energy. This paper presents the mechatronics of a six-legged insect-like robot Airbug with pneumatic muscles as actuators. Main focus lies on the control concept of the antagonistic actuators.

1 Motivation for the use of pneumatic muscles

Commonly walking and climbing robots are propelled using electrical motors combined with special gears and simple damper mechanisms. During locomotion of these machines unforeseen ground or obstacle contact of a foot could lead to a strong disturbance of the movement or damage parts of the mechanics.

Even if control concepts with active compliance are implemented it is very hard to cover these impact problems, especially in cases of fast movements. A second motivation for the use of other kind of actuators for walking machines is the possibility to redesign the locomotion apparatus of animals [11]. In both cases soft actuators like artificial muscles seem to solve the actuator problem. In literature there are three types of artificial muscles electrochemical, fluidic [7], and mechanical ones [8]. To build up a light weight machine, fluidic muscles are more powerful compared to the two other classes. Fluidic muscles have the advantage of high forces according to the size and the weight of the actuator, relatively high velocity for the contraction and a high-energy efficiency [4]. In this paper the insect-like robot Airbug is introduced. Main focus of the description is the control of the fluidic muscles to propell the machine.

2   Pneumatic Muscles as actuators for Walking Machines

Since the 60th research try to use pneumatic muscles as actuators for robots. E.g.
Prof. Morecki and his colleagues developed a robot arm actuated by air muscles. This
so-called biomanipulator was controlled by myopotentials of the muscles of a human
operator [9]. Since these days several research groups worldwide use fluidic actuators
for robot constructions. In most cases these muscles are self-made consisting of an
inner rubber liner, an outer containment layer of braided nylon and end caps that seal
the open ends of the muscles (see URL: http://www.shadow.org.uk/).

The fluidic muscles MAS-20 produced by the company FESTO are selected (for
more information see URL:http://www.festo.com/pneumatic/deu/muskel/index.htm and
http://www.festo.com/) as actuators for the Airbug machine. These membrane contraction systems contract under the pressure of 6 bar. Combining a fluidic tight, high load
bearing flexible tube and a strong fabric builds a three-dimensional compound structure. Embedding the fibres completely in the tube minimizes friction (fig. 1). Filling in
fluidic media (here: compressed air) forces the compound structure to deform in radial
direction. At the same time the actuator contracts in axial direction and the needed
tensile force arises. Without weight the actuator can reach a stroke of 25% of his maximum length.

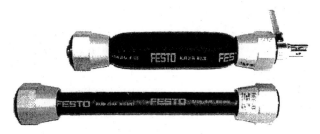

Figure 1: Design of the MAS-20 fluidic muscle

The tensile force depends on the actuator's contraction. The force of the MAS-20
decreases almost linear from its maximum of about 1700 N at its maximum length to
zero when the neutral angle is reached. The actuator's maximum force is only available
under static conditions (0 mm/s), and decreases with increasing velocity. In our case the
average velocity is about 300 mm/s giving us a waste of less than 9% of the actuator's
force.

3   Mechanics system of Airbug

Based on a prototype leg construction [3], which was used to optimise the leg geometry
and to develop first control concepts, the insect-like machine Airbug was designed (see
fig. 2). The leg of Airbug with a weight of 2.5kg is divided in three segments (coxa
94mm, femur 470mm, tibia 600mm) similar to the leg of a stick-insect (Fig.2). The
single legs are fixed on the body in an angle of 30°. For the drive of the leg altogether

eight FESTO MAS20 fluidic-muscles are used, thereof 2 for the control of the α-joint, 4 for the β-joint and two for the γ-joint.

Figure 2: Leg construction of Airbug

The angle range of α-joint is 78°, of the β-joint 94° and of γ-joint 78°. The maximum torque of α-joint and γ-joint is 59.5 Nm and that of the β-joint 98,6 Nm. At the begin of the contraction the max. torque is only available. It goes nearly linear to zero till the max. contraction.

The body is about 650mm in length, 600mm in width and 250mm in height. The final machine has a total weight of about 25 kg including the valves for the muscles and the embedded PC, which is located in the central body of the machine (see fig. 3).

4   Computer architecture and electronical components

The hierarchical modular control architecture of Airbug is similar to that of our walking machines LAURON III [6] and BISAM [2, 5]. A PC/104 system at the top level of the hierarchy performs the main robotic tasks e.g. path planning, calculation of foot trajectories and communication with the environment (Man Machine Interface). At the second level 80C167 microcontrollers are connected with the PC/104 system via CAN-Bus. They are installed in industrial controller boards (Phytec, miniMODUL-167) and execute the basic functions like close-loop joint control (valve control, recording the signals of joint encoders), sensor data acquisition and pre-computation. At the base level each sensor and actuator is connected directly to the microcontroller boards.

The modules of the different levels all have been developed observing the constraints which are important for mobile robots. The circuitry has been optimized for power consumption, less weight and small dimensions. Furthermore they were adapted for an easy integration in the entire system. In detail the control architecture consists of the following dedicated modules:

Figure 3: Central body of Airbug including controller cards for the muscles and a PC/104 system.

Valve driver board The microcontroller board is plugged on a PCB containing drivers for the control of up to 16 digital valves (see fig. 4). Each driver can deliver a maximum current of 1.1 Amps at 24V. A PWM channel of the 80C167 is used to generate a 10 bit PWM signal for the control of a proportional valve.

Pressure sensor The pressure inside the air muscles is measured for the closed-loop control of the joints. The measuring range is $-1 \ldots 10 bar$ with a linearity error of $\pm 0.3\%$ and covers the pressure range of the used air muscles.

Inclinometer To determine the inclination of the main body inclinometers are mounted on Airbug. They deliver information about the main body's inclination in two axis with a resolution of 1 degree and additional information about the acceleration (Analog Devices, Type ADXL202).

Measuring amplifier To amplify very low voltages e.g. produced by strain gauges a dual high precision amplifier was implemented. These amplifiers can be flexibly combined depending on the required measuring channels. Force sensors on the tendons and on the feet have been implemented in the test leg showing practical results.

Joint encoder For the measurement of the joint angles special shaft encoders have been developed. They are based on photo reflectors and provide the absolute position of a joint with a resolution of 0.4 degrees.

Power switch To control the power supply three different power circuits can be switched on or off by logic level signals. The semiconductor switches guarantee a secure soft start of the system and additionally deliver information about the power consumption of each of the three supply branches.

The use of different hardware and operating system units results in a three level system: C167, RT-Linux and Linux. In order to achieve a clear arrangement of those

Figure 4: The valve driver board with the plugged on micro-controller board.

levels the modular control architecture MCA is used [10]. This software architecture allows the programming of all levels in the same way. The C++-class library hides the communication between the levels. Hence, the system developers can focus on the development of methods while the communication is done automatically. Every method of the controller architecture is realised in a C++-class module. The modular architecture allows the manipulation of each parameter of every module via LAN while the system is running. This possibility results in small and fast development cycles.

The detachment of the user interface from the controlling PC disburdens the on board PC/104. For example, the user interface can include high-end graphic animations without straining the controlling mechanism too much.

5   Close loop control

As valves a 5/3 valve of the company Festo with a mechanical spring is used to fill and empty the muscle. This valve is slower than others (12/30 ms) but has a very low weight (50g).

The task to control one muscle-joint consists of the five sub-tasks of controlling the pressure and force of the protagonistic and the antagonistic activator and the joint angle position, which is controlled by the coupled actuator extension. The only variables of the control loop are the airflow into and out of the muscle in a trinary form (three valve states: in-close-out). Sensor input consists of the pressure in each muscle and the joint angle. The force of each muscle can be calculated with equation (1) with pressure and length.

$$F = K_p \cdot P \cdot (l - l_{min}) \tag{1}$$

The length of the muscle can be derived from the joint angle θ. l_{min} is the maximum contraction of the muscle and K_p is a constant for correlating the stiffness of the muscle

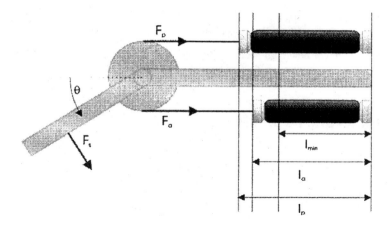

Figure 5: Mechanic structure of a joint

with the pressure. For a given joint-angle q both muscles will pull with the same force. This leads to the following equation (2):

$$F_a \equiv F_p, \quad p_a \cdot (l_a - l_{min}) \equiv p_p \cdot (l_p - l_{min}) \tag{2}$$

This correlation is only true as long as no external forces influence the joint. Since this is not the case in any real environment (2) is expanded in the following way:

$$F_a + F_S \equiv F_p \tag{3}$$

(3) shows, that with this system not only the position of the joint has to be given but also the overall force inside the whole system. In the currently implemented control system, we take not the force as input, but instead a maximum pressure p_s is set. With this knowledge a joint is controlled by the following scheme:

1. For a given θ and p_s the desired pressure for both muscles is set according to (4) and (5):

$$\theta \geq \theta_{med} : p_a = p_s \cdot \left(\frac{\theta_p - \theta}{\theta - \theta_a} \right), p_p = p_s \tag{4}$$

or for

$$\theta < \theta_{med} : p_p = p_s \cdot \left(\frac{\theta - \theta_a}{\theta_p - \theta} \right), p_a = p_s \tag{5}$$

While θ_p and θ_a are the minimum angles of the joint in protagonist or antagonist direction and therefore can be used to calculate $(l - l_{min})$.

2. The actual F_S is calculated with (6) and θ_n as the actual joint angle:

$$F_S = \frac{p_p \cdot (\theta_p - \theta_n) - p_a \cdot (\theta_n - \theta_a)}{(\theta_p - \theta)} \tag{6}$$

3. Recalculate the p_p and p_a according (4), (5) and (6). The pressure inside the muscles is than controlled by a classical PID controller. (see [1]).

At the moment new high speed switching valves (response time 2ms) are tested. Therefore, a test rig was build. Furthermore new control algorithms will be tested, which include an exact muscle model.

6   Experiments and further work

The first experiments on the assembled robot have shown the potential of the mechanical construction. With the described close-loop control it is possible to move the machine slowly in a pentapod, tetrapod and tripod gait. The transfer between the different gaits is always possible during locomotion. Based on an optimized muscle construction the weight of the robot will be reduced by 25%. Then much faster movements (0.5m/s) will be possible. A further study of the pressure-force relation in conjunction with the used pressure sensor will be done to evaluate the possibility of using a force observer for calculating the leg internal torques and forces. It is also planned to include the foot force sensor in control process so that a better adaptation to the ground condition is possible. Concerning the use of air muscles as actuator we have started to build two new robots. First a further six-legged machine which a compressed air tank onboard. With this tank it should be possible to run the machine for more than 20min without externeal power supply. Second a mammal-like machine is under development, for which a biological inspired leg was constructed. Aim of this reseach project is a four-legged robot able to walk in a statical and a dynamical stable way including balistic movements.

Acknowledgements

The reported here has been carried out at the Forschungszentrum Informatik (FZI) at the University of Karlsruhe. We want to thank Mr. Lorenz and Prof. Thalemer both from the company FESTO for the support of the project and for their technical help.

References

[1] K. Berns, J. Albiez, V. Kepplin, and C. Hillenbrand. Control of a six-legged robot using fluidic muscle. In *International Conference on Advanced Robotics*, Budapest, Hungary, August 2001.

[2] K. Berns, W. Ilg, M. Deck, J. Albiez, and R. Dillmann. Mechanical construction and computer architecture of the four-legged walking machine BISAM. *IEEE Transactions on Mechatronics*, 4(1):1–7, Mrz 1999.

[3] K. Berns, V. Kepplin, R. Mller, and M. Schmalenbach. Six-legged robot actuated by fluidic muscles. In *CLAWAR 2000 - Climbing and Walking Robots and the Support Technologies for Mobile Machines*, pages 545–550, october 2-4 2000.

[4] C. Chou and B. Hannaford. Measurement and modeling of mckibben pneumatic artificial muscles. *IEEE Transactions on Robotics and Automation*, 12(1), 1996.

[5] W. Ilg, J. Albiez, K. Berns, and R. Berns. Learning of posture controlmechanisms for the four-legged walking machine bisam. In *CLAWAR 2000*, 2000.

[6] V. Kepplin and K. Berns. Clawar 99: A concept for walking behaviour in rough terrain. In *Climbing and Walking Robots and the Support Technologies for Mobile Machines*, pages 509 – 516, september 13 - 15 1999.

[7] G. K. Klute, J. M. Czerniecki, and B. Hannaford. Pneumatic actuators with biomechanical intelligence. In *Proceedings of the IEEE/ASME International Conference on Advanced Intelligent Mechatronics*, Atlanta, September 1999.

[8] G. Mennitto, P. Gregoire, and M. Buehler. Carl: A compliant articulated robot leg for dynamic locomotion. In *IAS-4, Intelligent Autonomous Systems*, pages 702–707, Karlsruhe, Germany, 1995.

[9] A. Morecki. Polish artificial pneumatic muscles. In *4th International Conference on Climbing and Walking Robots*. Professional Engineering Publishing, september 24 -26 2001.

[10] K.-U. Scholl, V. Kepplin, and K. Berns. Steuerung eines mehrsegmentigen service roboters fuer die inspektion von abwasserkanaelen. In *it+ti Heft 1*, 2000.

[11] H. Witte, R. Hackert, K. Lilje, N. Schilling, D. Voges, G. Klauer, W. Ilg, J. Albiez, A. Seyfarth, D. Germann, M. Hiller, R. Dillmann, and M.S. Fischer. Transfer of biological priciples into the construction of quadruped walking machines. In *Second International Workshop On Robot Motion And Control*, Bukowy Dworek, Poland, 2001.

Intelligent Autonomous Systems 7
M. Gini et al. (Eds.)
IOS Press, 2002

Fusing Ladar and Color Image Information for Mobile Robot Feature Detection and Tracking

Tsai-Hong Hong, Christopher Rasmussen, Tommy Chang, and Michael Shneier

National Institute of Standards and Technology, Gaithersburg, MD
{hongt, crasmuss, tchang, michael.shneier}@nist.gov

Abstract. In an outdoor, off-road mobile robotics environment, it is important to identify objects that can affect the vehicle's ability to traverse its planned path, and to determine their three-dimensional characteristics. In this paper, a combination of three elements is used to accomplish this task. An imaging ladar collects range images of the scene. A color camera, whose position relative to the ladar is known, is used to gather color images. Information extracted from these sensors is used to build a world model, a representation of the current state of the world. The world model is used actively in the sensing to predict what should be visible in each of the sensors during the next imaging cycle. The paper explains how the combined use of these three types of information leads to a robust understanding of the local environment surrounding the robotic vehicle for two important tasks: puddle/pond avoidance and road sign detection. Applications of this approach to road detection are also discussed.

1 Introduction

An autonomous vehicle driving across unknown terrain must be able to detect potential obstacles and identify them accurately enough to determine if they can be traversed. This must be accomplished fairly speedily to ensure that the vehicle has sufficient time and space to avoid obstacles. The work described in this paper is part of the Army's Demo III project [1]. The requirements for the Experimental Unmanned Vehicle (XUV) developed for Demo III include the ability to drive autonomously at speeds of up to 60 kilometers per hour (km/h) on-road, 35 km/h off-road in daylight, and 15 km/h off-road at night or under bad weather conditions. The control system for the vehicle is designed in accordance with the 4D-Real-time Control System (RCS) architecture [2], which divides the system into perception, world modeling, and behavior generation subsystems.

The XUV, shown in Figure 1(a), has two principal sets of sensors for navigation. On the left, outlined in white, is a ladar system that produces range images at about 20 Hz. Mounted above the ladar is a color camera (not pictured) that produces images at up to 30 Hz. On the right are a pair of stereo color cameras, and a set of stereo FLIR cameras. The work described in this paper concerns the use of the ladar sensor and its associated color camera. The way each is used in conjunction with the other and with information stored in the vehicle's internal world model is the focus of the paper.

Given the need for relatively high speed driving, the sensory processing subsystem must be able to update the world model with current information as quickly as possible. It is not practical to process all images completely in the time available, so focusing attention on important regions is required. This is done by trying to predict which regions of future images will contain the most useful information based on the current

(a)   (b)

Figure 1: (a) The Demo III XUV; (b) Butterfly signs (see Section 3.1)

images and the current world model. Prediction is carried out between images, across images, and between the world model and each type of image.

Prediction and focus of attention are of special interest to robotic systems because they frequently have the capability to actively control their sensors [3]. The goal of focusing attention is to reduce the amount of processing necessary to understand an image in the context of a task. Usually, large regions either contain information of no interest for the task, or contain information that is unchanged from a previous view. If regions of interest can be isolated, special and perhaps expensive processing can be applied to them without exceeding the available computing resources.

Most focus of attention systems work by looking for features usually defined by some explicit or implicit model. The search may take many forms, from multi-resolution approaches that emulate human vision's peripheral and foveal vision, to target-recognition methods that use explicit templates for matching [4, 5, 6, 7]. Once a set of attention regions has been detected, a second stage of processing is often used to further process them, or to rank them. This processing may require more complex algorithms, but they are applied only to small regions of the image.

In this paper, we describe an approach to feature detection and tracking that falls within the above general description, but differs from previous approaches in using multiple sensor types that interact to locate and identify features. A world model containing the system's current best guess about the state of the world is used to predict where features should appear and how they should look to each of the sensors. The world model serves as a common coordinate system for comparing and integrating ladar and camera observations, as well as a framework for doing simple tracking of detected features. Judicious choice of which sensor to run initial perception routines on often significantly reduces computational burden, while combining the ladar and camera information tends to boost performance over either used singly.

2 Methods

2.1 The World Model

The World Model (WM) contains a 3-D, annotated representation of the current state of the terrain surrounding the vehicle and is updated continually by the sensors. We use a modified occupancy grid representation [8], with the vehicle centered on the grid, and

Figure 2: Camera-ladar registration. Darker laser pixels are more distant.

the grid tied to the world. The WM thus scrolls under the vehicle as the vehicle moves about in the world. The world model is the system's internal representation of the external world. It acts as a bridge between sensory processing and behavior generation by providing a central repository for storing sensory data in a unified representation, and decouples the real-time sensory updates from the rest of the system. The WM process has two primary functions:

Create a knowledge database (map) and keep it current and consistent by updating existing data in accordance with inputs from the sensors and deleting information no longer believed to be representative of the world. The WM also assigns confidence factors to all map data and adjusts them as new data are sensed. Types of information in the map include state variables (e.g., time, position, orientation), system parameters (e.g., coordinate transforms, sensor to vehicle offsets, etc.), and lists or classes of sensed objects. The world model process also provides functions to update and fuse data and to manage the map (e.g. scrolling and grouping objects.)

Generate predictions of expected sensory input based on the current state of the world and estimated future states of the world. For the Demo III off-road autonomous driving application, very little a priori information is available to support path planning between the vehicle's position and a final goal position. The world model therefore constructs and maintains all the information necessary for intelligent path planning [9].

Prediction is used to focus attention on regions that have previously been identified as interesting. It facilitates tracking, enables confidences in features to be updated, and allows information found in one sensor to influence processing in another. Prediction is mediated in our system by the world model. Since we use a grid representation fixed to the world, it is straightforward to project regions in the world model into each of the sensor coordinate systems. Currently, we only predict where a feature is expected to occur, not what it may look like.

2.2 Coordinate System Transformations

Since the sensors are mounted on a mobile platform, and the sensors themselves move, projections are not fixed, but must be computed each time they are needed. There are two kinds of projections: The ladar data are projected into the world model, and features identified in the ladar data are projected into the color image space.

Each sensor is at a known base position on the vehicle, and has a known sensor coordinate system. The vehicle is moving, however, and the WM maintains its representation in world coordinates, fixed on the ground. Thus, all coordinates must be converted from sensor to vehicle, and from vehicle to world. Some of the sensors also move relative to their base position. The ladar, for instance, may rotate about its horizontal axis (tilt). Finally, the sensors sample at different times, so a correction must be made for their relative positions in space when mapping between images.

The ladar-to-WM coordinate transformation includes the ladar-to-vehicle and vehicle-to-world-model transformations. The projection from the WM to the color camera image includes WM-to-ladar and ladar-to-image transformations. The ladar-to-image transformation is particularly important in order to achieve an accurate registration between ladar features and image features. This transformation is not invertible because of the lack of depth information in the camera image. In order to register the ladar and camera images, we first calibrated the camera's internal parameters using J. Bouguet's Matlab toolbox [10]. The external orientation between the camera and ladar was obtained by correlating corresponding points imaged by each device over a number of scenes and then computing a least-squares fit to the transformation according to the procedure described in [11]. Results are shown for a sample scene in Figure 2.

3 Feature Types

In this section we will discuss two examples of feature types that our system detects and tracks: puddles and road signs—specifically, signs marking an endangered butterfly sanctuary. Puddles, ponds, and mud are a serious problem for off-road mobile vehicles because of the danger they pose of the vehicle getting stuck in them, as well the possibility of water damage to the engine and/or critical electrical components. For our purposes, butterfly signs indicate the borders of an ecologically protected zone where we do testing that the vehicle must not enter, but human-readable signs might also mark minefields or contain other important information [12], making the ability to find them a critical one.

A third task that is ongoing work, road finding, is also briefly discussed. There has been some work on following marginal rural roads using color cues [13], but road *detection*, which is a vital skill for back-country navigation, has been less studied.

3.1 Butterfly signs

Butterfly signs are rectangular yellow placards mounted on six-foot wooden posts (painted orange on top) that delimit a "no driving zone," thus affecting the path-planning module of the XUV system. Two such signs are shown in Figure 1(b).

In the ladar domain, signs can be distinguished from the background because they frequently jut above surrounding foliage, with good depth contrast and nothing above them. When fixed, the limited vertical field of view of the ladar (15 degrees) tends to cut off the tops of the signs, so we use a sign-finding operator that simply searches for a vertical bar (i.e., not end-stopped) in the ladar range image at a scale corresponding to 5-10 meters distance to the sign.

The steps of the method are illustrated for a sample range image on the left hand side of Figure 3: first, two odd-phase Gabor filters [14] are run over the range image in Figure 3(a) to find left and right vertical edges, respectively, and the output of the filter in Figure 3(b) is thresholded to isolate strong edges. Second, we search for range image locations where left and right edges *co-occur*— that is, all (x, y) such that there is a left edge at $(x-\delta, y)$ and a right edge at $(x+\delta, y)$ for $\delta \leq 2$. This yields ladar-based hypotheses for sign locations, as shown in Figure 3(c).

Figure 3: **Butterfly sign detection steps**: (B1) Raw ladar range image containing signs, (B2) Gabor-filtered range image (right edges), (B3) Left and right ladar edge co-occurrences, (B4) Color confirmation matches; **Puddle detection steps**: (P1) Raw ladar image with puddles, (P2) Smoothed sky and puddles after segmenting voids, morphological closing, (P3) Puddles after sky removal, range calculation, (P4) Simultaneous color image with bounding boxes of projected puddles.

We can accumulate these hypotheses in the world model as the XUV drives; a representation of this part of the WM after the XUV has followed a road with ~ 15 signs along it for several hundred meters is given in Figure 4(a). The WM is shown at a 1-meter grid square resolution, with only squares that have had five or more sign hypotheses projected to them shown in gray; the vehicle path calculated from its inertial navigation system is indicated by a dark line. Nearly all of the signs are found, but at the cost of a number of false positives from vegetation. The spots inside the dotted rectangle, for example, are likely tree trunks.

To increase the accuracy of our sign-finder we add color, which has been shown to be a useful cue for sign detection [12]. A very simple method for modeling the sign

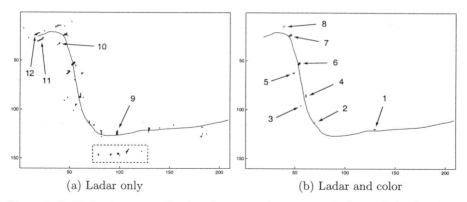

(a) Ladar only   (b) Ladar and color

Figure 4: Butterfly sign maps. Numbered arrows point to correctly-detected sign locations; the dotted box in (a) is a group of trees. Units are in meters.

color follows from sampling sign pixels over multiple images and performing principal components analysis on the pixel distribution to parametrize an ellipsoid in RGB space. Color similarity of an arbitrary pixel is then the Mahalanobis distance of the pixel color to the model ellipsoid's center. The ladar-based method above is easily extended by projecting ladar sign hypotheses (such as in Figure 3(c)) into the camera image, and then measuring the "yellowness/orangeness" of the local image neighborhood to check them. Specifically, we project each ladar sign hypothesis from the co-occurrence step to image coordinates (x, y) and compute the minimum Mahalanobis distance $d_{x,y}$ to the butterfly sign color over a 20×20 region about (x, y); if $d_{x,y}$ is less than a threshold then the hypothesis is *confirmed*. Bounding boxes on two clusters of confirmed ladar sign hypotheses from Figure 3(c) are shown in Figure 3(d). Color-confirmed ladar hypotheses for the road sequence are projected to the world model in Figure 4(b). This completely eliminates false positives, although a few actual sign locations are missed. A better color model would likely prevent this. Some signs were missed by both methods either because they were too far away to resolve or because foliage growing behind them eliminated depth contrast.

Focus of attention serves here to minimize computation: by searching first in the ladar domain, expensive image processing is limited to small neighborhoods around good candidates. By integrating sign detections over multiple frames, the world model throws out spurious sensor responses and betters the precision of the location estimates of the signs.

3.2 Puddles

In our standard ladar-based navigation system, we have found that puddles and other standing water appear as smooth, level surfaces. These qualities make such areas highly attractive to the motion planning system and therefore dangerous. We would like to detect puddles and flag them as "no go" or at least worthy of extra caution. Fortunately, a simple test follows from the optical properties of the ladar: laser beams hitting a puddle at an oblique angle are reflected away from the sensor, and result in no data being returned. Such points show up as *voids* in the ladar images, but are not the only source of missing data. Out-of-range depths are also recorded in any sky or otherwise distant regions in the ladar's field of view.

Our puddle detection algorithm thus looks for voids in the data, and then scans a region surrounding them. Puddle-derived voids are distinguished from sky by requiring that there be non-void pixels above every column in a connected component. Assuming that there are ground points somewhere adjacent to the puddle in the ladar image (rather than all pathological cases like overhanging limbs), we can obtain a reasonable estimate of the height of the water surface from the minimum height in the WM of over all points surrounding the puddle in the image. This allows us to solve for the missing range values in the puddle interior and thus properly place it in the map. These steps are illustrated in the right column of Figure 3. Without explicitly detecting puddles in this manner, height maps used for navigation have areas of missing data in them (see Figure 5(a)) that are similar to laser "shadows" behind obstacles and protruding objects. Puddle detection permits water hazards to be placed in the map for higher-level reasoning, as shown in Figure 5(b).

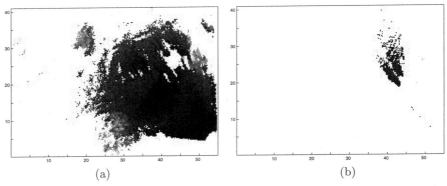

(a)   (b)

Figure 5: Mapping puddles. (a) Height map made from unprocessed ladar images; (b) Map of projected puddle regions with five or more hits per 0.25 meter grid square.

In the case of puddles, prediction plays an important role in ladar processing. As the vehicle approaches a puddle, the angle at which a laser ray hits the water gets steeper, and at a *critical angle* the sensor starts to record a return from the puddle. Without knowing that the region had already been identified as a puddle, the sensor would start to indicate that the region was traversable and smooth, which would make it a preferred location for the planner. Marking a region as already identified in the world model prevents this behavior. Note that puddles are unusual in that the confidence in most features increases through multiple views whereas using the ladar sensor to view puddles over time reduces their confidence. The behavior of the ladar sensor in the neighborhood of other features that produce voids, such as holes and occlusions, is very different from that around puddles. This enables a distinction to be made over time that might not be made from a single view.

Before the critical angle is reached, using the ladar alone is generally sufficient to correctly identify a puddle. After the critical angle the world model location of the puddle serves to guide the XUV away from water, but as it is unsupported by sensory data the reliability of the map decays over time. To maintain the reliability of the WM puddle locations even after the critical angle, we look in the color image for supporting evidence to reduce false positives. When a puddle is detected in the ladar data (with sufficient confidence in the map), a window is placed about the potential puddle region, and projected into the color image as in Figure 3. In the color domain, the system tries to determine if the region has a similar color to what is above it. Often, this will be the sky, so a blue color will mean a puddle. At other times, however, the puddle may reflect trees, grass, or clouds. The algorithm searches for a match, but may fail if the puddle is reflecting something not in the image. When the puddle is verified, color information from the puddle points can then be placed into the world model. By continually updating this color information while the ladar still sees a void, the system can smoothly transition to relying on color alone to segment a puddle even after the critical angle is reached. The confidence of puddle in the map is increased by a predefined value that depends on the robustness of the color classification algorithms.

3.3 Roads

Color from captured camera images can be combined with the 3-D information returned by the ladar range-finder in the world model. We do this by simply projecting ladar points into the current image and reading off the (R, G, B) values of the image pixels they land on, and then carrying that color information along when the ladar data is projected into the world map, averaging color per grid square. Example maps with fused height and color information created from two driving sequences are shown in Figures 6(a) and (b). The grid square size is 0.25 meters and every ladar image from each sequence is mapped (the sequence shown in (a) has 1072 ladar frames and (b) has 572). Height maps are displayed with the minimum height in the map as black and the maximum as white (unmapped squares are also white).

Observing that roads' structural and visual characteristics often differ from those of bordering areas in *height* (bushes, trees, and rocks tend to "stick up"), *smoothness* (roads are locally flat, while grass, etc. are bumpier), and *color* (asphalt, dirt, and gravel roads' hues are separable from those of vegetation and sky [13]), it is possible to formulate an objective function to distinguish roads in the ladar-color domain as contiguous world model regions with small variance in height and acceptable color distributions (brown, black, etc.). In Figure 6(a), the ladar data alone is sufficient to discriminate the road via height and smoothness features. When, however, the road and non-road are differentiated mainly by color as in Figure 6(b), color image information becomes critical. The ladar data remains useful, however, as a means of focusing attention by permitting obstacle regions such as trees and foliage to be masked out, reducing the map area searched for possible roads. We are currently investigating the use of GPS information from the XUV's navigational system and *a priori* map information to selectively cue evaluation of the objective function based on proximity.

4 Conclusion

By focusing on a few critical subtasks of the general off-road autonomous navigation problem, we have demonstrated the utility of combining information from a ladar and color camera for feature detection and tracking. The two sensors have strengths that are often complementary, and careful staging of algorithmic modules results in increased task performance without imposing the computational burden that simply analyzing or filtering both modalities and fusing them afterward would. Further, fusing data via a world model has proven a flexible way to integrate synchronized information from the two sensors while improving its quality over time, and being able to project the combined information into the sensor domain enables cheap prediction of what the sensors should see in subsequent views. By using both individual sensor characteristics

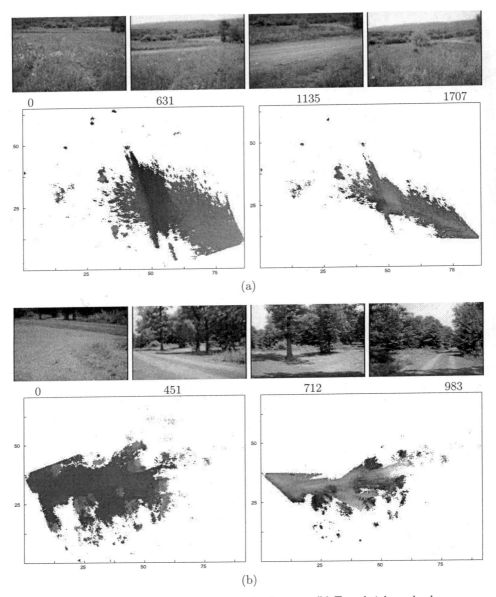

Figure 6: Road detection: (a) Field height and color map; (b) Trees height and color map.

and prediction, it is possible to focus attention on important features and to bring more sensor resources to bear on identifying them.

References

[1] C. Shoemaker and J. Bornstein, "The Demo3 UGV program: A testbed for autonomous navigation research," in *Proc. IEEE International Symposium on Intelligent Control*, 1998.

[2] J. Albus, "4-D/RCS: A reference model architecture for demo III," Tech. Rep. NISTIR 5994. 3-1-1997, National Institute of Standards and Technology, 1997.

[3] J. Clark and N. Ferrier, "Attentive visual servoing," in *Active Vision*, A. Blake and A. Yuille, Eds., pp. 137–154. MIT Press, 1992.

[4] K. Toyama and G. Hager, "Incremental focus of attention for robust vision-based tracking," *Int. J. Computer Vision*, vol. 35, no. 1, pp. 45–63, 1999.

[5] J. Eklundh, P. Nordlund, and T. Uhlin, "Issues in active vision: attention and cue integration/selection," in *Proc. British Machine Vision Conference*, 1996, pp. 1–12.

[6] C. Westin, C. Westelius, H. Knutsson, and G. Granlund, "Attention control for robot vision," in *Proc. Computer Vision and Pattern Recognition*, 1996, pp. 726–733.

[7] F. Ennesser and G. Medioni, "Finding Waldo, or focus of attention using local color information,," *IEEE Trans. Pattern Analysis and Machine Intelligence*, vol. 17, no. 8, pp. 805–809, 1995.

[8] D. Oskard, T. Hong, and C. Shaffer, "Real-time algorithms and data structures for under water mapping," in *SPIE Cambridge Symposium on Optical and Optoelectronic Engineering*, 1988.

[9] A. Lacaze, J. Albus, and A. Meystel, "Planning in the hierarchy of NIST-RCS for manufacturing," in *Proc. Int. Conf. on Intelligent Systems: A Semiotic Perspective*, 1996.

[10] J. Bouguet, "Camera Calibration Toolbox for Matlab," Available at www.vision.caltech.edu/ bouguetj/calib_doc. Accessed May 11, 2001.

[11] M. Elstrom, P. Smith, and M. Abidi, "Stereo-based registration of LADAR and color imagery," in *SPIE Conf. on Intelligent Robots and Computer Vision*, 1998, pp. 343–354.

[12] G. Piccioli, De Micheli, and M. Campani, "A robust method for road sign detection and recognition," in *Proc. European Conf. Computer Vision*, 1994, pp. 495–500.

[13] J. Fernandez and A. Casals, "Autonomous navigation in ill-structured outdoor environments," in *Proc. Int. Conf. Intelligent Robots and Systems*, 1997.

[14] T. Lee, "Image representation using 2D Gabor wavelets," *IEEE Trans. Pattern Analysis and Machine Intelligence*, vol. 18, no. 10, pp. 959–971, 1996.

Intelligent Autonomous Systems 7
M. Gini et al. (Eds.)
IOS Press, 2002

An Integrated Connectionist Approach to Reinforcement Learning for Robotic Control: Extension to Three Dimensions

Dean F. Hougen
School of Computer Science
University of Oklahoma, U.S.A.
hougen@ou.edu

Maria Gini and James Slagle
Department of Computer Science and Engineering
University of Minnesota, U.S.A.
gini—slagle@cs.umn.edu

Abstract. This paper explores the use of connectionism and reinforcement learning in robot control. It extends the authors' method to a problem with a three-dimensional input space. This is significant because it shows the method can handle maps equal in dimensionality to many of the sensor and motor maps found in animal brains. The approach is compared to a standard reinforcement-learning method and found to provide more rapid learning and a higher overall success rate. The trailer-backing task is used as an example problem.

1 Introduction

Reinforcement learning provides great potential for robots to improve their performance with experience. However, we face two obstacles when we apply reinforcement learning to robots: **Obstacle one** is the continuous nature of many robotic environments—the classic reinforcement model uses discrete environment states. **Obstacle two** is the large number of learning experiences that standard reinforcement-learning schemes require for performance to improve significantly.

We have addressed both obstacles previously, using an integrated connectionist approach, but only for two-dimensional input spaces [2]. An extension of this work to three-dimensions is important because animal brains have been found to use numerous two- and three-dimensional sensory and motor maps which are, presumably, serving useful purposes. It makes sense, then, to explore whether networks intended for robotic sensory and motor use can be constructed in both two- and three-dimensions. In this paper we revisit the trailer-backing problem that we previously solved in two-dimensions but add a second trailer. The angle between the trailers gives us our third dimension.

2 Related Work

Other authors have constructed learning systems for truck-backing using both supervised (e.g., [6]) and unsupervised methods (e.g., [17]). As we are interested in unsupervised robot learning, a closer look at the unsupervised systems seems warranted. Unfortunately, the methods used by these authors do not overcome the obstacles mentioned previously. From a learning systems perspective, then, it is more interesting to look at other research that attempts to overcome these two obstacles.

To overcome obstacle one, two basic approaches have been used. The first approach is to modify the model to accept continuous input. This has been done in several

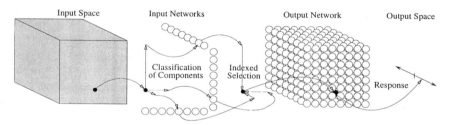

Figure 1: ICARL system for three-dimensional space and one-dimensional output spaces.

ways but always with difficulties, such as longer learning times (e.g., [1, 8]), poorer performance (e.g., [8]), and excessive complexity (e.g., [9]). The second approach is to discretize the input and use a standard model. Discretizing the input space by hand (e.g., [17]) requires either that the system designer have extensive knowledge of the environment in which the robot will act or that the partitioning be uniform and fine. Instead, systems may learn their own partitionings by successively dividing existing partitions regions (e.g., [14]), adjusting the boarders of partitioning regions (e.g., [7, 13]), or both (e.g., [10]). Our approach is a neighborhood-based, border-adjusting procedure.

To overcome obstacle two, approaches such as incorporating domain knowledge (e.g., [10]) or local control algorithms (e.g., [12]) have been used. Another approach has been to use a generalization scheme—treating similar states similarly. Generalization schemes have generally borrowed from methods used previously in supervised learning (e.g., [7]). Our approach is to use a neighborhood-based generalization scheme.

3   The Learning System

Rather than trying to surmount the obstacles separately, we take an integrated approach to both. We call this approach the Integrated Connectionist Approach to Reinforcement Learning (ICARL).[1] The overall learning system is shown in Figure 1.

To overcome obstacle one, we have the system learn a partitioning of the input space using self-organizing *input networks* that extract regularities from patterns of input data. Partitioning the input space into classes allows us to use mechanisms from traditional reinforcement learning. This step one in our integrated approach.

Step two uses the continuous nature of the input to overcome obstacle two. Because we know which classes are "near" one another in input space, and often we should respond similarly to similar situations, we use an experience involving one class to aid in determining responses to other "nearby" classes. This is done in the *output network*.

The third and final step in our integrated approach is the choice of method used. Both the input and output networks use laterally-connected networks of computational units. Different networks overcome each obstacle separately, but the overall working of the networks is similar. This parsimonious approach to both obstacles not only requires fewer total mechanisms in the system but allows for changes to the entire system to be made more easily. For example, to reduce the memory requirements of the system, both the input and output networks can be shrunk proportionally, as shown in Section 5.

These networks use a topological ordering of the units and associated neighbor-

---

[1]We previously called our method the Self-Organizing Neural Network with Eligibility Traces or SONNET. However, we found that Nigrin was referring to his Self-Organizing Neural NETwork as SONNET as well [11]. To avoid confusion with his unrelated systems, we have renamed our system.

hoods, as in Kohonen's Self-Organizing Topological Feature Maps [5]. Let D be the dimension of a network. Each unit is associated with a D-tuple that uniquely defines its coordinates in topology space. The existence of a network topology allows for the definition of a *distance* function d for the units—we use the maximum difference between coordinates of the units. E.g., for units (1,1,1) and (2,3,4) in a three-dimensional network, the distance between them is $max(|1-2|, |1-3|, |1-4|)$ or 3.

The distance function is used indirectly through the definition of *neighborhoods*. A neighborhood may have any width from zero to the maximum distance between units in topology space and varies with time, starting large and decreasing. Formally, if U is the set of units u in the network and W is a time-dependent function specifying the width of the neighborhood, then the neighborhood N of unit n at time t is defined as

$$N_n(t) = \{u \in U \ : \ d(n,u) \leq W(t)\} \tag{1}$$

Our system is trained in a series of *trials*. A trial is divided into discrete time steps, and lasts from an initial positioning of the controlled system until a success or failure signal is generated. During a *run* of multiple trials, the learning system progresses from random performance to competence.

3.1 Input-space partitioning

Input-space partitioning in ICARL systems uses self-organization principles introduced by Kohonen [5]. In this paper, three separate one-dimensional input networks are used, one for each dimension of the input space. On each time step a new input vector $\bar{x}$ is given to the system—one component x_i of this vector is given to each input network i.

Within each network, each unit has a weight to which the input value is compared. (The weights are given random values at the start of a run.) The unit s that has the weight most closely matching the input is said to be the *selected* unit. Formally,

$$\exists s[|w_s - x_i| \leq |w_u - x_i| : \forall u \in U, s \in U] \tag{2}$$

where w is a unit's weight. If more than one unit satisfies this equation, then the one with the lowest topological coordinate is selected. Once the trial has ended, the input vectors are re-presented to the input network and for each time step the weights of the selected unit and of all other units in its neighborhood are updated using

$$w^{new} = w^{old} + \alpha(t)\gamma(t_{final})(x_i - w^{old})f' \tag{3}$$

where f' is a feedback signal (+1 failure, 0 success), α and γ are functions that determine the influence of the input vector ($0 \leq \alpha, \gamma \leq 1$), and t_{final} is the total number of time steps in this trial. α starts near 1 and decreases with time to give the units independence as they become organized and γ ensures that all trials are weighted equally for purposes of partitioning the space, regardless of their length in time steps. Using Equation 3, there is competition among all the units in a network for selection and cooperation within a neighborhood as the units change their weights towards the same target. By presentations of successive data to the network, the network self-organizes so that closely neighboring units have similar values for their weights (respond to similar input) while units that, according to the network topology, are far from one another have very different values for their weights (respond to very different input).

The selection of one unit from each input network allows for the selection of a single corresponding unit from the output network. This correspondence is such that adjacent regions of the input space result in the selection of output units adjacent in their topology space. For this reason, the output network has the same dimensionality as the input space. In this paper, the one-dimensional topological coordinates of units in the three input networks are used as indices into a three-dimensional output network. The combination of all weights from all input networks, then, can be seen as a partitioning of the input space. This partitioning gives a discretization of the input space and allows for a single output response to be learned for each of the resulting discrete input regions.

3.2 Output-response learning

In the present application, each unit in the output network has one weight because the output space is one dimensional. These weights are also given random values at the start of a run. When a unit from the output network is selected as described above in Section 3.1, it produces a response based on the value of its weight. This is given as a control signal to the controlled system. The output weight updates use *eligibility traces*. The concept of the eligibility trace in brain function was presented by Klopf [4]. Eligibility traces have since been used in several reinforcement learning systems [15].

Eligibility traces allow for learning in domains in which performance is temporally dependent on responses and evaluations (such as terminal success and/or failure signals) are available. This eliminates the need to have a "teacher" that knows the correct control signal at each time step and thereby allows for unsupervised learning to occur.

At the start of a trial, all units have an eligibility value of zero. When an output unit is selected and responds (fires) it becomes amenable to change according to

$$e^{new} = e^{old} + I \tag{4}$$

where e is the eligibility and I is the initial additional eligibility just after firing. This value reduces with time but provides an opportunity for learning based on feedback received by the unit after its activity. At each time step, the eligibility of each output unit decays, regardless of whether that unit fired on that time step, according to

$$e(t + 1) = \delta e(t) \tag{5}$$

where δ is the rate of eligibility decay ($0 \leq \delta \leq 1$).

When success or failure occurs, it is likely that more recent control signals are more responsible than earlier control signals and, due to the eligibility trace, are rewarded or punished to a greater extent. When a success or failure signal is received by the output network, the weights of all of the output units are updated according to

$$v^{new} = \text{sign}(v^{old})(|v^{old}| + e\,\sigma(T)\,f) \tag{6}$$

where v is the weight, σ is a scaling function based on trial number T, and f is the feedback (+1 for success, -1 for failure). The function σ is used to allow for large changes to the weights in early training trials and smaller changes in subsequent trials.

The output network also uses inter-neural cooperation. After each trial each unit updates its weight a second time, this time using the weights of units in its neighborhood, according to

$$v_j = (1 - \beta(T))v_j + \beta(T) \sum_{n \in N_j} \frac{v_n}{m_j} \tag{7}$$

where N_j is the neighborhood of unit j, m_j is the number of units in that neighborhood, and β determines the degree to which a unit's value is "smoothed" with that of its neighbors ($0 \leq \beta \leq 1$). β starts near one and decreases with time. This means that each unit's value becomes more independent from those of its neighbors as time passes.

4　Application

As our application domain, we have chosen to look at the problem of learning to back a truck and trailer rig to a goal. The rig consists of a truck with two trailers. While non-learning controllers for backing such a rig can be devised (e.g., [16]), learning to control these movements provides a very interesting benchmark task nonetheless.

The control system has three inputs: The angle of hitch 1 between the truck and the first trailer, the angle of hitch 2 between the first and second trailers, and the angle between the spine of the second trailer and the goal. There is a single output dimension (wheel angle) and the single response value given is thresholded at zero, to give either a turn-left or turn-right control signal to the truck. The initial position of the rig at the start of each trial is chosen randomly with a uniform distribution over the range $-6°$ to $+6°$ for each angle.

While the maximum initial angles specified may seem small, it should be noted that the inherent instability in backing with passive trailers means that, for two trailer systems like the one described, large differences in angles inevitably lead to failure, regardless of the control signal given. This is because, if the signs of the angles between adjacent components of the rig are opposite, then the active truck cab must swing out in arcs wider than those traced by the first trailer in order to move it around into position to correct the trajectory of the second trailer. However, in trying to bring the truck into such a position, the angle between the trailers and that between the second trailer and the goal are increased in magnitude as the entire rig backs up. In order for the system to be able to physically maneuver to the goal with larger possible initial angles, it would be necessary for the system to utilize forward motion, as well as backing.

5　Experiments and Results

To investigate the flexibility of the system with regard to resources provided, we constructed eight ICARL systems with networks of sizes varying from 3 to 10 units per dimension of the input space. This gave us networks ranging in overall size from 9 input/27 output units to 30 input/1000 output units. We compared the performance of these eight ICARL systems to eight baseline reinforcement-learning systems. These baseline systems used eligibility traces but did not learn the input space partitioning nor use inter-neural information sharing in the output learning. Instead, the input space was simply divided into equal-sized regions in each dimension and each output unit learned its response individually. The baseline systems, therefore, used Equations 2, 4, 5, and 6 but not 1, 3, or 7.

We conducted experiments in simulation and results are shown in Figure 2. Each run was 1000 trials in length and each graph is made by averaging 100 runs together. The graphs show success rates.

While we could have compared our results to those obtained using other standard reinforcement learning methods, such as Adaptive Heuristic Critics, TD(λ), or Q-Learning (see, for example, [3]), such methods require exponential decays for all regions to be computed on each time step, regardless of which region contains the new input. This greatly increases the computational requirements of the algorithm. Perhaps, as computing speed continues to grow, such a computational cost will be minimal and such methods will become practical for learning on board real robots. However, it should be noted that robots will likely to be asked to learn many things at once, as people do, so minimizing the computation required for each task may still be necessary.

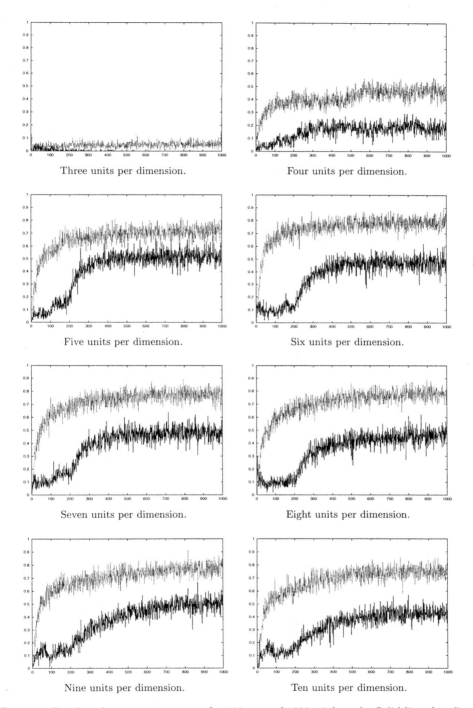

Three units per dimension.

Four units per dimension.

Five units per dimension.

Six units per dimension.

Seven units per dimension.

Eight units per dimension.

Nine units per dimension.

Ten units per dimension.

Figure 2: Results. Average success rate for 100 runs of 1000 trials each. Solid line: baseline system. Dotted line: ICARL system. (The dotted line is the upper line in all graphs.)

As Figure 2 makes clear, the ICARL system greatly outperformed the baseline system for all network sizes. At size three (3 units per dimension), the ICARL system achieved only a 5% success rate yet it still outperformed the baseline system which never moved beyond random performance. At size four, the ICARL system reached approximately 45% success; at size five and higher its performance was relatively constant at between 70% and 80%. The baseline system came in approximately 20% to 30% lower in these cases. The ICARL system learned much more quickly as well—note the initial slope for each system.

It is interesting to note that the baseline system achieved its best performance, both in terms of learning rate and ultimate success rate, at size five. At smaller sizes it likely did not have sufficient resources available to it. At larger sizes it most likely fell victim to an increasingly difficult credit assignment problem—it needed to determine output values for more units but had no more experiences on which to base its determinations.

The ICARL system, on the other hand, had consistent top results at sizes six through ten, and only slightly worse results at size five. Not only was it able to make better use of the resources available to it at small sizes by adjusting its partitioning of the input space, it also was able to overcome much of the spatial credit assignment problem through information sharing in the output network.

Also interesting to note is the fact that the baseline system did not show a uniform drop-off in performance as it increased in size above size five. At size nine, for example, it had slightly better learning and ultimate success rates than it did at either sizes eight or ten. This is likely due to the appropriateness of the uniform partitioning in each case. The ICARL system, on the other hand, was much more consistent as size varied.

While no systematic effort was made to test either the baseline or ICARL systems for sensitivity to learning parameters (such as decay rates), neither were these parameters tuned for these systems on this particular task. Instead, several values of these parameters were tried for each system with no noticeable effect on learning or success rates; a more detailed analysis of the effect of these parameters is left for a future date.

6 Conclusions

Reinforcement learning allows systems to autonomously improve their performance through interactions with their environment but obstacles have remained. We have addressed the obstacle of continuous input by having the system learn its partitioning. We have addressed the obstacle of long learning times through cooperative response learning. Our system overcomes both obstacles using an integrated approach.

Our system improves upon a baseline method that uses eligibility traces. If greater computational resources become available to robotic agents, making the use of Adaptive Heuristic Critics, TD(λ), and Q-Learning practical for time-critical systems, we could likewise build on any or all of those methods using our integrated approach for input space partitioning and cooperative response learning.

By showing that our system works well for three-dimensional input spaces, we have made a significant new demonstration of our system, as many of the sensory-motor maps found in animals are known to be three-dimensional. While it might be interesting to investigate ICARL systems with higher numbers of dimensions in the input space, we are more interested in exploring ways in which two- and three-dimensional maps can be combined to accomplish complex tasks. After all, brains seem to accomplish a great deal through the combination of many such sensory and motor maps. We are also interested in demonstrating the performance of this system in other application areas.

References

[1] C. W. Anderson. Learning to control an inverted pendulum using neural networks. *IEEE Control Systems Magazine*, 9(3):31–37, 1989.

[2] D. F. Hougen, M. Gini, and J. Slagle. An integrated connectionist approach to reinforcement learning for robotic control: The advantages of indexed partitioning. In *Proceedings of the International Conference on Machine Learning*, pages 383–390, June 2000.

[3] L. P. Kaelbling, M. L. Littman, and A. W. Moore. Reinforcement learning: A survey. *Journal of Artificial Intelligence Research*, 4:237–285, 1996.

[4] A. Klopf. Brain function and adaptive systems—a heterostatic theory. In *Proceedings of the International Conference on Systems, Man, and Cybernetics*, 1974.

[5] T. K. Kohonen. *Self-organizing and associative memory*. Springer-Verlag, Berlin, 3rd edition, 1989.

[6] S.-G. Kong and B. Kosko. Adaptive fuzzy systems for backing up a truck-and-trailer. *IEEE Transactions on Neural Networks*, 3(2):211–223, 1992.

[7] R. M. Kretchmar and C. W. Anderson. Comparison of CMACs and radial basis functions for local function approximators in reinforcement learning. In *Proceedings of the IEEE International Conference on Neural Networks*, pages 834–837, Houston, TX, 1997.

[8] B. Maričić. Genetically programmed neural network for solving pole-balancing problem. In *Proceedings of the 1991 International Conference on Artificial Neural Networks*, pages 1273–1276, Espoo, Finland, 1991.

[9] P. Martín and J. del R. Millán. Learning reaching strategies through reinforcement for a sensor-based manipulator. *Neural Networks*, 11(2):359–376, 1998.

[10] J. del R. Millán. Rapid, safe, and incremental learning of navigation strategies. *IEEE Transactions on Systems, Man, and Cybernetics*, 26, Part B(3), 1996.

[11] A. Nigrin. SONNET: a self-organizing neural network that classifies multiple patterns simultaneously. In *Proceedings of the International Joint Conference on Neural Networks*, volume 2, pages 313–318, 1990.

[12] J. Randløv, A. G. Barto, and M. T. Rosenstein. Combining reinforcement learning with a local control algorithm. In *Proceedings of the International Conference on Machine Learning*, pages 775–782, 2000.

[13] B. E. Rosen, J. M. Goodwin, and J. J. Vidal. Process control with adaptive range coding. *Biological Cybernetics*, 66(5):419–428, 1992.

[14] M. Salganicoff, L. H. Ungar, and R. Bajcsy. Active learning for vision-based robot grasping. *Machine Learning*, 23:251–278, 1996.

[15] S. P. Singh and R. S. Sutton. Reinforcement learning with replacing eligibility traces. *Machine Learning*, 22:123–158, 1996.

[16] M. Vendittelli, G. Oriolo, and J.-P. Laumond. Steering nonholonomic systems via nilpotent approximations: The general two-trailer system. In *Proceedings of the IEEE International Conference on Robotics and Automation*, pages 823–829, 1999.

[17] N. Woodcock, N. J. Hallam, and P. D. Picton. Fuzzy BOXES as an alternative to neural networks for difficult control problems. In *Applications of Artificial Intelligence in Engineering VI*, pages 903–919, Oxford, UK, July 1991.

Intelligent Autonomous Systems 7
M. Gini et al. (Eds.)
IOS Press, 2002

Biologically Inspired Visual Odometer for Navigation of a Flying Robot

Fumiya Iida
AILab, Department of Information Technology
University of Zurich, *Switzerland*
iida@ifi.unizh.ch

Abstract. The ability to navigate in a complex environment is crucial for both animals and robots. Particularly flying insects are capable of surprisingly good navigation despite the small size and relative simplicity of their brains. Recent experimental research in biology has uncovered a number of different ways in which insects use cues derived from optical flow for navigational purposes, such as safe landing, obstacle avoidance and dead reckoning. In this study, we use a synthetic methodology to gain additional insights into the navigation behavior of bees, specifically focus on the mechanisms of course stabilization behavior and visually mediated odometer by using a biology model of motion detector for the purpose of long distance goal-directed navigation, and evaluate its performance with the blimp type flying robot platform in uncontrolled indoor environments. The result shows that the proposed mechanism can be used to provide an estimate of the distance traveled, but the performance depends on the route the robot follows. Further analysis is also conducted in order to improve the performance of the proposed method.

Keywords. Biorobotics, visual navigation, Visual odometry, Optical flow

1 Introduction

Navigation in complex 3-D environments is very challenging for both animals and robots. Because of the limited weight capacity and potentially hazardous conditions, the control systems of flying machines are mostly highly complex, where a lot of sensory systems are generally required, such as GPSs, gyroscopes, compasses, ultrasonic sensors, inclinometers, accelerometers, and laser rangefinders [1][3][5]. Compare to flying machines, however, flying insects solve this task in a robust and adaptive manner despite their tiny brain. Surprisingly, behavioral studies with insects have revealed that a number of important navigational abilities rely mainly on visual information, although visual processing is generally regarded as a computationally demanding task. Due to the fact that image motion induced by ego-motion plays a crucial role in insects, navigation using biologically inspired optical flow has been investigated mainly on land-based agents. The basic behaviors observed in flying insects, e.g. obstacle avoidance, fixation behaviors, were demonstrated with relatively simple mechanisms [4][8]. Owing to its simplicity, such mechanisms have been incorporated in a robot exclusively using analog hardware [4]; a VLSI implementation has been also realized [7]. In a similar way, simulated flying agents were used for altitude control and obstacle avoidance [9][10][11], and a robotic gantry demonstrated the landing behavior of flies [16].

In addition to such basic behaviors, however, mechanisms for long distance navigation are crucial for such insects as honeybees, which navigate accurately and repeatedly from their hive to a food source. This type of navigation normally requires that both directional and distance information be available. Whereas in direction estimation, there is a lot of evidence that celestial cues and especially the polarization pattern of the sky play crucial role [6], the way that bees gain information about the distance traveled has been a point of dispute for many years. The behavior studies with honeybees have recently uncovered a mechanism behind visually mediated odometer, in which the primary cue is the integral, over time, of the image motion that is experienced en route. Given these biological backgrounds, in this paper, we propose a biologically inspired navigation method using visual odometry model in combination with a course stabilization mechanism, and test its performance by using a freely flying robot we developed in uncontrolled environments.

In the following section, we introduce navigation mechanisms of flying insects in detail. In section 3, we propose a visual odometry model and explain the experiments with an autonomous flying robot in section 4. Further issues are discussed in section 5.

2   Insect Navigation

The vision systems of flying insects are exquisitely sensitive to motion, because visual motion induced by ego-motion can tell the animal much about its own motion and also about the structure of its environment. Behavior experiments with flies and bees show a number of different ways in which insects use cues derived from optical flow for navigational purposes (for review, see [14]). Early studies showed that a tethered fly inside a striped drum tends to turn in the direction in which the drum is rotated [13]. This reaction, so-called optomotor response, serves to help the insect maintain a straight course by compensating for undesired deviations.

For the purpose of long distance goal-directed navigation, recent studies of bees' behavior suggested that the amount of image motion plays an important role in estimating the distance traveled [15]. Early studies suggested that the distance is gauged in terms of total energy consumption during a foraging journey, but recent studies of the bees' behavior questioned this hypothesis, and suggested that visual cues, more specifically the amount of image motion, play an important role on estimating the distance traveled [15]. In these studies, bees were initially trained to search for the food source in the vertically striped tunnel. Then the searching behavior of the trained bees was observed when the pattern of stripes on the walls was changed, by using different stripe periods, different width between right and left walls, and walls with horizontal stripes. With horizontal stripes, bees tended to fail searching for the food source at the right location, therefore it was concluded that vertical stripes are used for the visual odometer. With different period of stripes, bees succeeded in finding the food source position, which implied that bees measure the distance independent of the spatial frequency of the stimulus in the environment. However, when the distance to both walls was increased or decreased compared to the training situation, bees tended to fail in searching at the right location. The main conclusion from these studies was that the bee's visual odometer provides an estimate of distance that is independent of the spatial frequency of the visual stimulus, i.e. it only depends on the angular velocity of the image projected on the insect's retina.

On the basis of above mentioned behavioral experiments as well as electrophysiological studies, a model of motion detection in the insect's nervous system, the Elementary Motion Detector (EMD), has been proposed (for review, see [2]). A well-known model of the EMD is the so-called Reichardt detector, which belongs to a class of correlation-type detectors, shown in Figure 1. Two adjacent photoreceptors send their outputs to temporal high-pass filters that remove constant illumination containing no motion information. These signals are then "delayed" by exploiting the phase lag inherent in a first order temporal low-pass filter. While not a true time delay, the low-pass filter is a good approximation that biology appears to use. Delayed channels are then correlated with adjacent, non-delayed channels by means of a multiplication operation. Finally the outputs of two opponent EMDs are subtracted to yield a strongly direction-sensitive response. Although the nature of the neural mechanisms and the location in the visual pathway remains to be elucidated, some behaviors of the motion sensitive

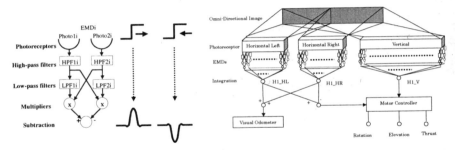

Figure 1. Left: The Reichardt model of elementary motion detection. Photoreceptors, high-pass filters, low-pass filters, multipliers, and the subtraction module are wired in series. The output of the last step (subtraction) is an estimate of direction sensitive image speed. **Right**: The controller circuit for the flying robot and visual odometer. The gray rectangle denotes the panoramic image extracted from the omni-directional camera on the robot. The image is given to both horizontally and vertically arranged photoreceptor arrays of EMDs. After the integrators and comparator, the outputs are given to motor controller and visual odometry modules.

Figure 2 Top: The autonomous flying robot, Melissa and its gondola, on the bottom of which a miniature panoramic camera attached. **Bottom:** An image obtained by the panoramic vision system (left) and its log-polar transformed image (right), which is used in the experiments.

neurons in insects can be well characterized by this motion detector model [14]. The salient properties of the movement-sensitive mechanism underlying these responses are that it is directional, and it does not encode the speed of the moving image, rather it is sensitive to the temporal frequency of intensity fluctuations generated by the moving image, and therefore confounds the speed of the image with its spatial structure.

3   Navigation models of a freely flying robot

On the basis of the biology studies, this section explains models of course stabilization and visual odometer for a freely flying robot. Although, in nature, insects are believed to use celestial cues together with visual odometry, in this paper, a robot takes advantage of a model of course stabilization behavior together with a visual odometry model for a long-range navigation.

3.1   Course Stabilization Model

Inspired from the flying insects' course stabilization behavior, which is observed as a fixation behavior using optomotor response, we propose a course stabilization model for a flying robot. Figure 1 illustrates the control procedure: At first, image from panoramic vision system (see next section for the details) are log-polar transformed, whose intensity information is given to three 2-dimensional EMD arrays, i.e. Horizontal Left (HL), Horizontal Right (HR), and Vertical (V). These arrays extract horizontal motion from the left and the right lateral images, and vertical motion from the whole image, respectively. The outputs of each EMD array are then integrated in order to simulate the wide field motion sensitive interneurons H1:

$$H1\_HL(t) = \sum_i \sum_j EMD\_HL_{ij}(t) \tag{1}$$

$$H1\_HR(t) = \sum_i \sum_j EMD\_HR_{ij}(t) \tag{2}$$

$$S_{rotation}(t) = H1\_HL(t) - H1\_HR(t) \tag{3}$$

To estimate the rotational ego-motion of the robot, the right and left horizontal motions are compared by a simple subtraction. Vertical motion, i.e. changes in height, is estimated by the vertical 2-dimensional EMD array, in which EMDs are vertically arranged. The outputs are integrated as $S_{altitude}$:

$$S_{altitude}(t) = H1\_V(t) = \sum_i \sum_j EMD\_V_{ij}(t) \tag{4}$$

These rotation and height sensory information induced by ego-motion is then given to a motor controller module. The three parameters are required for the control of the flying robot (see the next section for the details), i.e. rotation, elevation and thrust motor outputs, M_R, M_E and M_T, respectively, which are determined by the following equation.

$$\begin{bmatrix} M_R \\ M_E \\ M_T \end{bmatrix} = \begin{bmatrix} W_{RS} & 0 & 0 \\ 0 & W_{ES} & 0 \\ 0 & 0 & W_{bias} \end{bmatrix} \cdot \begin{bmatrix} S_{rotation} \\ S_{altitude} \\ 1 \end{bmatrix} \tag{5}$$

In this motor controller, thrust motors are driven at constant speed.

3.2 Visual Odometry Model

The same horizontal EMD arrays that are used for course stabilization control are again used for the visual odometry (Figure 1). Since the EMD response provides an angular velocity signal (more exactly, a spatio-temporal signal that depends on the speed of the stimulus as well as its spatial frequency), the distance traveled could be estimated by integrating the EMD outputs over time. The responses from both right and left horizontal EMD arrays are integrated every time step, which is given to the visual odometer module, then accumulated over time. Namely the visual odometer response, VO, is calculated by integrating the H1_HL, H1_HR output over time:

$$VO = \sum_t H1\_HR(t) + H1\_HL(t) \tag{6}$$

4 Experiment

This section presents a series of navigation experiments by using a flying robot we developed. The models explained in the previous section are implemented to the robot, and then tested in uncontrolled indoor environments.

4.1 Method

To evaluate the performance of our model, we developed an autonomous flying robot, shown in Figure 2. The flying robot Melissa is a blimp-like flying robot, which consists of a helium balloon, a gondola hosting the onboard electronics, and a host computer. The balloon is 2.3m long and has a lift capacity of approximately 500g. Inside the gondola, there are 3 motors for elevation and thrust control (a motor for rotation control is attached directly to the balloon), a

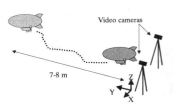   

Figure 3 Left: Experimental setup for the navigation experiments. The experiments start with the same initial conditions, i.e. initial robot position and orientation. Two video cameras are installed to record the absolute trajectory of the robot for later analysis. **Right:** Pictures of three different uncontrolled environments, which are tested in the experiments. (EXP 0, 1, 2, from left to right)

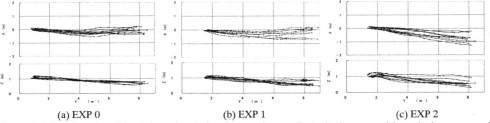

(a) EXP 0   (b) EXP 1   (c) EXP 2

Figure 4. 3-D trajectories of the flying robot during the experiments. Each plot is extracted from the images recorded with the stereo video camera. The plots denotes the position of the robot at one second time-step, and each graph shows the trajectories of 10 trials.

four-channel radio link, a miniature panoramic vision system, and the batteries. The panoramic mirror has a hyperbolic surface that provides a visual field of 360 degrees on the horizontal plane and 260 degrees vertically. The control process of Melissa can be decomposed to three basic steps. First, the video signal from the CCD camera attached to the gondola is transmitted to the host computer via a wireless video link. Second, the images are then digitized on the host computer, which also performs the image processing in order to determine the target motor command. And third, the motor command is sent to the gondola also via radio transmission. The frame rate of the visual processing is set to a constant speed of 10.0 fps. The robot uses 180 (horizontal) x 60 (vertical) pixels (90 x 30 EMDs), which covers 360 degrees along horizontal plane and 120 degrees along vertical plane in the panoramic image homogeneously, i.e. angular distances between photoreceptors, the sampling bases, are 1.0 degree. The W matrix of the equation (5) is heuristically determined before experiments, since it is strongly dependent on the hardware conditions (floating balance of the robot etc.) and environment factors (air currents etc.). Three sets of experiments are conducted in three different uncontrolled indoor locations (EXP0, EXP1 and EXP2 shown in Figure 3), where we installed two video cameras to track and record the absolute trajectory of the robot for later analysis (Figure 3). Each set of experiment consists of 10 flight trials: in one trial, the robot starts controlling at the same initial conditions, i.e. initial positions and initial orientations, and stops after the same time duration. Due to the limitation of the experimental environments, the time durations are set to 25 seconds.

4.2 Results

In Figure 4, the plots show 3-D coordinates of the robot which are extracted from two external video camera images in one second step. In these graphs, each trial starts from left toward right. In these experiments, the robot simply balances the right and left lateral image speeds to maintain a straight route, a small difference in the initial orientations of the robot results in a relatively large deviations of the goal position, as shown in the X-Y plane figures of EXP 1 and

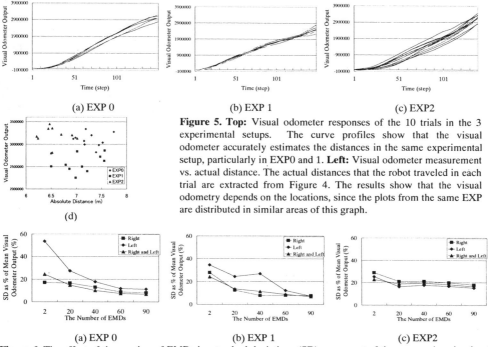

(a) EXP 0   (b) EXP 1   (c) EXP2

Figure 5. Top: Visual odometer responses of the 10 trials in the 3 experimental setups. The curve profiles show that the visual odometer accurately estimates the distances in the same experimental setup, particularly in EXP0 and 1. **Left:** Visual odometer measurement vs. actual distance. The actual distances that the robot traveled in each trial are extracted from Figure 4. The results show that the visual odometry depends on the locations, since the plots from the same EXP are distributed in similar areas of this graph.

(d)

(a) EXP 0   (b) EXP 1   (c) EXP2

Figure 6. The effect of the number of EMDs by standard deviations (SD) as percent of the mean values in visual odometry measurement. The graphs show that the number of EMDs changes the accuracy of visual odometry measurements.

2. However, most of the trajectories show that the robot maintains straight routes. For the height control, the robot tends to lose the height at the beginning of experiments, but eventually maintains the height in a certain range in all of the trials.

Figure 5 (a), (b) and (c) show the visual odometer responses that the robot experienced during each trial. The proposed visual odometry model measures almost the same distances in each experimental setup, EXP 0, 1, and 2. However, despite the same flight duration of 25 sec, there are differences in the final values of visual odometry, i.e. the outputs in EXP1 are smaller than those of EXP0 and 1. In addition, the visual odometer responses are plotted against the corresponding actual distances the robot flies in Figure 5(d), where the actual distances are estimated from the externally observed 3D trajectories shown in Figure 4. The results from each EXP0, 1 and 2 tend to distribute in a similar area in this figure, i.e. the EXP 0 plots distribute in the upper right, EXP 1 in the bottom, EXP 2 in the upper left area. A main conclusion of this analysis is that deviations of the visual odometer in a route rely on the route the robot follows, which is biologically plausible since natural insects tend to adopt a fixed route during foraging.

Further analyses are conducted with respect to the effects of the number of EMDs. In this analyses, from the stored data of EMD outputs, we re-calculated the visual odometer outputs by using the different number of EMDs: 1, 10, 20, 30, and 45 EMDs are selected over each left and right laterally distributed EMDs, then integrated their outputs over time. These results are again normalized by the corresponding actual distances, and their standard deviations are calculated, in which, first, the odometry output of each trial is divided by the corresponding actual distance, then second, the standard deviations (SDs) are calculated as percent of mean values (Figure 6) There are two implications that should be mentioned. Firstly, in principle, one EMD is essentially the minimum requirement of the proposed visual odometry model, although redundancy improves the measurement errors for all EXP 0, 1, and 2. Secondly, the numbers of EMDs required for the visual odometer depend on the environment, since the curve profiles of SDs are different in each experiment, as shown in Figure 6. For example, in EXP1 and EXP2, the standard deviations of both right and left visual fields are significantly larger than either right or left ones. Moreover, in EXP 2, the performance is not significantly improved in the case of more than 20 EMDs. From these results, depending on the environment, there would be specific visual fields that are able to estimate the distance more reliably than others, which suggests that there is a possibility to increase the performance of the proposed visual odometer model by adaptively changing areas of EMDs used for the visual odometer.

5 Discussion

Given the experimental results, in this section, we discuss general properties of the proposed navigation model and possible solutions to improve the performance.

In our experiments, we take advantage of the optomotor control for the yaw rotational control, instead of yaw, roll, and pitch controls of flying insects. However the mechanism we employed can be applicable to all three rotational controls in case of the other robotic platforms with an omni-directional vision. For the altitude controls, in this paper, we employed the vertical EMDs to measure the vertical image movement. It is relatively easy in our robotic platform, because, in principle, it has almost no motion induced by roll and pitch rotations. In the case of other robotic platforms, it requires a compensation mechanism for these roll and pitch motions, because they also produce vertical motions. However such motions can be distinguished from the vertical translation motion with direction-sensitive motion detectors, since the opposite side of visual fields has always motions in the other direction: e.g. for roll motion, right motion is going down, but left motion going up. Measuring the ground speed could be an alternative mechanism to control altitude. For example, the robot can exploit the physical constraints of the slower apparent ground speed when the robot flies at higher altitude. It has been reported that flies actually use such a constraint to control altitude and speed at the same time in the landing process [16].

There are some potential solutions to improve the performance of the proposed visual odometer model. Since the above mentioned experimental results show that the performance of the visual odometer relies significantly on the spatial structure experienced during the navigation, and this course stabilization mechanism is potentially vulnerable against unanticipated deviations, one solution would be to minimize route deviations. Considering that celestial cues play an important role as a global compass in natural systems, compass information would be biologically plausible and it could improve the performance. According to the results above, however, the combination of course stabilization behavior and visual odometer would be a redundant backup system. In another solution, as suggested in Figure 6, the EMD locations of the visual field need to be carefully considered for the better performance of the visual odometer, especially in unstructured environments. Alternatively, according to a number of experiments with bees, visual odometry would be based on a pure speed detector, rather than spatio-temporal filters such as the EMDs employed in this paper, although mechanisms of such pure speed detectors are still remained to be elucidated.

6 Conclusion

This paper presents a biologically inspired visual odometer mechanism in combination with course stabilization behavior based on an EMD model in the context of goal-directed navigation framework of bees. We performed experiments using a blimp-type robot platform in unstructured indoor environments, and the combination of course stabilization and visual odometry models successfully demonstrates a long-range navigation. In addition, the further analysis shows that the redundancy in sensory system plays an important role for the robust performance of the visual odometry measurement, which leads to some possible solutions for the improvement of the proposed model. Although the control of our blimp-type robotic platform is by far simpler than those of other platforms such as helicopters, by enhancing the "cheap vision" approach [12], it would be possible to realize more sophisticated controls for more demanding situations with a simpler architecture, as the natural evolution has found a solution for flying insects.

Acknowledgements

This work is supported by the Swiss National Science Foundation, grant no 2000-061372.00, and the Swiss Federal Office for Education and Science (VIRGO TMR network, BBW-No. 96.0148).

References

[1] Amidi O., Kanade T., and Fujita K., "A Visual Odometer for Autonomous Helicopter Fight," Intelligent Autonomous Systems, Y. Kakazu et al. (Eds.), IOS Press, pp.123-130, 1998.

[2] Borst A., Egelhaaf M., "Detecting visual motion: Theory and models, Visual Motion and its Role in the Stabilization of Gaze," Eds. F.A. Miles and J. Wallman, Elsevier Science, pp. 3-27, 1993

[3] Fagg A. H., Lewis M. A., Montgomery J. F., Bekey, G. A., "The USC Autonomous Flying Vehicle: an Experiment in Real-Time Behavior-Based Control," IEEE/RSJ International Conference on Intelligent Robots and Systems (IROS), Yokohama, Japan, pp. 1173-80, 1993

[4] Franceschini N., Pichon J. M., Blanes, C., "From insect vision to robot vision," Phil. Trans. R. Soc. Lond. B, 337, pp. 283-294, 1992

[5] Fürst S., Dickmanns E. D., "A vision based navigation system for autonomous aircraft," Intelligent Autonomous Systems, Y. Kakazu et al. (Eds.), IOS Press, pp.765-774, 1998

[6] Gould, J. L., Gould, C. G., The Honey Bee, Scientific American Library, New York, 1988

[7] Harrison R. R., Koch C., "A neuromorphic visual motion sensor for real-world robots," Workshop on Defining the Future of Biomorphic Robotics, IROS'98, 1998

[8] Huber S. A., Franz M. O., Bülthoff H. H., "On robots and flies: Modeling the visual orientation behavior of flies, " Robotics and Autonomous Systems 29, Elsevier, pp.227-242, 1999.

[9] Mura F., Franceschini N., "Visual control of altitude and speed in a flying agent," Proceedings of 3rd international conference on Simulation of Adaptive Behavior: From Animal to Animats III, pp.91-99, 1994

[10] Netter T., and Franceschini N., "Towards nap-of-the-earth flight using optical flow, " Proceedings of ECAL99, pp. 334-338, 1999

[11] Neumann T. R., Bülthoff H. H., "Insect Inspired Visual Control of Translatory Flight," J. Kelemen and P. Posik (Eds) ECAL2001 LNAI2159 Springer Verlag , pp. 627-636, 2001

[12] Pfeifer R., Lambrinos D., "Cheap Vision - Exploiting Ecological Niche and Morphology", Theory and practice of informatics: SOFSEM 2000, 27th Conference on Current Trends in Theory and Practice of Informatics, Milovy, Czech Republic; Vaclav Hlavac [et al.] (Eds.), pp. 202-226, 2000.

[13] Reichardt W., "Movement perception in insects," In W. Reichardt (Eds.), Processing of optical data by organisms and machines, pp.465-493, New York: Academic, 1969.

[14] Srinivasan M. V., Poteser M., Kral K., "Motion detection in insect orientation and navigation," Vision Research 39, pp. 2749-2766, 1999

[15] Srinivasan M. V., Zhang S., Altwein M., and Tautz J., "Honeybee Navigation: Nature and Calibration of the "Odometer"," Science, vol. **287**, pp. 851-853, 2000.

[16] Srinivasan M. V., Zhang S. W., Chahl J. S., Barth E., Venkatesh S., "How honeybees make grazing landings on flat surfaces," Biol. Cybern. 83, pp.171-183, 2000.

Intelligent Autonomous Systems 7
M. Gini et al. (Eds.)
IOS Press, 2002

CPG Model for Autonomous Decentralized Multi-Legged Robot System
–Generation & Transition of Oscillation Patterns and Dynamics of Oscillators–

Shinkichi Inagaki
The University of Tokyo
Tokyo, Japan
inagaki@prince.pe.u-tokyo.ac.jp

Hideo Yuasa
The University of Tokyo & RIKEN–BMC
Tokyo, Japan & Nagoya, Japan
yuasa@prince.pe.u-tokyo.ac.jp

Tamio Arai
The University of Tokyo
Tokyo, Japan
arai@prince.pe.u-tokyo.ac.jp

Abstract. A central pattern generator (CPG) model is proposed for a walk pattern generation mechanism of an autonomous decentralized multi legged robot system. The topological structure of the CPG is represented as a graph, on which two time evolution systems, Hamilton system and a gradient system, are introduced. The CPG model can generate oscillation patterns depending only on the network topology and bifurcate different oscillation patterns according to the network energy. It means that the robot can generate gait patterns only by connecting legs and transit gait patterns according to a parameter such as the desired speed.

1 Introduction

Legged animals can generate suitable walk patterns (gaits) for the walk speed. This gait generating system is achieved by the central pattern generator (CPG), which is a group of neurons located in a central nervous system[1]. The CPG is modeled as a system of coupled nonlinear oscillators because of the periodical output (oscillation patterns), and it is used to coordinate leg movements (Figure 1) [2, 3, 4, 5].

An autonomous decentralized multi-legged robot system (ADMRS) has a leg as an autonomous partial system (subsystem) (Figure 1)[6]. This robot system achieves the function of a CPG as the whole system by coupling with neighbor oscillators, that is, coupling locally. Each subsystem is assumed to be unable to get information about the total number of subsystems or its position in the network. On the contrary, this robot

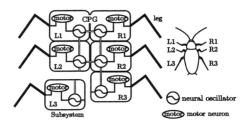

Figure 1: Schema of CPG: each leg is coordinated by the neural oscillator output via the motor neuron.

Figure 2: Autonomous Decentralized Multi-Legged Robot (under development): W 0.32, H 0.23, L 0.37[m].

system is expected to have properties such as the change of the number of legs, easy maintenance, failure-tolerance, and environmental adaptability.

Many CPG models were proposed in previous research. However, they couldn't deal with the change of the number of oscillators, and the number of legs. This is because they are divided into the next two types[4]: one is what phase relation of common gait patterns is required according to the number of oscillators[2, 3]. The other is required to analyze on how to change the parameters and inputs in various cases[4, 5].

On the other hand, our research purpose is the construction of the CPG model to realize the following features in ADMRS:

1. Gait patterns can be generated automatically according to the number of legs without embedded gait patterns.

2. Transitions of gait patterns can be generated only by changing certain parameter, e.g., the network energy.

In this paper, the method to construct a CPG model is proposed from the viewpoint of the graph expression of the oscillator network and the oscillator dynamics. The oscillator and the interaction between others correspond to a vertex and an edge respectively (Figure 3). Thereby, the oscillation patterns result in the eigenvalue problem of the graph and the transition is achieved only by changing the oscillation energy (network energy). Furthermore, the locality of the subsystem is realized due to the locality of the oscillator dynamics.

The organization of this paper is as follows. In Sec.2, a graph and a functional space will be defined. In Sec.3, Hamilton system will be introduced on the graph and obtain inherently multiple natural modes of oscillation. In Sec.4, the construction method of a CPG model will be shown by using a simple example (hexapod). In Sec.5, Hamilton system will be synthesized with a gradient system to select and transit the modes. In Sec.6, computer simulations of a hexapodal CPG model will be demonstrated.

2   Definition of a graph and function spaces

This section starts with some definitions about a complex function on a graph. Let a set of vertices be V, a set of edges be E, then a graph G can be defined as a set of these two terms $G = (V, E)$ (Figure 3).

Generally, verteces in the graph correspond to the agents of an autonomous decentralized system. In the case of the CPG model, they correspond to the oscillators.

It is assumed that G is a finite and oriented graph with N vertices. Each edge connects from the initial vertex to the terminal one. Note that this direction is not related with the information flow, that is, the interaction of each vertex is bi-direction[7].

Each vertex $u \in V$ has a complex valiable $\psi(u) = q(u) + ip(u)$ for $q(u), p(u) \in \mathbf{R}$. Let $C(V)$ be the set of all complex (or real) functions on V , then $\boldsymbol{\psi} = (\psi(u)_{u \in V}) \in C(V)$ is defined as the group of $\psi(u)$ $\forall u \in V$. $\boldsymbol{\psi}$ is also defined as $\boldsymbol{\psi} = \boldsymbol{q} + i\boldsymbol{p}$ for $[\boldsymbol{q}, \boldsymbol{p}]^T = [(q(u)_{u \in V}), (p(u)_{u \in V})]^T$.

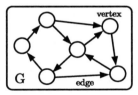

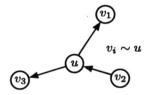

Figure 3: Graph expression of a system   Figure 4: Vertex u and its neighbor vertices v

Laplacian operator Δ_A on a finite graph G is defined for $\psi \in C(V)$ as

$$(\Delta_A \psi)(u) = \sum_{v \sim u} (\psi(u) - \psi(v)). \tag{1}$$

$v \sim u$ shows the set of verteces in which vertex v connects to vertex u(Figure 4). Note that Δ_A is calculated as $\Delta_A = A_a A_a^T$, where A_a is an incidence matrix of G. This incidence matrix A_a is determined by connections of vertices.

3   Hamilton system on a graph

3.1   Wave equation as Hamilton system

Schrödinger's wave equation is introduced on graph G, which is one of Hamilton system. Let $\psi(t) = (\psi(u, t))_{u \in V} \in C(V \times R)$ be a wave function on graph G. Then, Schrödinger's wave equation in graph G is defined as

$$ih\frac{\partial \psi(t)}{\partial t} = \mathcal{H}\psi(t), \tag{2}$$

where h plays the role of Planck's constant which relates Hamiltonian and the angular velocity. Hamiltonian operator $\mathcal{H}$ is defined as $\mathcal{H} \equiv \Delta_A$.

Then, each dynamics of Eq.(2) at each vertex is expressed as

$$\frac{\partial q(u, t)}{\partial t} = \frac{1}{h}\sum_{v \sim u}(p(u, t) - p(v, t)), \tag{3}$$

$$\frac{\partial p(u, t)}{\partial t} = -\frac{1}{h}\sum_{v \sim u}(q(u, t) - q(v, t)). \tag{4}$$

This equation shows that the dynamics at each vertex is determined by valiables of neighbor vertices $v \sim u$ and itself u (Figure 4).

Total Hamiltonian H of this Hamiltonian operator $\mathcal{H}$ shown in Eq.(5) is defined as the energy of this network (H is called 'Total Hamiltonian'). Total Hamiltonian H can be decomposed as local value $H(u)$ at each vertex u.

$$H = \psi^*(t) \cdot \mathcal{H}\psi(t) = \sum_{u \in V} H(u), \tag{5}$$

$$H(u) = \sum_{v \sim u} \left(q^2(u, t) + p^2(u, t) - q(u, t)p(v, t) - p(u, t)q(v, t) \right), \tag{6}$$

where, $\psi^*(t) = (\psi^*(u, t))_{u \in V}$ is complex conjugate of $\psi(t)$, and $\psi_1(t) \cdot \psi_2(t)$ is an inner product of $\psi_1(t)$ and $\psi_2(t)$. $H(u)$ is also determined by valiables of neighbor vertices and itself, so $H(u)$ is called 'Local Hamiltonian' of u.

3.2   Natural frequency mode of wave

Natural frequency modes of the wave Eq.(2) are solutions of the following equation,

$$\mathcal{H}\psi(t) = H\psi(t). \tag{7}$$

That is, the natural frequency modes $\psi_m(t)$ ($m = 0, 1, 2, \cdots, N - 1$) are provided as eigenfunctions of Hamiltonian operator $\mathcal{H}$. Then total Hamiltonian H_m is given as the eigenvalue of Hamiltonian operator $\mathcal{H}$.

Table 1: The solution of the eigenvalue ploblem and hexapodal gaits (phase relations) according to each mode : one cycle is set to be 1 on the basis of the right rear leg.

m	0	1	2	3	4	5
H_m	0	1	1	3	3	4
φ_m	$\frac{\sqrt{6}}{6}\begin{bmatrix}1\\1\\1\\1\\1\\1\end{bmatrix}$	$\frac{1}{2}\begin{bmatrix}-1\\0\\-1\\1\\0\\1\end{bmatrix}$	$\frac{1}{2}\begin{bmatrix}0\\-1\\1\\-1\\1\\0\end{bmatrix}$	$\frac{1}{2}\begin{bmatrix}1\\0\\-1\\-1\\0\\1\end{bmatrix}$	$\frac{1}{2}\begin{bmatrix}0\\1\\-1\\-1\\1\\0\end{bmatrix}$	$\frac{\sqrt{6}}{6}\begin{bmatrix}-1\\1\\1\\-1\\-1\\1\end{bmatrix}$
gait	0 L1--R1 0 0 L2--R2 0 0 L3--R3 0 (a)stand	5/6 L1--R1 2/6 4/6 L2--R2 1/6 3/6 L3--R3 0 (b)metachronal tripod		1/6 L1--R1 4/6 5/6 L2--R2 2/6 3/6 L3--R3 0 (c)rolling tripod		1/2 L1--R1 0 0 L2--R2 1/2 1/2 L3--R3 0 (d)tripod

With the separation of variables, these natural frequency modes of Eq.(7) can be expressed by the spatio-function $\varphi_m = (\varphi_m(u))_{u \in V} \in C(V)$ where $\|\varphi_m\| = 1$, and the temporal-function $\phi_m(t) = e^{-i\omega_m t} \in \mathbf{R}$.

$$\psi_m(t) = \varphi_m \phi_m(t) = \varphi_m e^{-i\omega_m t}, \tag{8}$$

where ω_m is the natural angular velocity. It is related with total Hamiltonian H_m as $H_m = h\omega_m$. In this Hamilton system, an arbitrary wave function $\psi(t)$ is expressed by the linear sum of the natural frequency modes

$$\psi(t) = \sum_{m=0}^{N-1} k_m \psi_m(t), \tag{9}$$

where $k_m \in \mathbf{C}$ are mode coefficients.

4   Example : graph with six vertices

4.1   The natural oscillation modes that appear in a graph of six vertices

In this section, we consider the graph composed of six vertices as shown in Figure 5. In this case, the incidence matrix A_a is as follows:

$$A_a = \begin{bmatrix} -1 & 1 & 0 & 0 & 0 & 0 \\ 1 & 0 & -1 & 0 & 0 & 0 \\ 0 & -1 & 0 & 1 & 0 & 0 \\ 0 & 0 & 1 & 0 & -1 & 0 \\ 0 & 0 & 0 & -1 & 0 & 1 \\ 0 & 0 & 0 & 0 & 1 & -1 \end{bmatrix}. \tag{10}$$

Eigenfunctions $\varphi_m = [\varphi_m(0), \varphi_m(1), \cdots, \varphi_m(5)]^T$ and total Hamiltonians H_m are obtained from Eq.(7) as shown in Table 1.

4.2   Construction of the CPG model

Here, a N-legged robot (N is an even number) has the CPG model composed of $2N$ oscillators is shown in Figure 6 [6, 8]. Two networks are symmetrically arranged, and oscillators 0, 2, $\cdots$, $N-2$ and 0′, 2′, $\cdots$, $(N-2)′$ control timings of the legs L1, L2, $\cdots$, $L\frac{N}{2}$ and R1, R2, $\cdots$, $R\frac{N}{2}$ respectively. Here, it is assumed that the phases of the wave in one side are half period earlier (or later) than the other's. Then, common hexapodal

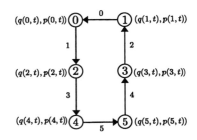

Figure 5: Graph of 6 vertices

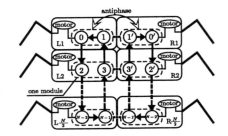

Figure 6: The structure of a N legged robot: a region in a broken oval is treated as one module

gait patterns of insects can be applied in (Table 1)[6]. The reason for employing the structure is

① the generation of same-directive traveling waves in both sides, and

② the locality of the connections between oscillators.

For example, the CPG model can generate 'metachronal tripod' in hexapodal gaits and 'walk'[1] in quadrupedal gaits.

The natural oscillation modes obtained from Figure 5 are matched with hexapodal gaits as shown in Table 1. The natural modes with the same eigenvalue form one gait by exchanging themselves through one oscillation cycle. This oscillation pattern is called 'exchange-pattern', e.g., 'exchange-pattern 1&2' or 'exchange-pattern 3&4'. Interestingly, the larger total Hamiltonian H_m becomes, the more rapid these gaits in real insects are.

5   Synthsis of Hamilton system and a gradient system

Hamilton system merely produces a linear sum of the oscillating modes as we demonstrated in Sec.3. The dynamics (a), (b) is realized via synthesis of Hamilton system and a gradient system.

(a) One (or two) specific natural oscillation mode(or modes) appears alternatively out of multiple natural oscillation modes.

(b) Changes between gait patterns are performed properly by changing certain parameter, such as the walk speed.

The prosedures to realize (a), (b) are considered as follows:

(a) The variable $\psi(u,t)$ of each vertex converges to a unit circular orbital in a complex plane. Then the natural oscillation mode appears, whose set of local Hamiltonian is the closest to the local Hamiltonian target value H_T.

(b) When the target value H_T of local Hamiltonian $H(u)$ increases (or decreases) continuously, the variable of each vertex follows it by enlarging (or lessening) the orbital radius at first. When the amount of the radius change becomes even larger, the variables of vertices change (bifurcate) the natural oscillation mode, and continue the follow-up to the target value H_T (Figure 7).

---

[1]The step order is left-forward, right-hind, right-forward, left-hind.

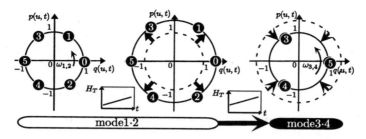

Figure 7: Transition between natural oscillation modes with a target value H_T

Let $[x_1(t), x_2(t)]^T = \left[(x_1(u,t))_{u\in V}, (x_2(u,t))_{u\in V}\right]^T$ be the function in graph G. If a potential function is represented as $P(x_1(t), x_2(t))$, the dynamics of the gradient system[7] can be expressed at each vertex $u \in V$ as

$$\frac{\partial x_1(u,t)}{\partial t} = -\frac{\partial P(x_1(t), x_2(t))}{\partial x_1(u,t)}, \tag{11}$$

$$\frac{\partial x_2(u,t)}{\partial t} = -\frac{\partial P(x_1(t).x_2(t))}{\partial x_2(u,t)}. \tag{12}$$

This dynamics is effectual in converging $[x_1(t), x_2(t)]^T$ to the state in which $P(x_1(t), x_2(t))$ can be minimized. The potential function $P(x_1(t), x_2(t))$ is given as the sum of local potential functions

$$P(x_1(t), x_2(t)) = \sum_{u\in V} P(u), \tag{13}$$

where $P(u)$ is determined by valiables of neighbor vertices $v \sim u$ and itself u.

From Eq.(3)(4) and Eq.(11)(12), each vertex dynamics of the synthesized system can be expressed as

$$\frac{\partial q(u,t)}{\partial t} = \frac{1}{h}\sum_{v\sim u}(p(u,t) - p(v,t)) - \frac{\partial P(q(t), p(t))}{\partial q(u,t)}, \tag{14}$$

$$\frac{\partial p(u,t)}{\partial t} = -\frac{1}{h}\sum_{v\sim u}(q(u,t) - q(v,t)) - \frac{\partial P(q(t), p(t))}{\partial p(u,t)}. \tag{15}$$

In this research, the potential function is chosen from (a), (b) as

$$P(u) = \alpha P_0(u) + \beta P_1(u) + \gamma P_2(u), \tag{16}$$

$$P_0(u) = q^2(u,t) + p^2(u,t) + \frac{1}{q^2(u,t) + p^2(u,t)}, \tag{17}$$

$$P_1(u) = (H(u) - H_T)^2, \tag{18}$$

$$P_2(u) = P_2(u') = (q(u,t) + q(u',t))^2 + (p(u,t) + p(u',t))^2, \tag{19}$$

where α, β and γ are real constants, and H_T is the target value of local Hamiltonian $H(u)$. $P_0(u)$ works so that the variable of each vertex converges to a circular orbital with radius 1 in the complex plane. $P_1(u)$ works so that the local Hamiltonian $H(u)$ of each vertex follows the target value H_T. $P_2(u)$ works so that waves in left and right networks become antiphase.

The method of constructing the gradient system in the graph is the same as the one in Yuasa et al.[7]. The significant difference is that the method of Yuasa et al.[7] needs to embed oscillation patterns beforehand (as phase differences of the oscillators), and this proposed method is to obtain oscillation patterns which are generated from the connecting way of the oscillator network. This is due to the introduction of Hamilton system.

6   Gaits generation and transition in computer simulations

We simulated generation and transition of hexapodal gaits with the connection as shown in Figure 6. The parameters in Eq.(16) were $\alpha = 1.0$, $\beta = 0.1$, $\gamma = 0.05$ and the initial variables $q(0)$, $p(0)$ were generated respectively by uniform random number of $[-1, 1]$. The target value of local Hamiltonian H_T was

$$H_T = 6\left(1 - \cos\left(2\pi t/400\right)\right) \qquad (20)$$

as shown in Figure 8 (the dashed line). The result of this simulation ($t = 0 \sim 400$) is shown in Figure 9 and Figure 8.

Figure 8 shows the changes of total Hamiltonian H (the solid line) and the target value of local Hamiltonian H_T (the broken line). H_T is given to all vertices as the convergence target value of local Hamiltonian $H(u)$. H is calculated by Eq.(5). H dosen't absolutely follow H_T because the system follows the change of H_T by changing the oscillation modes (Sec.5 (a),(b)). As shown in Figure 8, H changed dramatically between the exchange-patterns 1&2 and 3&4.

Figure 9 is a plot of real parts of $\psi(u, t)$ ($u = 0, 2, 4$ and $0', 2', 4'$) at $t = 380 \sim 400$. The oscillation pattern transited from exchange-pattern 3&4 to 1&2 at around $t = 390$. The number above each curve expresses the number of the corresponding oscillator. The former and latter oscillation patterns correspond to rolling tripod and metachronal tripod respectively from Table 1.

It turns out that the proposed CPG model can achieve transitions between oscillating patterns against the change of H_T. Furthermore, if the target value of local Hamiltonian H_T is matched with the desired walk speed, the gait patterns would be changeable to it.

Extraction and transition of modes are also observed in the cases of 4-legged and 8-legged CPG models with same parameters. However, they are depended on the initial condition in some degree. This is because the potential functions Eq.(16)~(19) were chosen with the heuristics method to the procedures(a)(b). We are now establishing a methodology for designing a suitable potential function.

7   Discussion

In this paper, the method to construct a mathematical CPG model was proposed. The dynamics of each oscillator is determined only by information of itself and neighbor oscillators. Nevertheless, the total system has the natural oscillation modes as the

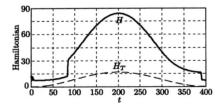

Figure 8: Total Hamiltonian H of the left network of Figure 6 and the target value of local Hamiltonian H_T

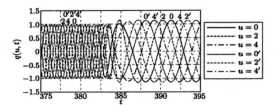

Figure 9: The wave forms of $p(u, t)(u = 0, 2, 4, 0', 2', 4')$

total-order (gait pattern). The order depends on a scale of the total system, that is, the number of oscillators. As a result, we can produce the CPG model adequate to the actual robot ADMRS (Sec.1–1,2).

The influence to the CPG model from the body characteristic is important when the robot walks actually[9, 10]. In regard to this, the feedback to the CPG model from the robot's sensors is required. However, this paper is not treating the feedback. Our research builds the open loop system first:

Input target speed (desired value of a part Hamiltonian),

State wave function,

Output oscillating pattern.

We take the standpoint at which we will introduce the feedback to the CPG model from sensors. Concretely, we consider

- the correspondence of the actual walk speed and the target value of local Hamiltonian H_T,

- the change of Hamiltonian operator $\mathcal{H}$ of Eq.(2) to the sensor information and

- the addition of new potential functions to Eq.(16).

Moreover, we will consider how each leg moves against the changing variables of the oscillator. The correspondence is required to carry the CPG model in the actual robot. We are now installing the CPG model to the actual multi-legged robot (Fig.2).

References

[1] K. Pearson. The control of walking. *Scientific American*, 235-6:72–86, 1976.

[2] G. Schöner, W. Y. Jiang, and J. A. S. Kelso. A synergetic theory of quadruped gaits and gait transitions. *Journal of theoretical biology*, 142:359–391, 1990.

[3] H. Yuasa and M. Ito. Coordination of many oscillators and generation of locomotory patterns. *Biological Cybernetics*, 63:177–184, 1990.

[4] J. J. Collins and S. A. Richmond. Hard–wired central pattern generators for quadrupedal locomotion. *Biological Cybernetics*, 71:375–385, 1994.

[5] C. C. Canavier, R. J. Butera, R. O. Dror, D. A. Baxter, J. W. Clark, and J. H. Byrne. Phase response characteristics of model neurons determine which patterns are expressed in a ring circuit model of gait generation. *Biological Cybernetics*, 77:367–380, 1997.

[6] T. Odashima, H. Yuasa, Z. W. Luo, and M. Ito. Emergent generation of gait pattern for a myriapod robot system based on energy consumption. *Journal of the Robotics Society of Japan (in Japanese)*, 17-8:1149–1157, 1999.

[7] H. Yuasa and M. Ito. Autonomous decentralized systems and reaction-diffusion equation on a graph. *Transactions of the Society of Instrument and Control Engineers (in Japanese)*, 35:1447–1453, 1999.

[8] M. Golubitsky, I. Stewart, P. L. Buono, and J. J. Collins. A modular network for legged locomotion. *PhysicaD*, 115:56–72, 1998.

[9] M. Yano, M. Tokiwa, and Y. Makino. Simultaneous self-organization of optimal gait pattern of insect and trajectry to its destination. *The Fourth IFAC Symposium on Intelligent Autonomous Vehicles*, pages 261–264, 2001.

[10] K. Tujita, K. Tuchiya, A. Onat, S. Aoi, and M. Kawakami. Locomotion control of a multipod locomotion robot with CPG principles. *Proc. of The Sixth International Symposium of Artifical Life and Robotics*, 2:421–426, 2001.

Intelligent Autonomous Systems 7
M. Gini et al. (Eds.)
IOS Press, 2002

Interactive Evolutionary Computation for Real Robot from a Viewpoint of Observation

Daisuke Katagami and Seiji Yamada
CISS, IGSSE, Tokyo Institute of Technology, JAPAN
{katagami, yamada}@ymd.dis.titech.ac.jp

Abstract. In this paper, we describe investigation on a viewpoint of observation in an interactive evolutionary robotics system. We propose a behavior learning system ICS (Interactive Classifier System) using interactive evolutionary computation and a mobile robot is able to quickly learn rules by direct teaching of an operator. Also ICS is a novel evolutionary robotics approach using an adaptive classifier system to environmental changes. We classify teaching methods as internal observation and external one, and investigate the relationship between the observation methods and the results. We have two experiments based on our teaching methods on a real world.

1 Introduction

In previous robot learning studies, optimization of control parameters has been applied to acquire suitable behaviors in an real environment. Also in most of such researches, a model of human evaluation has been used for validation of learned behaviors. However, since it is very difficult to build a human evaluation function and adjust control parameters, a system hardly learns behaviors intended by a human operator. In contrast with modeling human evaluation analytically, we introduce another approach in which a system learns suitable behaviors using human direct evaluation without its modeling. Such an interactive method with *Evolutionary Computation* (EC) as a search algorithm is called Interactive EC (IEC) [1], and a lot of researches on it have been done thus far [2] [3]. Additionally reinforcement learning has been applied to robot learning in a real environment[4]. Unfortunately the learning takes pretty much time to converge. Furthermore, when a robot hardly gets first reward because of no priori knowledge, the learning becomes far slower.

To solve these problems, we have been proposed the framework of Interactive Evolutionary Robotics (IER)[5][6]. It is an interactive EC learning method for the purpose of designing a robot using EC methods like *genetic algorithm*, *genetic programming* and *evolutionary strategy*. We can expect IER to perform high emergent property of ER and subjective adaptability of IEC. This method quickly learns effective rules by simple instructions of a human operator. The objective of IER is to make initial learning more efficient and learn behaviors that a human operator intended through interaction with him/her. Though [7] is a research that a robot learns from interaction with human, it is significantly different from our work using interactive experiences with human to evolutionary learning.

First of all, we developed a learning system based on *classifier system*[8] on IER framework, which is able to adapted to multiplicity of an environment and a variable

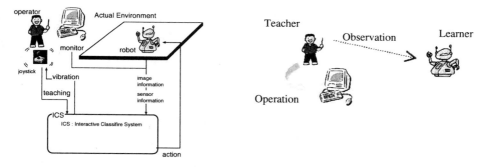

Figure 1: Teaching environment   Figure 2: Teacher view

dynamic state. we call it as *Interactive Classifier System* (ICS). Fig.1 give an environment of teaching using ICS. The difference between ICS and an usual learning classifier system is to introduce an interactive method. Accordingly, we expect that the system performs efficient initial learning in an actual environment and can operate concentrative incremental learning. However there is few framework that an operator observes from robot's view. Therefore, the system can not make the best use of the learning. In this paper, we propose interactive method based on internal observation in order to solve the problem.

2 Teaching Method Based on the Viewpoint of Observation

A robot hardly knows how the robot modifies behaviors of itself for a task and recognizes that the task is in the process of achieving. Besides, a robot, which observes through an inside viewpoint of a system, may be unable to recognize that a task is realizing, or rather an operator which observes through the outside viewpoint of a system can recognize. However the outside information of a system differ from the inside one. Thus a robot hardly learns from only the outside information. We, then, had a significant difference between cognition of an operator and one of a robot what is called a perceptual aliasing problem. It hence become a subject of discussion when a operator teach skills for a task to a robot.

In this paper, we prepare the simple setting based on the observation to examine how the difference influence to acquired rules by teaching. We call the method which observe through an outside viewpoint of a system as *teacher view* (Fig.2), and the method which observe through the inside viewpoint of a system as *learner view* (Fig.3).

We examine the difference by IER based on teaching with this *teacher view* and *learner view*. To realize this IER, we applied these methods to developed ICS, which is a robot learning system based on interactive EC.

3 Interactive Classifier System

3.1 System overview

ICS applies XCS[9], which is a kind of Learning Classifier System (LCS), as evolutionary computation and equipped a interactive function. XCS equipped a function which preserve classifiers from overgeneralization makes system's performance worse. Moreover, XCS applies restricted mating which is a kind of strategy based on *genetic*

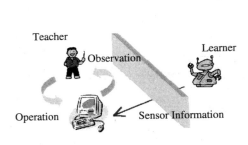

Figure 3: Learner view

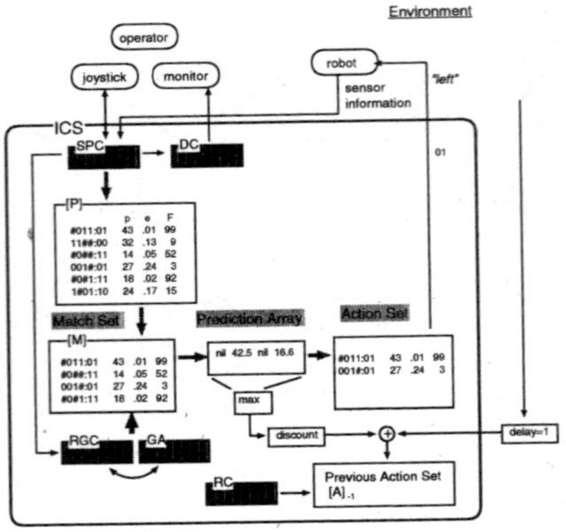

Figure 4: Overview of Interactive Classifier System

argorithms. For these reason, XCS improves learning performance of traditional CS. It is constructed as a robot learning model that can not only learn through teaching but also learn autonomously using XCS.

ICS mainly consist of a rule generation component (RGC), a sensor processing component (SPC), a display component (DC) and a reinforcement component (RC). It was developed with C language and GTK+ on Linux. It utilizes Video4Linux for image processing. The rule generation component makes a new classifier from teaching by the operator. The SPC processes each information of some sensors and camera, and through it for the RGC. The DC displays by GUI interface and processes the input from a joystick. Finally the RC performs learning by updating parameters in ICS. Fig.5 shows the developed interface of the system.

The experiments are made with a standard miniature mobile robot Khepera (Fig.6). The mobile robot has a cylinder shape, a diameter of 6 cm and a height of 5 cm. It possesses two motors and on-board power supply. The motors can be independently controlled by a PID controller. The eight infrared sensors are distributed around the robot in a circular pattern. They emit infrared light, receive the reflected light and measure distances in a short range: 2-5 cm. The robot is also equipped with a Motorola 68331 micro-controller which can be connected to a computer via serial cable. Moreover, the system utilizes SONY analog controller DUALSHOCK as a joystick. Fig.6 shows them respectively.

We describe a learning procedure in ICS as follows.

1. At first, a human operates robot with a joystick by viewing sensor information displayed on GUI, and the DC processes it.

2. Next, the SPC gets operator's instruction and robot's sensor information.

3. The RGC makes new rules from them and adds them into a rule list. When nothing is input from the operator, a mobile robot executes autonomous behaviors from interaction.

4. Finally, the RC reinforces the classifiers by updating their parameters in the actions which were previously executed.

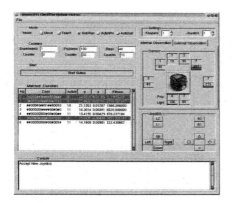

Figure 5: User interface

Figure 6: A mobile robot Khepera and a joystick

In traditional works of robot learning in a real environment, a learning takes pretty much time to converge because of learning by trial and error. In this work, we consider that this learning by trial and error is a problem in a real environment. However, it hardly prepare suitable apriori knowledge in an environment. For this reason, the ICS generate initial individuals by teaching from human-robot interaction. We can perform efficiently initial learning in this way.

4   Experiments

4.1   Experiment A with Cognitive Observation

We experimented in a real world to investigate difference in two teaching methods: *teacher view* and *learner view*. Fig.7 shows an experimental environment. As an experimental task ICS reduces the number of steps from any start points to a light source which set up as a goal in a field surrounded with white plastic plates. We compared two teaching methods with a traditional method in which a robot autonomously learns by simple EC.

An encoding of a classifier is as follows. A classifier is the twenty bit string "#000#10 0000###100#:01". The robot's condition is the left sixteen bit string, it represented eight infrared proximity (left eight bit) and light (right eight bit) sensors around the robot in a circular pattern respectively (Fig.6). The bit is "1" if a input sensor value larger than a threshold, or else "0". "#" is a "don't care" symbol which *classifier system* employed. The robot's previous action is the next two bit string, represented as forward "11", left-turn "01", right-turn "10 " and back "00". The robot's current action is the two bit string similarly.

A fitness function of ICS defined as follows. Reward F is computed by the sum of eight light sensors through a sigmoid function.

$$u = \sum_{i=0}^{7} light_i \frac{Payment\,Range}{light_{MAX} \times 8} \qquad F = \frac{1}{1 + exp(-u)}$$

We consider fourty steps as a trial and begin teaching or autonomous exploration for five trials at randomly start points, and test a trial at each of five start points

Table 1: Experimental Parameters

Parameters	Value
number of problems in one experiment	30
number of expriments	1
maximum size of the population	300
probability to do crossover	0.8
probability of mutating one bit	0.04

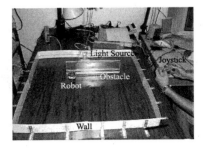

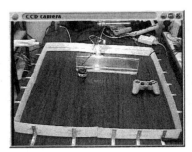

Figure 7: Experimental Environment   Figure 8: External Observation Settings

as examination of created rules by the teachings or the exploration. It performs this procedure six times, and consequently, we have thirty trials as examination. Table 1 shows experimental parameters.

In the case of *teacher view*, the operator can not recognize a small obstacles as the robot can perceive although he looks a whole environment. To represent *teacher view*, an operator teach to a robot using the camera which looks a whole environment. Fig.8 shows information through GUI by a camera which sets up the environment.

In the case of *learner view*, an operator can not look a whole environment although it can recognize directly a small obstacles and a recognition error which the robot can perceive. To implement *learner view*, ICS uses GUI which represent some sensor values. Since an operator hardly understands although one looks only numerical values of sensor data, we developed GUI interface which can represent sensor values by graphs. Fig.9 shows sensor information which represented by graphs through GUI interface.

4.2   Experimental Results of Examination A with Cognitive Observation

In this experiments, we examined the number of steps to a light source, system error, and fitness. Fig.10 shows the average of the number of steps at test trials from five start points. System error(Fig.11) is the absolute difference between the system prediction for the chosen action by a system and the actual external payoff. Fitness (Fig.12) which is used in *genetic algorithms* for evaluation.

Teacher view improves a robot's learning in the simple environment which cognitive difference is little by teaching is easy because an operator looks a whole environment. However, the difference is not so large. There was no difference in both observations regard for system error and fitness. We found out that the learning improves without they effect on system error or fitness by using two kind of teaching methods.

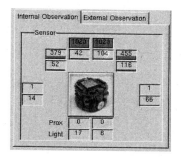

Figure 9: Internal Observation Settings

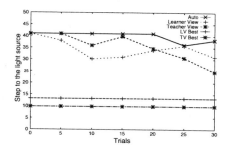

Figure 10: Step to Light Source

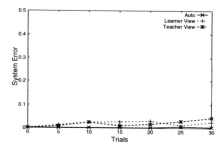

Figure 11: System Error

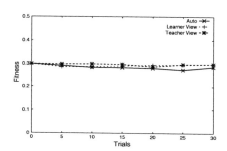

Figure 12: Fitness

4.3 Experiment B with Cognitive Observation

We introduced an obstacle in the environment of experiment A to investigate the effect by a more difficult task. Fig.7 shows the experimental environment. The obstacle was made of the transparent plastic board because of perception a direction of a light source. A robot must reach to the light source as avoiding the obstacle. We compared a teaching method by *teacher view* with *learner view* in the same way as experiment A.

We consider twenty steps as a trial and begin teaching for twenty trials at randomly start points. It is different from experiment A because of simplifying an experiment. We test a single trial at a start point every one trial as examination of created rules. The experimental parameters is as same as Table 1.

4.4 Experimental Results of Examination B with Cognitive Observation

In experiment B, we examined the number of steps to a light source. The number of steps to a light source shows Fig.13.

Seeing from a best value of teaching (TV Best and LV Best), we see that *teacher view* improves teaching likewise experiment A because of looking a whole environment. However, we can see that *learner view* outperformed *teacher view* about steps to a light source in contrast with experiment A. Since ICS hardly uses information of proximity sensors and can acquire effective rules by using only information of light sensors in the environment, the difference of cognition between an operator and a robot is a little. For this reason, the difference of two teaching methods was not so large.

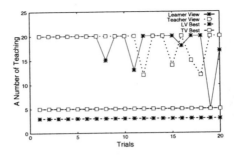

Figure 13: Step to Light Source

Table 2: Experimental Results of Exploit

	Step to Light Source	Reach to Goal
times	1 2 3 4 5	total
teacher view	- - - - -	0
learner view	6 9 - - -	2

Table 3: Created Rules by *learner view* Method

Condition	Action	Prediction	Teach
0#0#0#######0#10##1#	10	413.7	2
0010###0##000#1###	01	364.0	1
00###0#000#1#0#01#	11	292.0	4
###000000101001011	11	256.0	1
0#0#0#######0#10#01#	10	280.3	3
00#000001001##0##1	10	244.5	1
00###0#000#1#0##11	11	215.0	1
#000#100000###100#	01	101.5	3
#000##001000##0001	11	100.8	6
#000#0001000##0001	11	99.4	10

ICS however hardly learns from teaching of an operator in case of experiment B because there is the difference between the situation of the robot which an operator estimated by observation from the outside and the one of the robot in the real world. *Leaner view* actually creates effective rules, because an operator performed teaching as verifying robot's internal status.

After we have experiments *teacher view* and *learner view* each twenty trials, we test a trial at each of five start points as examination using each created rules. Table 2 shows the experimental results. The robot can not reach to a light source at any start points of five in the examination of *teacher view* because teaching does not improve. On the other hand, the robot reached to a light source by six steps and nine steps respectively at two of five start points in the examination of *learner view*. We can see that ICS can create rules which effective and do not depend on start points since *learner view* improves the robot's learning from teaching.

Table 3 shows ten rules which have best values of the system prediction in created ones after twenty trials with *learner view*. ICS created effective and common-sense rules like as the robot moves forward when the light source faces in front of the robot, and it moves right when the light source faces right. We can see teaching improves very well because all the rules which the system prediction is high is created from teaching or its offspring.

Though an operator did not teach a robot to go back in the both experiments, the robot goes back for avoiding an obstacle when the robot colides with a wall and reaches to a light source. Because ICS created these rules by which a robot works in cooperation with human.

5 Conclusion

In this paper, we proposed a novel interactive method from the viewpoint of observation and investigated its effects in a real world experiments. It was found that internal observation increase the effects according to arise the difference between the robot's recognition and operator's one in a complex environment.

This study examined that it is possible to perform a fast learning of a robot by human-robot interaction in a real world. Besides, our system learns and automatically acquires a complex robot's program which human hardly describes by the way that an operator only teachs to a robot simply. Our future direction will be one that acquires the unconscious information of an operator like reactive behaviors of human and skills of an operator.

References

[1] R. Dawkins. *The Blind Watchmaker.* Longman, Essex, 1986.

[2] Y.. Nakanishi. Capturing Preference into a Function Using Interactions with a Manual Evolutionary Design Aid System. Genetic Programming, pages 133–140, 1996.

[3] T. Unemi. A Design of Multi-Field User Interface for Sumulated Brdding. Asian Fuzzy System Symposium (AFSS'98), pages 489–494, 1998.

[4] E. Uchibe, M. Asada, and K. Hosoda. Behavior coordination for a mobile robot using modular reinforcement learning. In IEEE/RSJ International Conference on Intelligent Robots and Systems 1996 (IROS96), pages 1329–1336, 1996.

[5] D. Katagami and S. Yamada. Interactive classifier system for real robot learning. In *IEEE International Wortkshop on Robot and Human Interaction(ROMAN-2000)*, pages 258–263, 2000.

[6] D. Katagami and S. Yamada. Real robot learning with human teaching. In *The Fourth Japan-Australia Joint Workshop on Intelligent and Evolutionary Systems*, pages 263–270, 2000.

[7] M. Nicolescu and M. J Mataric. Learning and Interacting in Human-Robot Domains. In IEEE Transactions on Systems, Man, Cybernetics, special issue on "Socially Intelligent Agents - The Human in the Loop", to appear in 2001.

[8] J. H. Holland and J. S. Reitman. Cognitive systems based on adaptive algorithms. In Donald A. Waterman and Frederick Hayes-Roth, editors, *Pattern-Directed Inference Systems*, pages 313–329, Orlando, 1978. Academic Press.

[9] S.W. Wilson. Classifier fitness based on accuracy. Evolutionary Computation, 3(2):149–175, 1995.

Intelligent Autonomous Systems 7
M. Gini et al. (Eds.)
IOS Press, 2002

A Study on Low-invasive Measurement System from Motor and Somato-sensory Cortex of Rat for Prosthetic Application

Takahiro Kawashima, Masahiro Oikawa, Hiroshi Yokoi and Yukinori Kakazu
Autonomous Systems Engineering, Hokkaido University
N13-W8, Kita-ku, Sapporo, Hokkaido, 060-8628, Japan

Abstract. This paper aims to develop a low-invasive measurement system from a brain for prosthetic application. For the purpose, corresponding to the difference of individuals' neural activity patterns and responsiveness is required. Therefore, recorded signals are classified by using the kohonen's Self-Organizing Maps. First, a reactive signal from a barrel cortex in a somato-sensory cortex is recorded in order to simulate the low-invasiveness. Next, a signal from a rat which is fastened on a rotation table is measured using electrodes fixed to a cranial bone of another rat in order to realize stable measurement, and classified with the same method as the barrel. The result shows that the map is divided into two states of move and stop, respectively.

1 Introduction

If a path from a motor cortex as an output of central nerve system (CNS) to extremities is damaged by an accident or disease, then it becomes the cause of losing a motor function. In the case of serious illness, they suffer severe disability. Therefore, to develop an adaptive prosthetics, which is controlled by user's intention, leads these persons to the more comfortable activity of daily living.

Recently, the computational approaches to reconstruct or substitute the lost motor function are studied. One of these studies is the development of interface with the computer using the human electroencephalogram for communication [1]. Moreover, measuring the activities in the cerebral cortex directly and classifying of a signal in motor area to activate extremities are expected to identify the signals more correctly. From this viewpoint, approaches using minute electrodes such as glass electrode[2] or metal [3] implanted in the cerebral cortex are reported. The progress of these studies enables the patients with the motor function disease to do the activity of daily living, for example, communicating, walking or eating by themselves. The key point to realize this requirement lies in how to make use of the remaining function effectively. And problems of these studies are how to construct the low- or non-invasive measurement system in terms of the safety, real time processing, to decrease the learning time of the patients, and the development of prosthetics easy to use.

In this study, we focus on the electrocorticogram measured at the surface of the cerebral cortex. The amplitude property of the recorded signal is larger than EEG. This method is not non-invasive but low-invasive. We try to measure from the surface of CNS in order to realize such a low-invasive measurement system. Moreover, corresponding to difference of individuals' nervous activity patterns and responsiveness is necessary. Therefore, we classify the signals by using Fast Fourier Transform (FFT) as a pre-process and Self-Organizing Maps (SOM), which maps higher dimension data to lower dimension. SOM is possible to observe control signals for the motor system by measuring in the cerebral cortex directly [4]. The barrel data is investigated in order to evaluate the classification possibility of SOM.

2 Measurement

2.1 Measurement from Somato-sensory Cortex

We focus on the barrel cortex, which is specifically responding the stimulation for the mystacial vibrissae in the rat somato-sensory cortex, and measure the response for its stimulation. In the somato-sensory cortex of the rodent, module structures which process the information corresponding to the response of the mystacial vibrissae respectively exist. This area is named the barrel cortex. The projection map for the sensory cortex of the rat brain is the formation like the reversely standing. The composite map of the rat cutaneous representation is indicated [5]. According to this map, the center position is 2mm posterior and 6mm lateral from the bregma and the barrel cortex exists in the square of 2-3mm. The processing for the whisker tactual stimulation is occurred in the barrel area of its opposite cerebral hemisphere.

The rat is anaesthetized with urethane, and is fixed with a brain stereotaxis apparatus. After incising its scalp, the cranial bone around the area including the barrel cortex is drilled till the cranial dura mater appears. In the position of the barrel which is 2mm posterior and 5.5mm lateral from bregma, the electrode is contacted at the surface of the cerebral cortex. Then, we measure the signal from the surface of the cortex for the stimulation of the rat whisker and classify. The stimulus patterns are from up to down (UD, Fig.1(I)), from down to up(DU, Fig.1(II)), from forward to backward(FB, Fig.1(III)), and from backward to forward(BF, Fig.1(IV)). And the arrow in the Fig.1 represents the stimulus direction. We stimulate the whisker by hands. One pattern is repeatedly stimulated 20 times at intervals of a second as a trial. These four trials are inputted in turn and a set of these patterns repeats five times. The sampling rate on the A/D converter is 20 kHz.

2.2 Measurement from Motor Cortex

Two electrodes are fixed to a cranial bone of a rat. The measurement position is 1.8mm posterior and 2.5mm lateral from bregma at the both hemisphere. This position is corresponding to the hind-limb motor cortex and is decided based on the brain map of a rat [6]. Moreover, we measure and classify the signal from the rat which is fastened on a rotation table and is moving freely. If the rat moves, then the table rotates. This classification is off-line. In order to evaluate the movement of the rat hind-limb and extract the signal on move and on stop, respectively, we use tracking vision system. Tracking speed of this system is 30-frame per a second. Sampling rate is 10kHz.

3 Classification

Generally, FFT is used as the methodology for the time series analysis. In this study, FFT is applied as the means of extracting the frequency characteristics from the time series data. The recorded signal is converted into the frequency range using FFT as preprocessing. This converted data is a set of the input codebook vector of SOM.

In this study, the update function of SOM codebook is shown in Eq.(2).

$$m_i(t) = m_i(t-1) + \lambda_i(t)\Phi(||r_i - r_{i(x(t))}||)(x(t) - m_i) \tag{1}$$

$$\Phi(p) = exp(-\frac{p^2}{2\sigma^2(t)}) \tag{2}$$

$$\sigma(t) = (10 - (9/time)*t) \tag{3}$$

$$\lambda_i(t) = 0.9*(1 - t/time) \tag{4}$$

Where, m_i is a codebook vector, r_i is a coordinate on the lattice of map, $i(x(t))$ represents the closest codebook to the input vector, and t is time. $x(t)$ is an input vector. Eq.2 is the neighborhood function decreasing on p. This function satisfies following relations: $\Phi(0) = 1$, $\Phi(\infty) = 0$. The function $\sigma(t)$ and $\lambda_i(t)$ depending on t is given by Eq.3 and Eq.4, respectively. *time* means total time of the learning. Moreover, the map shape is a hexagon. Evaluation between input and codebook is executed by an inner product. Under these conditions, the classification of the recorded signal is performed. The results of computation are shown in the next section.

4 Results

Barrel Cortex: In the Fig.2, the sampled data at the position of the surface in every stimulus of one time is transformed by FFT at 2^{14} (=16384) data size as one input data, and the total 216 input data (4 patterns × 54 data/pattern) is prepared. We train SOM map 5000 times by inputting the 216 data repeatedly, projecting all the input patterns lastly on the map. The maximum of the bandwidth of frequency data converted by FFT is the 10kHz. The input data size is very large. Therefore, on an average of every one hundred data, 100-dimension vectors are inputted SOM. The map size of SOM is 30 × 30. The symbol $\diamond$ represents the stimulation UD, $\square$ is DU, $\triangle$ is FB, and $\bigcirc$ is BF.

Motor Cortex: The data size converted by FFT is 256. The number of SOM character vector is 128. The SOM map size is 25 × 25. Fig.3 represents the SOM map on the signal recorded from the left hemisphere when right hind-limb moving. Fig.4 is the map on stop.

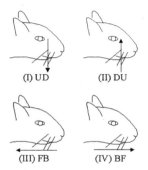

(I) UD   (II) DU

(III) FB   (IV) BF

Figure 1: Stimulus patterns

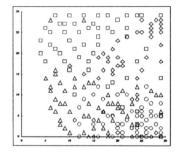

Figure 2: Result of the barrel surface

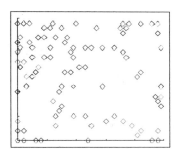

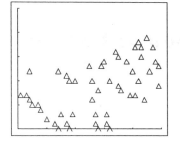

Figure 3: Map on right hind-limb movement   Figure 4: Map on stop

5   Discussion and Future Works

The barrel cortex result (Fig.2) shows the possibility of classification to surface-recorded signal. The results of the motor cortex (Fig.3 and Fig.4) show the differences of the spacial distribution between these patterns. This means it is possible to classify the two patterns, stop and right hind-limb move. However, the details of the state of the limb movement are not clear in current experimental environment. In this experiment, we evaluate only two values, which is whether the hind-limb moved or not. Therefore, It is not clear how many states of hind-limb sequential movement can be classified from the obtained brain signal. Moreover, it is also necessary to verify the difference between right and left hind-limb movement. These are future works about the experiment of the motor cortex which will have the necessity of carrying out from now on.

Acknowledgements

This work was supported by National Institute of Bioscience in Human-Technology on the measurement in the rat barrel cortex. We thank Dr. H. Kaneko and Dr. S. Suzuki.

References

[1] N. Birbaumer et al. A spelling device for the paralysed. *Nature*, 398:297–298, 1999.

[2] P.R. Kennedy and R.A.E. Bakay. Restoration of neural output from a paralyzed patient by a direct brain connection. *NeuroReport*, 9:1707–1711, 1998.

[3] J.K. Chapin et al. Real-time control of a robot arm using simultaneously recorded neurons in the motor cortex. em Nature Neuroscience, 2:664–670, 1999.

[4] S. Lin, J. Si, A.B. Schwartz. Self-organization of firing activities in monkey's motor cortex: trajectory computation from spike signals. *Neural Computation*, 9:607–621, 1997.

[5] J.K. Chapin and C-S. Lin. The sensory cortex of the rat. In: *The Cerebral Cortex of the Rat* (B. Kolb, R.C. Tees Ed.), MIT Press, 1990.

[6] G. Paxinos and C. Watson. The Rat Brain in Stereotaxic Coordinates, Compact 3rd Edition, Academic Press, 1997.

Intelligent Autonomous Systems 7
M. Gini et al. (Eds.)
IOS Press, 2002

Modular Fuzzy-Reinforcement Learning in Multi-Agent Systems

Mehmet KAYA   İrfan GÜLTEKİN   Ahmet ARSLAN

Fırat University, Department of Computer Engineering, 23119 Elazığ, TURKEY
e-mail:{mekaya, igultekin, aarslan}@firat.edu.tr

Abstract: In multi-agent systems, the state space to be handled constitutes a major problem efficiently in learning of agents. This paper presents a novel approach to overcome this problem. The approach uses together the advantages of the modular architecture and fuzzy logic in multi-agent systems. Fuzzy logic maps the input fuzzy sets, representing state space of each learning module, to the output fuzzy sets representing the action space. The fuzzy rule base of each module is built through the Q-learning, which is one of the reinforcement learning schemes. Experimental results done on pursuit domain show the effectiveness and applicability of the proposed approach.

1. Introduction

In recent years there has been a considerable amount of interest in multi-agent systems. One approach to modeling multi-agent learning is to augment the state of each agent with the information about other agents [8, 11, 12]. However, as the number of agents increases in a multi-agent environment, the state space of each agent grows exponentially. Thus, even simple multi-agent learning problems are computationally intractable by standard reinforcement learning approaches. In order to remedy the problem of combinatorial explosion in multi-agent reinforcement learning, Whitehead [14] proposed a modular approach. He considered a variant of the pursuit problem as a multi-agent learning problem suffering from the combinatorial explosion, and showed how successfully modular Q-learning prey-pursuing agents synthesize coordinated decision policies needed to capture a randomly-moving prey agent.

In another work done on modular Q-learning for a multi-agent environment, Park et al. [10] proposed an action selection mechanism among the robots in a robot soccer game and the effectiveness of their scheme was demonstrated through real robot soccer experiments.

In our previous work [5], we grouped the state space of each hunter agent acting in a pursuit domain to different fuzzy labels. Thus, we decreased the size of the problem space and obtained faster convergence to the near optimal solution.

In this paper, we combine the advantages of the modular architecture and fuzzy logic in a multi-agent system and attempt to additionally validate the effectiveness of our approach through the application to a variant of the well-known pursuit problem, which is a typical testbed for multi-agent cooperation algorithms.

The rest of this paper is organized as follows. Section 2 presents a variant of the pursuit problem, to be used as a platform for experiments throughout this study. In Section 3, we describe our modular fuzzy-reinforcement learning approach. In section 4, we give the experimental results done for this approach and in Section 5 we will discuss the results we have obtained.

2. Problem Domain

In order to evaluate our learning approach, we consider a variant of the pursuit problem defined as follows:

- The environment consists of a 12x12 grid. Edges are connected (torus).
- Initial position of each agent is determined randomly. There are four hunter agents and a prey agent in the environment.
- At each time step, agents synchronously execute one out of five actions: moving up, down, left, or right from the current position, or staying at the current position. More than one hunter agent can share the same grid. However, hunter agents cannot share with the prey. Therefore, hunter agents that try to move to the grid already occupied by the prey, cannot move and must stay at the their current positions. The prey agent selects its own action so as to maximize the total sum of Manhattan-distance to the located hunters.
- Every agent can see objects at a certain distance. This distance and its cells are called the visual depth and the visual environment of that agent, respectively. Each agent is assigned an identifier. A hunter agent can recognize the relative position and identifier of any other agents in its sight.
- The prey is captured, when the hunters occupy all of its four neighbor positions, as shown in Figure 1(b). Then all of the prey and the hunter agents are relocated at new random positions in the grid world and the next trial starts.

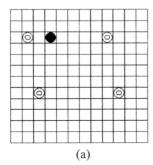   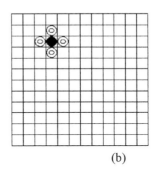

(a)   (b)

Figure 1: (a) An initial position (b) A goal state

3. Modular Fuzzy-Reinforcement Learning Approach

Q-learning is a recently explored reinforcement learning algorithm that does not need a model for its application and can be used online [13]. Q-learning algorithms store the expected reinforcement value associated with each situation-action pair, usually in a look-up table.

Q-learning is an incremental reinforcement learning method. According to Q-learning, the agent selects an action based on an action-value function, called the Q-function, $Q(s,a)$, which defines the expected sum of the discounted reward attained by executing action $a(\in A)$ in state $s(\in S)$, and determining the subsequent actions by the current policy π. The Q-function is updated using the agent's experience. The learning is as follows:

1. observe the current state s
2. select an action a_i with the probability:

$$\pi[a_i \mid s] = \frac{e^{Q(s,a_i)/\tau}}{\sum_{a_k \in A} e^{Q(s,a_k)/\tau}}$$

where τ is called the temperature, which determine the randomness of the stochastic policy. Although several methods for selecting an action have been proposed, we adopt this Boltzmann selection method.
3. observe the new state s'
4. receive a reward r from the environment
5. update the corresponding Q value for state s and action a according to the following formula:

$$Q(s,a) = (1 - \beta)Q(s,a) + \beta(r + \gamma \max_{a' \in A} Q(s',a'))$$

where β $(0 \le \beta < 1)$ is the learning rate and γ $(0 \le \gamma \le 1)$ is the discounted parameter.
6. If the new state s' satisfies a terminal condition, then the single trial ends. Otherwise let $s' \to s$ and go back to step 1.

However, in applying Q-learning to the multi-agent system, there are some difficult problems because the dimension of state space for each learning agent grows exponentially in the number of its partners. For example, for the pursuit domain shown in Figure 1 the state of a hunter is determined by the positions of the other agents currently inside its visual environment. The state can be represented by an n-tuple where n is one less than the number of agents in the world. Each element of this state tuple consists of the coordinates of the corresponding agent relative to the hunters standing point. If the visual depth of a hunter is d, the number of possible locations of the other agents is $(2d+1)^2+1$. Note that there are $(2d+1)^2$ cells in the visual environment and the other is for indicating out of sight. Also, because there are one prey and four hunters in our problem domain, each hunter may be observing one of the $4m^2 + 6(m^2 - 1)^2 + 2(m^2 - 1)(m^2 - 2)^2 + (m^2 - 1)(m^2 - 2)(m^2 - 3)^2/6$, where $m=2d+1$, distinct states. If we set each hunter's visual depth to value of 3, the number of states amounts to 1,021,700.
In order to overcome the state space with high size, Whitehead [14] implemented a modular Q-learning architecture as shown in Figure 2. The architecture consists of three learning modules and single mediator-module. Each learning-module focuses on specific attributes of the current perceptual input and performs Q-learning. The i-th learning module of the architecture, L_i, receives only the relative position of the prey and that of the i-th partner that have been located, and ignores any information concerning the other partners.
Its state is represented by a combination of these two relative positions and the state space size equals m^4+1. This size of state space is computationally tractable by the standard Q-learning algorithm when d is relatively small. On the other hand, a mediator-module, without learning, combines learning modules' decision policies using a simple heuristic decision procedure and selects a final decision by the corresponding agent.
Agents decide their action by using *the greatest mass* strategy, proposed by Whitehead [14],

$$\arg\max_{a \in A} \sum_{i=1}^{i=3} Q_i(s_i, a)$$

where Q_i denotes the action-value function maintained by the learning-module L_i. Modules do not refer to other modules' Q-tables and each updates its own Q-table individually.

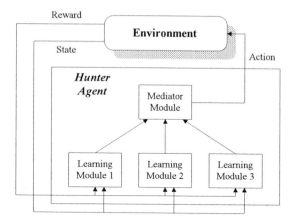

Figure 2: Modular Reinforcement Learning Architecture

In modular fuzzy-reinforcement learning approach we propose, the state space of each hunter whose visual depth is 3 can be represented using fuzzy sets as shown in Figure 3.
The fuzzy variables shown in figure constitute the input of fuzzy system. The state of a hunter is determined by 3 different membership functions according to the x and y positions of other agents. For instance, if a fuzzy input at x-variable belonging to the prey is at interval of [-3, 0], the prey is at left-hand side of the hunter. Similarly, if a fuzzy input at y-variable for the prey is at interval of [0, 3], the prey is above from the hunter. It should be noted that if one of the agents is outside the visual environment of the corresponding hunter, that agent hasn't the fuzzy inputs.

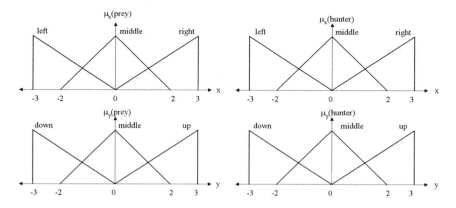

Figure 3: The input variables of the fuzzy system representing the state of a hunter.

The output of the fuzzy system that represents the action space of the hunters is shown in Figure 4.

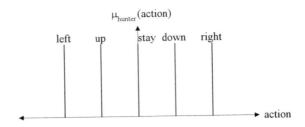

Figure 4: The output variable of the fuzzy system representing the action space of a hunter.

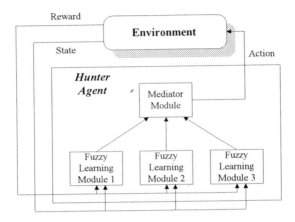

Figure 5: Modular Fuzzy-Reinforcement Learning Architecture

Figure 5 shows the architecture of modular fuzzy-reinforcement learning proposed here. The rule base of the fuzzy system in each block is found by Q-learning. Thus, the number of the state needed to find the rule base in each fuzzy learning-module is decreased to 9*9+9+9+1 (i.e., 100) where 9*9 is the number of the state in which each agent is observed by the corresponding agent in visual environment, 9+9 is the number of the state in which only one agent is observed, and 1 is the number of the state in which no agent is perceived.

If there is more than one membership function intersecting with the fuzzy input (for instance, x=-1 and y=-2 for prey) while updating Q-values, then all of Q-values related to these membership functions are updated.

After the hunter agent under consideration observes the current state, its mediator module makes the final decision and selects the most suitable action based on the Q-value received from each fuzzy-learning module. The mediator module makes this selection by considering *the greatest mass* strategy mentioned earlier.

4. Experimental Results

In our experiments, the learning consists of a series of trials, which begins with a single prey and four hunter agents placed at random positions, and ends when the prey is captured or at 2000 time steps. Upon capturing the prey, individual hunters immediately receive a reward of

1.0, and accordingly all of its component fuzzy learning-modules uniformly receive the same reward regardless of what decision policies they have. Hunters receive a reward of –0.1 for each move when they do not capture the prey.

Parameters for Q-learning are as follows. The learning rate is $\beta=0.8$, the discount factor is $\gamma=0.9$, and the initial value of the Q-value is 0.1. The visual depth of the hunters and the prey is 3. The results are average values over 10 distinct runs.

At first of the experiments, both of x and y positions used for a hunter to sense the state of the other agents were represented by 3 membership functions. In the next experiment, the number of membership functions was increased from 3 to 5. In this case, the number of state hold by each fuzzy learning-module is equal to 676.

Figure 6 shows the experimental results done by modular fuzzy-reinforcement learning vs. modular reinforcement learning.

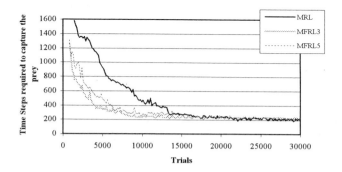

Figure 6: Comparison of MRL and MFRL on the pursuit domain.

As can be seen from Figure 6, Modular Fuzzy-Reinforcement Learning with 3 membership functions (MFRL3) and Modular Fuzzy-Reinforcement Learning with 5 membership functions (MFRL5) have obtained faster convergence than MRL to the near optimal solution. Also, it has been demonstrated that MFRL3 have reached solution faster than MFRL5, because the number of state of MFRL3 is less than that of MFRL5.

5. Conclusions

In order to remedy the problem of combinatorial explosion in multi-agent reinforcement learning, in this paper, we combined the advantages of modular architecture and fuzzy logic in a multi-agent system. Thus, we decreased more the number of state spaces of learning agents. We additionally investigated the effectiveness of the idea. For this purpose, we considered a variant of the pursuit problem and made some experiments on this domain. Experimental results showed that the proposed learning approach could be used more effectively to achieve quality optimal solution similar to modular reinforcement learning but faster than it.

References

1. H. Berenji and D. Vengerov, "Cooperation and coordination between fuzzy reinforcement learning agents in continuous state partially observable markov decision processes", *In Proceedings of the 8th IEEE International Conference on Fuzzy Systems (FUZZ-IEEE'99)* 1999.

processes", *In Proceedings of the 8th IEEE International Conference on Fuzzy Systems (FUZZ-IEEE'99)* 1999.

2. H. Berenji and D. Vengerov, "Advantage of cooperation between reinforcement learning agents in difficult stochastic problems", *In Proceedings of the 9th IEEE International Conference on Fuzzy Systems (FUZZ-IEEE'00)* 2000.

3. P. Y. Glorennec, "Fuzzy Q-learning and Evolutionary strategy for adaptive fuzzy control", In Proceedings of EUFIT'94 ELITE Foundation, Aachen, pp:35-40, 1994.

4. J. Hu and M. P. Wellman, "Multiagent reinforcement learning: theoretical framework and an algorithm" In Proceedings of the Fifteenth International Conference on Machine Learning (ICML-98), pp:242-250, 1998.

5. M. Kaya and A. Kılıç. "Fuzzy-Reinforcement Learning in Cooperative Multi-Agent Systems", *International Symposium on Computer and Information Sciences (ISCIS 2001),* Turkey, November 5-7, 2001.

6. T. Kohri, K. Matsubayashi, and M. Tokoro. "An adaptive architecture for modular Q-learning", *In Proceedings of the Fifteenth International Joint Conference on Artificial Intelligence (IJCAI-97)*, pages 820--825, 1997.

7. L. J. Lin. "Self-improving reactive agents based on reinforcement learning, planning and teaching", *Machine Learning*, Vol: 8, pp: 293-321, 1992.

8. M. L. Littman. "Markov games as a framework for multi agent reinforcement learning", *Proceedings of the Eleventh International Conference on Machine Learning*, pp. 157-163. San Francisco, CA 1994.

9. N. Ono and K. Fukomoto, "Multi-agent reinforcement learning: A modular approach". *In Proceedings of the Second International Conference on Multi-Agent Systems (ICMAS96),* pp: 252-258, 1996.

10. K. H. Park, Y. J. Kim and J. H. Kim. "Modular Q-learning based multi-agent cooperation for robot soccer", *Robotics and Autonomous Systems* Vol:35, pp:109-122, 2001.

11. T .W. Sandholm and R. H. Crites. "Multi agent reinforcement learning in the Iterated Prisoner's Dilemma", *Biosystems*, 37:147-166, 1995.

12. M. Tan, "Multi-agent reinforcement learning: independent vs. cooperative agents*" In Proceedings of the Tenth International Conference on Machine Learning (ICML-93),* pp:330-337, 1993.

13. C. J. C. H. Watkins and P. Dayan. "Technical Note: Q-Learning" *Machine Learning,* 8:279-292, 1992.

14. S. Whitehead, J. Karlsson, and J. Tenenberg. "Learning Multiple Goal Behavior via Task Decomposition and Dynamic Policy Merging", *Robot Learning*, Kluwer Academic Press, 1993.

Intelligent Autonomous Systems 7
M. Gini et al. (Eds.)
IOS Press, 2002

Finding Sub-optimal Policies Faster in Multi-Agent Systems: FQ-Learning

Alper KILIÇ   Mehmet KAYA   Ahmet ARSLAN

Fırat University, Department of Computer Engineering, 23119, Elazığ, TURKEY
e-mail: {alperkilic, mekaya, aarslan}@firat.edu.tr

Abstract: Recently, delayed reinforcement learning (RL) has been proposed as a strong method for learning in multi-agent systems (MASs). In this method, agents are concerned with the problem of discovering an optimal policy, a function mapping states to actions. The most popular RL technique, Q-learning, has been proven to produce an optimal policy under certain conditions. In this paper, we present a new method called FQ-learning for more quickly learning agents acting in MAS. In experimental results done on the pursuit domain, we demonstrate the superiority of the proposed algorithm over standard Q-learning method in terms of convergence speed and number of convergence steps.

1. Introduction

MASs are a particular type of distributed artificial intelligence systems. Many real world problems are best modeled by using a set of agents instead of a single agent. Learning in MASs is an active research field with many open issues. In such systems, RL is the most promosing paradigms among the other machine learning methods, because RL is based on the interactions of the entities with the environment in which they operate.

Recent developments in the area of reinforcement learning have yielded a number of new algorithms for the prediction and control of Markov environments. These algorithms, including the TD(λ) algorithm of Sutton [15] and the standard Q-learning algorithm of Watkins [20], can be motivated heuristically as approximations to dynamic programming.

Some works have been done for accelerating the learning of agents acting in a multi-agent environment. However, most attempts done in this field focus on several aspects. The first of these aspects is concerned with the state space of MASs. Kuter [7] has partially improved it with the S-Learning algorithm he developed for a specific multi-agent environment. It has been shown in experiments that S-Learning gives very satisfactory results in a dynamic and non-deterministic simulated environment compared to the other RL-based algorithms. Second, a teacher is employed to improve learning in MASs. That is, graded lessons are used starting near the rewards and then backing away, and examples of good behavior are used [14]. In this manner, agents are able to learn so quickly. The third aspect is to form a hierarchy of agent actions. Agents learn primitive actions first, freeze the useful sequences into macros, and then learn how to use the macros [1]. The final aspect is our previous work. In that study, we proposed a new method using Q-learning for faster learning agents acting in MASs [6].

In this study, a robust method of Q-learning is presented. The advantage of this learning approach is that it becomes independent from the multi-agent environment and is easily applicable to any multi-agent system.

The rest of the paper is organized as follows. Markov decision process, the proposed Q-learning approach vs. standard Q-learning method is described in Section 2. The pursuit domain is presented in Section 3. The experiments and the results achieved for the pursuit domain are given in Section 4. Conclusions are included in Section5.

2.1 Markov decision process and Q-Learning

In this study, learning method is based on Q-learning. Q-learning was originally defined to deal with an MDP, which is a model of an agent interacting with a stationary environment.
An MDP is described by the a 4-tuple $<S, A, T, R>$, where S is the set of finite states of the environment, A is the set of finite actions, T is the state-transition function, and R is the reward function. At each time step, an agent observes a state $s(\in S)$, executes an action $a(\in A)$, and receives a reward r from the environment. The dynamics of the environment is modeled by the state-transition function, $T:S \times A \rightarrow \pi(S)$ is the *transition function*, giving for each state and agent action, a probability distribution over states, $T(s,a, s')$ is the probability of ending in state s', given that the agent starts in state s and takes action a;
The reward r, which is a random variable, is determined by the reward function $R:SxA \rightarrow r$.
Q-learning is an incremental reinforcement learning method. According to Q-learning, the agent selects an action based on an action-value function, called the Q-function, $Q(s,a)$, which defines the expected sum of the discounted reward attained by executing action $a(\in A)$ in state $s(\in S)$, and determining the subsequent actions by the current policy π. The Q-function is updated using the agent's experience. The learning is as follows:

1. observe the current state s
2. select an action a_i with the probability:

$$\pi[a_i \mid s] = \frac{e^{Q(s,a_i)/\tau}}{\sum_{a_k \in A} e^{Q(s,a_k)/\tau}}$$

where τ is called the temperature, which determine the randomness of the stochastic policy. Although several methods for selecting an action have been proposed, we adopt this Boltzmann selection method.
3. observe the new state s'
4. receive a reward r from the environment
5. update the corresponding Q value for state s and action a according to the following formula:

$$Q(s,a) = (1 - \beta)Q(s,a) + \beta(r + \gamma \max_{a' \in A} Q(s',a'))$$

where β $(0\leq\beta<1)$ is the learning rate and γ $(0\leq\gamma\leq1)$ is the discounted parameter.
6. If the new state s' satisfies a terminal condition, then the single episode ends. Otherwise let $s' \rightarrow s$ and go back to step 1.

2.2 The Proposed FQ-Learning Approach

RL is a problem of learning from interactions in an environment to achieve some goals. An entity interacting with the environment through actions is called an agent. At each time step, an agent observes its environment and selects an action based on that observation. In the next step, the agent obtains a new observation that may reflect the effects of its previous action and a payoff value indicating the quality of the selected action.
In this study, each RL agent uses the one-step Q-learning algorithm. Its learned decision policy is determined by the state/action value function, Q, which estimates long-term discounted rewards for each state/action pair. Therefore, the Q-Learning algorithm suggests applying an action that is selected against each observed state and updating only the Q value

of this action (Standard Q-Learning or SQL). This is because whatever behavior the agent does in its environment, it is experimenting with only this behavior [8].

In the standard Q-learning, Q(s,a) is defined as the maximum discounted cumulative reward that can be obtained by choosing the action a while in state s, where $s \in S$ and $a \in A$. The Q values are held in a table where each state and action pair has a distinct entry.

However, if an agent is killed when it moves to any state, it should learn not to approach that state. If only the behavior to move to this deadly state were updated, the agent would exhibit behaviors to avoid that state [6].

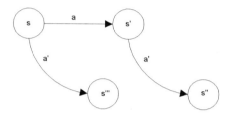

Figure 1: Schematic diagram of T (s, a', s''')

From this perspective, it can be said that the experiment obtained by a behavior constitues an important effect for other behaviors in the repertory of an intelligent agent. With the proposed Q-Learning approach (FQL) in this paper, this influence relation among behaviors of agents is established.

As seen in Figure 1, if $a' \in A(s)$, where A(s) is set of the actions in state s, there exists a $T(s,\ a',\ s''')$ transition function. Consequently, when the agent execute a policy π that maximizes $\sum_a \pi(s,a)Q^\pi(s,a)$, there exist $Q^*(s,\ a')$ that converges to optimal policy π called sub-optimal policy. In this paper, the updating rule of FQL as apart from SQL is as follows: if $a' \in A(s)$ then

$$Q(s,a') = k[(1-\beta)Q(s,a') + \beta(r + \gamma \max_{a' \in A} Q(s',a'))],$$

where s is the current state, a is an action, k $(0<k\leq1)$ is an convergence parameter such that

$$k = \begin{cases} 1 & \text{if a is the selected action} \\ (0,1] & a' \in A(s) \end{cases}$$

When the state space in a RL problem is large, learning separate Q-values for every state-action pair is very inefficient. If there is more than one updating process at any action case, the updating of unselected action a^l accelerates the learning speed.

The experiment obtained by a selected behavior is reflected on other behaviors at the rate of influence coefficient k. That is, after normally updating the Q value of the selected action, the values of Q for the other actions, which are not active in this state, update with influence coefficient k times the Q value of the corresponding action.

The second step of the algorithm is the selection of an action for the current state. Boltzman distribution mentioned earlier was used for this selection.

3. Pursuit Domain

In order to evaluate our learning approach, we consider a variant of the pursuit problem defined as follows:

- The environment consists of a 12x12 grid. Edges are connected (torus).
- Initial position of each agent is determined randomly as shown in Figure 2(a). There are four hunter agents and one prey agent in the environment.
- At each time step, agents synchronously execute one out of five actions: moving up, down, left, or right from the current position, or staying at the current position. More than one hunter agent can share the same grid. However, hunter agents cannot share with the prey. Therefore, hunter agents that try to move to the grid already occupied by the prey, cannot move and must stay at the their current positions. The prey agent selects its own action so as to maximize the total sum of Manhattan-distance to the located hunters.
- Every agent can see objects at a certain distance. This distance and its cells are called the visual depth and the visual environment of that agent, respectively. Each agent is assigned an identifier. A hunter agent can recognize the relative position and identifier of any other agents in its sight.
- The prey is captured, when the hunters occupy all of its four neighbor positions as shown in Figure 2(b). Then all of the prey and hunter agents are relocated at new random positions in the grid world and the next trial starts.

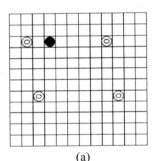 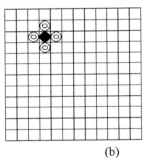

(a)   (b)

Figure 2: (a) An initial position (b) A goal state

4. Experiments

In our experiments, the learning consists of a series of trials, which begins with a single prey and four hunter agents placed at random positions, and ends when the prey is captured or at 2000 time steps. Upon capturing the prey, individual hunters immediately receive a reward of 1.0. Hunters receive a reward of –0.1 for each move when they do not capture the prey.

Parameters for the proposed FQ-learning are as follows. The learning rate is $\beta=0.8$, the discount factor is $\gamma=0.9$, and the initial value of the FQ-value is 0.1. The visual depth of the hunters and the prey is 3. The results are average values over 10 distinct runs.

Figure 3 shows the experimental results done by the proposed FQ-learning vs. standard Q-learning with two different k values.

As can be seen from Figure 3, the proposed learning approach has obtained faster conversions than SQL to the near-optimal solutions.

When the prey sees a hunter in this experiment, because of escaping by using the Manhattan-distance algorithm, some Q entries in the lookup tables of hunters with SQL remain empty.

This is the result of the algorithm the prey uses. In this case, the hunter cannot exactly learn for those states. In particular, when the prey starts moving around using a different algorithm, the hunters entering those states can be at a loss for what to do and move randomly.

This problem has disappeared in hunters using FQL. Hunters appropriately update every state/action pair in every state. This is because there are no Q entries remaining empty in the lookup tables. Thus, more effective learning is observed.

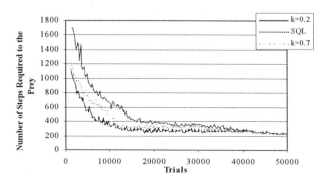

Figure 3: Comparison of SQL and FQL

5. Conclusions

In the multi-agent environments, agents are concerned with the problem of discovering an optimal policy. The most popular RL technique, Q-learning, has been proven to produce an optimal policy under certain conditions. In this paper, we proposed a new approach called FQ-learning instead of standard Q-learning, when the state-space is large. We considered a variant of the pursuit problem and made some experiments on this domain. Experimental results showed that the proposed learning approach could be used more effectively to achieve quality optimal solution using sub-optimal policies. Consequently, we accelerated the learning speed updating sub-optimal actions. This influence relation among behaviors of agents is established with experimental results.

References

1. Andre D., "Learning Hierarchical Behaviors", Draft, Submitted to AHRL Workshop, 1998.
2. Barto, A. G., Sutton, R. S. & Watkins, C. J. C. H., "Learning and sequential decision making", In: Gabriel, M. & Moore, J. (Eds.), Learning and computational neuroscience: foundations of adaptive networks, MIT Press, Cambridge, MA, Tech. Rep. 89-95, Department of Information Science, University of Massachusetts, Amherst, MA, 1989.
3. Bertsekas, D. P., "Dynamic Programming: deterministic and stochastic models", Prentice Hall, Englewood Cliffs, NJ.,1987
4. Howard, R. A., "Dynamic Programming and Markov processes", MIT Press, Cambridge, MA.,1960
5. Jaakkola, T., Jordan, M. I. & Singh, S. P., "On the convergence of stochastic iterative dynamic programming algorithms.", Neural Comput. 6(6), 1185-1201.,1994.
6. Kaya, M., Arslan, A., "Learning Faster in Cooperative Multi-Agent Systems", Proceeding of the AISB'01 Symposium on Adaptive Agents and Multi-agent Systems, pp:85-89, UK, March, 2001

7. Kuter U., "S-Learning: A Multi-Agent Reinforcement Learning Method", MS Thesis, Middle East Technical University, 2000.

8. Lin L. J., "Self-improving reactive agents based on reinforcement learning, planning and teaching", Machine Learning, Vol: 8, pp: 293-321, 1992.

9. Littman M. L., "Markov games as a framework for multi agent reinforcement learning", Proceedings of the Eleventh International Conference on Machine Learning, pp. 157-163. San Francisco, CA 1994.

10. Littman, M. L., "Value-function reinforcement learning in Markov Games", Journal of Cognitive Systems Research 2, pp:55-66, 2001

11. Puterman, M. L., "Markov decision processes-discreate stochastic dynamic programming", John Wiley, New York, 1994

12. Russel S. J. and Norvig P., "Artificial Intelligence: A Modern Approach", Prentice-Hall, 1995. (5)

13. Sandholm T. W., Crites R. H., "Multi agent reinforcement learning in the Iterated Prisoner's Dilemma" 1995.

14. Sheppard J. W., "Multi agent reinforcement learning in Markov Games", PhD thesis, John Hopkins University 1997.

15. Sutton, R.S. "Learning to predict by the methods of temporal differences", Machine Learning, 3,9-44

16. Szepesvari, C. & Littman, M. L., "A unified analysis of value-function-based reinforcement-learning algorithms." Neural Comput. 11(8), 2017-2059, 1999

17. Tan M., "Multi-agent reinforcement learning: independent vs. cooperative agents", In Proceedings of the Tenth International Conference on Machine Learning, Amherst, MA 1993.

18. Tsitsiklis, J. N., "Asynchronous stochastic approximation and Q-learning.", Machine Learning, 16(3), 185-202, 1994

19. Watkins C. J. C. H. and Dayan P., "Technical Note: Q-Learning" Machine Learning, 8 pp: 279-292, 1992.

20. Watkins, C.J.C.H., "Learning from delayed rewards", PhD Thesis, University of Cambridge, England, 1989

Intelligent Autonomous Systems 7
M. Gini et al. (Eds.)
IOS Press, 2002

Autonomous Behavior of Collective Block Agents

Masahiro KINOSHITA, Takashi KAWAKAMI*, Michiko WATANABE** and Yukinori KAKAZU
Autonomous System Engineering, Complex Systems, Hokkaido University, Japan
{kino,kakazu}@complex.eng.hokudai.ac.jp
* *Dept. of Industrial Engineering, Hokkaido Institute of Technology, Japan*
kawakami@hit.ac.jp
***Asahikawa National College of Technology*
michy@asahikawa-nct.ac.jp

Abstract. This paper describes the autonomous behavior of block agents existing in a spatially constrained environment. The difficulty of this problem is the determination of the behavior of a block agent within an environment requiring mutual action of many block agents. This difficulty is caused by dynamic obstacle avoidance problem resulting from physical collisions among the autonomous motion of block agents. The objective of this research is to build an adaptive decision mechanism of behavior for autonomous block agents when they are given a task. Specifically, this paper shows one case study of the proposed mechanism for the problem of removing blocks from a container. Our approach uses Classifier System based architecture. The results of simulation experiments are presented which indicate the possibility of this architecture in allowing block agents to adapt to dynamic environments.

1. Introduction

The automatic warehouse is one of the most important technologies for an advanced automated factory. The crates in the warehouse often must be moved while minimizing waste. Also, optimizations such as minimization of waste area or maximization of free space are required in the final configuration of the crates. Such kinds of problems have appeared in the field of Operations Research, known as Trim problems or Cutting Stock problems. When designing a fully automated and flexible manufacturing system or factory, we have to solve several difficult problems. One of the difficult problems is the solution of the robot navigation task. In this task, an optimal solution of the path planning must be calculated effectively. For example, a mobile robot can move in a factory according to instructions about its movement from a navigation system (e.g., a human supervisor) that determines an optimal path from a starting position to a destination position while avoiding ill-known obstacles. In particular, if there are multiple autonomous robots in the navigation area, then it is more difficult to solve a navigation task that involves the problem of collision avoidance among the robots. There are many conventional approaches to this problem. For example, there are approaches from production rule-based ones to fuzzy controlled ones[1]. However, all of these approaches adopt predetermined rules or theories that are peculiar to the problem being solved.

2. The Block Removal Problem as a Multi-Agent Problem

Removing blocks is one of the task which block agents may perform. Given that there is a container with at least one exit, and in that container are several blocks, the problem is to remove a portion of those blocks from the container. The considerations are: 1) maximize the number of blocks removed from the container, 2) minimize the time required to remove the blocks from the container, and 3) avoid collision while the blocks are moving. An adequate solution for this problem would be one which satisfies all of the above considerations. Thus, a satisfactory solution to block removal problem would be one that makes the following determinations: 1) the path for the blocks, and 2) the sequence of block movement. To treat the geometrical block removal, this navigation area is defined as a 2-dimensional container.

We describe a Block Removal Problem(BRP) as follows;

$$BRP = (Q, \Delta, q_0, q_G, R) \tag{1}$$

where,

> Q:the set of finite states
> Δ:the state transition function
> q_0:initial state
> q_G:goal state
> R:evaluation from the environment

The state transition function Δ maps state q_0 to q_G.

$$q(t+1) = \Delta(q(t)) \quad (q \in Q, t \in I) \tag{2}$$
$$A = \{a_i; i = 1, 2, \cdots, n_b\} \tag{3}$$
$$q = (u_i; i = 1, 2, \cdots, n_b) \tag{4}$$
$$u_i = (a_i, w_i, h_i, G_i, \theta_i, goal(G_i)) \tag{5}$$

where, A is a set of block agents, n_b is a number of blocks, u_i is the state of ith block agent, w_i, h_i, G_i, θ_i are width, height, center of gravity and angle of the block agent respectively. The function $goal(\cdot)$ represents achievement of goal state of block agent.

$$goal(G_i) = \begin{cases} 1 : \text{goal attained} \\ 0 : \text{otherwise} \end{cases} \tag{6}$$

Here, the function Δ can written as:

$$\Delta = (\Psi, \Sigma) \tag{7}$$

where Ψ decides a block agent which should be move and Σ decides a place where block agent would move in next time step. When the time step is t=n, the state transition function are applying n times. Therefore, the solutions to the problem can be obtained by finding Δ such that

$$q_G = \Delta^n(q_0) \quad (n < L), \ q(n) = q_G, \ q(0) = q_0 \tag{8}$$

3. Implementing the Classifier System

In order to solve the block removal problem mentioned above, the classifier system is implemented as a navigation system of a corresponding block. In the classifier systems and the production rule-based systems, a strategy is identically performed by the activated rules that satisfy the environmental conditions. Classifier systems depart from representational difficulties by restricting a rule to fixed length representation using alphabets or number. By this restriction, string rules can be operated easily. In the multi-agent environment, each autonomous block has such a classifier system individually.

To achieve this reactive planning process, through the detector as a perceptive organ of the classifier system, information from the environment is encoded to the finite length message. After the condition matching, a winner classifier of the rule competition instructs a motion direction of the agent according to its action part. As the useful classifiers for the given task tend to higher credit values, the credit assignment is performed as follows;

P1)   if($Ar(a_i,t) \cap Ar(a_j,t) \neq \varnothing$ $(\exists\, a_j \in A, i \neq j)$)   then R<0   (9)

P2)   if($G_i(t+1) = G_i(\tau)$ $(\exists \tau \leq t)$)   then R<0   (10)

P3)   if($d(a_i,t) \geq d(a_i,t)$)   then R<0   (11)

P4)   if($G_i(t) = dp_i$)   then R>0   (12)

where R is an environmental reward, a function $Ar(a_i,t)$ represents the sensor range of the block agent, and a function $d(a_i,t)$ is for distance. In these reinforcement plans, P1 means that if there is a contacting other block on the indicated position, in order to avoid deadlock, then give a negative reward, P2 means that if the indicated position is a previously passed position, then give a negative reward, P3 means that if the distance between the center of gravity of block and center of the exit of the container became larger, then give a negative reward, P4 means that a block attain the goal, receives a positive reward. Since a current situation is made by the chain of the selected classifiers from a starting position, an environmental reward is distributed to later classifiers of the chain . In this study, we applied the profit sharing rules. As a rule reinforcement algorithm, the environmental reward R at time t is distributed to three classifiers that are later members of the searching chain. As a result of learning, one chain of appropriate classifiers that links from a starting position to a destination position is realized.

4. Computer Simulations

Based on the proposed method, the block removal simulator is constructed and some computational experiments are carried out. Our one of major objective is avoiding such deadlocks under the condition of the autonomous behavior of the blocks. The number of classifier rules is fixed to 47200. The initial population of classifiers is generalized randomly, and every strength value of initial classifiers is set equivalently. First, to determine the effective reinforcement scheme, we performed primary experiments using the simple and same blocks, and all of the blocks are able to attain the goal position.

That is, there are no block which dimensions are large enough compare with the width of the exit of the container. In this simulation, the number of blocks is 12, its dimension is 40x80, width of the exit of the container is 70, dimension of the container is 500x470. Even this situation, the deadlocks arise from interaction of blocks easily on the near side of the exit of the container. Because autonomous block agents have no priority for moving, each block agent becomes to wish to approach the exit simultaneously, according to increasing learning. Then we adjusted the environmental reward and observed the behavior of the block agents. In the reinforcement plan P1, each block obtains the negative rewards when contacting other blocks for avoid-ing the deadlocks. Figure 1 shows the trajectories of blocks. Block configurations and a navigation area given as an environment are shown as Figure 1. In this figure, the destination position of each block is the center of the exit of the container. As a result, all of the blocks are removed from the container. And the cooperative behavior appeared when several blocks gathered in the same places. Figure 2 shows the learning curve of the maximum number of steps required for all of the blocks to exit from the container. The number of blocks are 12 and 5. We assumed that the system could not find a solution when some blocks could not attain the goal within 4000 steps. Thus, deadlocks occurred during the learning process. In the early stage of the learning, we can see such kind of phenomena. Figure 3 shows a state of strengths in the Reward field after 100 times learning process. The place of highest strength indicates goal position. We can see the constructed potential field in the cause of applying classifier systems to the block removal process. Figure 4 shows vectors transform from the selected actions on each place in the environment.

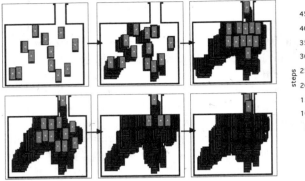

Figure 1 The Trajectories of Simple Shape Blocks

Figure 2 The Learning Curves

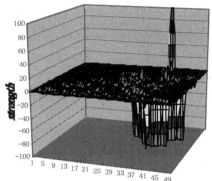

Figure 3 An aspect of the strength

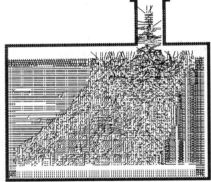

Figure 4 Vectors of the selected action

5. Conclusions

We proposed a two-dimensional problem which we termed the Block Removal Problem and suggested that an adaptive removing method using a Classifier Systems would produce a near optimal solution. The collective behavior appeared while the block agents were moving in order to adapt to the environment. This simple simulation should be followed up by more complicated simulations, with further research to investigate more efficient solutions.

References

[1] B.Beaufre and S Zeghloul; A mobile Robot navigation method using a fuzzy logic approach, Robotica, 13(1995), 437..
[2] Roberts, S.G. and Turega, M., Evolving Neural Network Structures: An Evaluation of Encoding Techniques, Artificial Neural Nets and Genetic Algorithms, Springer, (1995), 96-99..
[3] Wilson,S.W. and Goldberg D.E., A Critical Review of Classifier Systems. Proc. of 3rd International Conference on Genetic Algorithms, Morgan Kaufmann Publishers, (1989),244-255..
[4] M.Kinoshita, T.Kawakami and Y.Kakazu, Taking Block Out Problem by Adaptive Packing Method, Intelligent Engineering Systems Through Artificial Neural Networks,7(1997), 389-394..

Intelligent Autonomous Systems 7
M. Gini et al. (Eds.)
IOS Press, 2002

A BIOMIMETIC APPROACH TOWARDS AUTONOMOUS REAL WORLD ROBOTS

Frank Kirchner
GMD, AiS-BAR, St. Augustin, Germany, Frank.Kirchner@gmd.de
Also affiliated with
Northeastern University, Boston, MA, USA, fkirchne@lynx.dac.neu.edu
Dirk Spenneberg
GMD, AiS-BAR, St. Augustin, Germany Dirk.Spenneberg@gmd.de.

Abstract: We present an eigth-legged walking robot, which is controlled using a biomimetic approach to ambulation control. The approach is based on two biological control primitives: Central Pattern Generators and Reflexes. Using this approach omnidirectional walking and smooth gait transitions can be achieved. Additionally the posture of the robot can be changed while walking. The robots were successfully tested in rough terrain, with obstacles as high as the robots and different terrains like sand, grass, concrete and rock piles.

1 INTRODUCTION

The focus in this project is the development of an eight legged robot (see Figure 1). The length of the robot is 65 cm and the weight of the fully equipped robot (including 3.8 Ah accumulators, communication equipment and sensors) is 9.8kg.

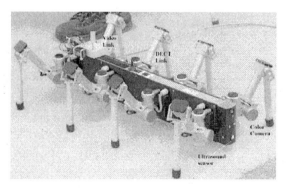

Figure 1: The 8-legged Robot 'Scorpion'.

2 MECHATRONICS

The most challenging parts of a walking robot are the legs (figure 2). The leg design presented here provides 3 degrees of freedom, which is the minimum needed for a robust, outdoor walking robot, e.g. it provides the possibility to walk omnidirectional in narrow environments. The leg consist of a thoracical joint for protraction and retraction, a basalar joint for elevation and depression and a distal joint for extension and flexion of the leg (see also figure 2). The joints are actuated by standard 6 Watt 24V DC-Motors with high gear transmission ratio for sufficient lifting capacity.

An important constraint in the development was the outdoor capability. Therefore a good trade-off between making the leg as light as possible to improve its lifting capacity and shielding it against the environmental influences like dust and water had to be found.

We achieved a weight of 950 grams and a weight to lifting-capacity ratio of 1:8. This is a prerequisite to walk up steep rises or to walk over obstacles higher than the robot itself. Another challenge was to integrate compliant elements in the design in order to make the robot robust enough to withstand the mechanical stress in an outdoor terrain.

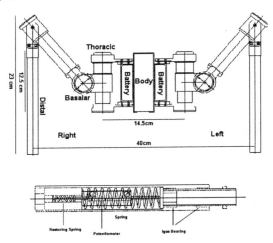

Figure 2: The mechanical design of the Scorpion legs. This front view of the robot shows left and right side legs with the body in the center. Each leg consists of 3 parts: 1) thoracic joint, 2) basalar joint and 3) distal joint. (right) The distal segment contains a spring damped compliant element with a built in potentiometer to measure contact and load on individual legs. The most energy absorbing part in our design is a spring element integrated in the distal segment of the leg (see figure 2). The distal spring element is also used for measuring the ground contact force by an integrated linear potentiometer. From this the robot can compute the load for each leg. This enables us to make use of the principle of "Early Retraction Acceleration" which is described in (2, 13).

2.1 The Sensors

The robot is equipped with the following proprioceptive sensors:

- Motor Encoders for each motor to measure the relative joint angle
- Hall-Effect Motor Current Sensors for each motor
- The analog load/pressure sensor in each the foot tip
- The Power-Management sensors, providing current battery voltage and current power drain

- Three dimensional inclinometers (pitch, roll and yaw)

The following exteroceptive sensors are integrated:
- Utrasound distance sensor for obstacle avoidance
- Compass sensor for heading control
- Contact/pressure sensors at the foot tip

It is important to note that the legs itself can be used as exteroceptive sensors. One can use the current sensors of the joint motors during the movement as a tactile sensor during movement. In order to allow an operator to communicate with the robot or to take data samples during a test run, the robot is equipped with an wireless 28K Baud bi-directional communication link and a PAL CCD Camera with a 5GHz video/audio link for video transmission. So it is possible to use the robot as a semiautonomous system.
The Operator can control it via high-level commands like walk forward, left, right, go up, go down, move sideward, turn etc. To supervise the system all relevant sensor data is send back from the robot to the operator.

2.2   The processing hardware

A network of Infineon C167 and C164 Microcontroller Derivates are used that consists of one Master-Controller (C167) which functionally contains the higher behavioral level, the communication to the operator and the data processing of the exterioceptive sensors.   The master controller is connected via a CAN-Bus network with the Leg-Controllers (C164). The leg-controller contains the local control functions like the CPG and the local reflexes and processing the proprioceptive sensors and controlling the DC-Motors. In a new version of the hardware we will use a MPC555 for the higher level control and a FPGA for the local control of the legs.

3   AMBULATION CONTROL

Our architecture (14) is based on two approaches to robust and flexible real world locomotion in biological systems, which seem to be contradictory at first sight. These are the Central Pattern Generator (CPG) model and the pure reflex driven approach (1, 6, 7).
A CPG is able to produce a rhythmic motor pattern even in the complete absence of sensory feedback.   The general model of a CPG has been identified in nearly every species even though the specific instantiations vary among the species to reflect the individual kinematical characteristics in the animals.
The idea seems to be very promising as a concept to stabilize locomotion in kinematically complex robotic systems, see figure 3 and 4. As it resembles the divide and conquer strategies that are reflected in nearly all solutions to complex control problems (4).

Figure 3: The scorpion robot during an autonomous exploration into a sand bed. The beach like sand bed was 3m wide and 9m long. The robots feet penetrated the sand for aprox. 3-5cm. A reflex mechanism helped to overcome the obstacle.

Another model for the support of robust locomotion is also provided by evolution in the animal kingdom. This is the concept of reflex based control (7). A reflex can be viewed as a closed loop control system with fixed input/output characteristics. In some animals, like the locust, this concept is said to actually perform all of the locomotion control and no further levels of control, like the CPG, are involved (6).

Whether or not complex motion control can be achieved only via reflex systems is subject to further discussion, however, the concept of a set of fixed wired reactions to sensory stimuli is of high interest to roboticists who aim to gain stability in the systems locomotion.

The design of the control architecture described here was thus driven by these two concepts. The CPG approach appeared to be interesting to generate rhythmic walking patterns which can be implemented computationally efficient, while the reflex driven approach seemed to provide a simple way to stabilize these walking patterns by providing 1) a set of fixed situation-reactions rules to external disturbances and 2) as a way to bias leg coordination among multiple independent legs. (6). Figure 4 outlines the general idea.

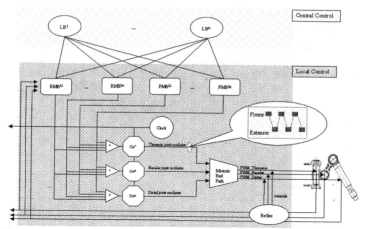

Figure 4: The overall architecture for low level actuation is depicted here.

On the global level(light gray area) we have implemented Locomotion Behaviours (LB's), typically (Forward, Backward and Lateral locomotion). These global behaviours are connected to all local leg controllers and activate (with continuous strength) the local (per leg) motion behaviours. At the same time they implement the inter leg phase relation by setting/resetting the local clocks. The local level (dark gray area) implements Rhytmic Motion Behaviours (RMB's) and Postural Motion Behaviours (PMB's). These behaviours simultaneously influence the amplitude and frequency (see figure 4 and 5) parameters of three oscilating networks (OS^T, OS^B and OS^D). The oscillators are connected to a common clock which is used for local and global (in relation to other legs) synchronization purposes. The oscillators output is a rhythmic, alternating flexor and extensor, stimulation signal (see callout box in figure 4) which is implemented as splined sine waves. This activation signal represents the desired behavioural locomotion, which is translated into PWM signals via the motoric end path. Inline with the output of the motoric end path are a set of perturbation specific reflexes, which are implemented as 'watchdogs'. They override the signals on the end path with precompiled activation signals if the sensor information from the physical joints meets a set of defined criteria.

4 RESULTS

This approach was implemented using inter leg coordination data as observed in real scorpions (3) and successfully tested on our 'SCORPION' robot study (see figure 5). In the figures 5 A through C, data of the performance of one leg is shown. The solid line is the real angle of the leg, measured with the motor encoders. The angle for the distal and the basalar joints increases during elevation, while the angle for the thoracic joint increases during protraction. The frequency was set to 1.3 Hz (19 time units on the x-axis). The data was taken every 1/25 sec. and the curves are directly computed from the raw data. The mean starting position is at 37 degree for the thoracic joint, 121 degree for the basilar joint, and ca. 30 degree for the distal joint. At first only a local FORWARD behaviour is stimulated (until $t = 1375$), then the LATERAL behaviour is activated simultaneously. Because of the equal strength of the activation the system now tries to walk forward as well as laterally, which results in a diagonal walking.

Approximately at time $t = 1460$ the activation of the FORWARD behaviour is set to 0 which leaves only the LATERAL behaviour to influence the oscillator networks (see figure 4 and figure 5 A-C). Thus the system walks laterally, which can be observed from the data as the amplitude of the thoracic joint is 0 while the basalar and especially the distal joint perform large amplitude oscillations. Subsequently the described process is reversed.

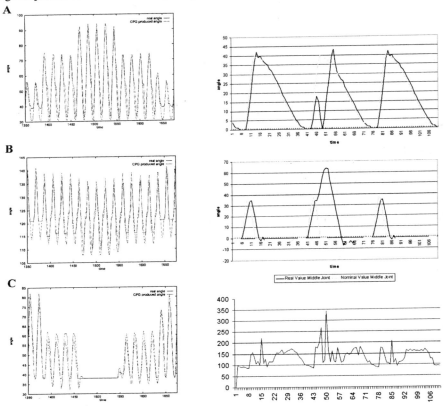

Figure 5: Traces A through C show the movements of the thoracic (A), the basalar (B) and the distal (C) joints during a transition from forward walking (pure FORWARD activation), to diagonal (equal activation of FORWARD and LATERAL), to lateral walking (pure LATERAL activation) and back to forward walking. See figure 4 for the pathways of activation.

A reflex initiated at a leg during a course through a rock bed. The current in the thoracic joint (Trace C) increased as a result of the obstacle blocking the way. At the same time the angular displacement error (Trace A) in the thoracic joint increased, indicating an exception in the regular swing cycle. As a result of these factors the basalar joint controller initiated the reflex (Trace B) that elevated the leg further, thereby overcoming the obstacle.

4.1   Local reflexes

The approach described here for the generation of rhythmic motion deals very well with plain surfaces without obstacles. However, in the case of uneven ground poor results would be expected.

Our approach to deal with uneven terrain was to implement a set of reflexes in parallel to the motor end path (see figure 5, right side), which override, for a short and predefined period of time, the rhythmic activity of the oscilaltors. E.g. (see figure 5, right side), if the current values of the thoracic joint increase steep and a significant angular displacement error is detected at the same time, it is assumed that the 'planed' trajectory is blocked. This triggers a reflex, which moves the leg backward and upward (via joint activity in the thoracic and basalar joints) and then forward at max. speed. This reflex is illustrated in figure 5, right side, (start point at t = 46, stop at t=58). The reaction-time of the reflexes are as fast as 1/100sec. because they are directly in line with the motor control signals. The 3 pictures also illustrate how fast the motor controller returns back to the pattern given by the oscillator, after the reflex is no longer active. It is important to notice that the action of the reflex does infact sit right on top of the ongoing rhythmic activity. As can be seen in Figure 5 (right hand traces) the oscillatory activity is always present in the background (light, grey lines), as soon as the reflex is terminated, the locomotion returns to the oscillation.

5   OUTLOOK

We intend to develop the system, which was presented here, further with respect to the mechanism, electronics and also the software architecture. In the field of mechanics we are currently in the process of redesign of the legs. We intend to further decrease the weight (aprox. 150g/leg) whereby increasing the weight/lift ratio to 1:9 without loosing stability or protection against environmental influences. This will be achieved by means of optimizing the design as well as material changes. In the electronics we are also in the process of redesign. The new design will incorporate the Motorola MPC555 as the master controller and an FPGA, which will be programmed to perform the tasks of the 8 local (C164) micro-controllers. This design will improve the motor control, as it allows for higher sampling speed and more sophisticated controller design, e.g., pid instead of only proportional controllers. On the MPC555 we will be able to explore the use of more complicated algorithms and even learning and optimization procedures due to the integrated FPU and the higher clock speed. The biggest effort however, has to be put on the autonomous operation of the system, which has to go beyond low-level locomotion. Here we are planning for high-level task planning and scheduling algorithms as well as on-line learning and adaptation methods.

We intend to incorporate a separate vision-processing module, which will analyze the video taken onboard, and evaluate it according to object recognition and classification procedures. These high level data will then be transferred to the master controller (MPC555), which can incorporate the information in the higher-level task planning and navigation procedures.

REFERENCES

(1) Ayers, J. , Kashin, S. ,Blidberg, D.R. and Massa, D. : Biologically-Based Underwater Robots; in Unmanned Systems, vol. 12, 30-36; 1994.

(2)Blickhan, R. & Full, R. J. 1998 Similarity in multilegged locomotion, bouncing like a monopode. J. Comp. Physiol. A173, 509-517.

(3) Bowerman, R. F., The Control of Walking in the Scorpion,
I. Leg Movement during Normal Walking, J. comp. Physiol.,
Vol. 100, 183-196, 1975

(4)Brooks, R. A. : A robust layered control system for a mobile robot; in IEEE Journal of Robotics and Automation, vol. 2, 14-23; 1986.

(5)Calvitti, A. and Beer, R.D. (2000). Analysis of a distributed model of leg coordination. I. Individual coordination mechanisms. Biological Cybernetics 82:197-2

(6)Cruse,H. et al.: Walknet- a biologically inspired network to control six-legged walking; Neural Networks. 11, 1435-1447; 1998

(7)Delcomyn, F. : Neural Basis of Rhythmic Behavior in Animals; in: Sciene, vol. 210, 492-498; 1980.

(8)R.S. Fearing, K.H. Chiang, M. Dickinson, D.L. Pick, M. Sitti, and J. Yan, Wing Transmission for a Micromechanical Flying Insect, IEEE Int. Conf. on Robotics and Automation, April, 2000.

(9)Hebb, D. : The Organization of Behavior (A Neuropsychological Theory), Wiley, New York; 1949

(10)Kirchner, F., Automatic Decomposition of Reinforcement Learning Tasks, In Proceedings of the AAAI 95 Fall Symposium Series on Active Learning, November 10-12, Massachusetts Institute of Technology, Cambridge, Massachusetts, USA 1995. Page 56-59. Eds. D. Cohn, D. Lewis, AAAI Press.

(11)Kirchner, F. Q-Learning of complex behaviors on a six-legged walking Machine; in Journal of Robotics and Autonomous systems, vol. 25, 256-263, Elsevier Science Publishers, Amsterdam Netherlands, 1998.

(12)Kupferman, I. and Weiss, K. R. : The command neuron concept; Behav. Brain Sci., vol. 1, 3-39; 1978

(13)Stein,P.S.G: Motor Systems with specific reference to the control of locomotion. Ann.Rev. Neuroscience,1:61-81; 1978

(14)Spenneberg D. and Kirchner F.: Omnidirectional Walking in an Eight Legged Robot ,Proc. International Symposium of Robotics and Automation (ISRA2000), Monterrey, Mexico; 2000

(15)Whitehead, S. Active Perception and Reinforcement Learning, Proc. Of the 7th int. conference on Machine Learning, 1990.

Intelligent Autonomous Systems 7
M. Gini et al. (Eds.)
IOS Press, 2002

Towards an optimal scoring policy for simulated soccer agents

Jelle Kok   Remco de Boer   Nikos Vlassis

Computer Science Institute, University of Amsterdam

Kruislaan 403, 1098 SJ Amsterdam, The Netherlands

{jellekok,remdboer,vlassis}@science.uva.nl

Abstract. This paper describes the scoring policy used by the agents of our simulation robot soccer team. In a given situation this policy enables an agent to determine the best shooting point in the goal, together with an associated probability of scoring when the ball is shot to this point. The ball motion can be regarded as a geometrically constrained continuous-time Markov process. Our main contribution is an approximate method for learning relevant statistics of such a process.

1   Introduction

RoboCup is an attempt to foster AI and intelligent robotics research by providing a standard problem where a wide range of technologies can be integrated and examined. Since the main purpose of a soccer game is to score goals, it is important for a robotic soccer agent to have a clear policy about whether he should attempt to score in a given situation, and if so, which point in the goal he should aim for. In this paper we describe the implementation of a scoring policy that was used by the agents of our *UvA Trilearn 2001* team, which reached fourth place in the *RoboCup-2001* simulation world cup.

An interesting aspect of the *soccer server* simulator is that, although the motion noise added to the ball is known, an analytical solution of the corresponding diffusion process (position of the ball in each time step) is difficult for two reasons: (1) the noise added by the server is by construction non-white, and (2) the process is geometrically constrained (the ball must end up inside the goal). We propose an approximate solution to the problem of learning the statistics of such a geometrically-constrained continuous-time Markov process, which we believe can also be useful in other applications.

2   The Optimal Scoring Problem

The optimal scoring problem can be stated as follows: find the point in the goal where the probability of scoring is the highest when the ball is shot to this point in a given situation. This problem can be decomposed into two independent subproblems:

1. Determine the probability that the ball will enter the goal when shot to a specific point in the goal from a given position.

2. Determine the probability of passing the goalkeeper in a given situation.

Since the two subproblems are independent, the probability of scoring when shooting at a certain point in the goal is equal to the *product* of these two probabilities.

2.1 Subproblem 1: Probability that the Ball Enters the Goal

When the ball is shot to a point somewhere inside the goal, it can miss the goal due to motion noise (introduced by the server). We are interested in the probability that the ball will end up *somewhere* inside the goal when shot at a *specific* point. To this end we need to compute the deviation of the ball from the aiming point. This deviation is caused by the noise which is added to the ball velocity in each simulation cycle.[1] The complication arises from the fact that the added noise in each cycle depends on the speed of the ball in the previous cycle, making the noise non-white.

Treating the ball motion as a continuous-time Markov process, computing exact statistics for each time step would require the solution of a corresponding Fokker-Planck equation [2], further complicated by the fact that the motion noise is non-white. Moreover, the solution will not be generic but depends on the current values of the server parameters. To avoid both problems, we propose to *learn* the statistics of the ball motion directly from experiments, and to compute the required probabilities from these statistics.

We estimated the *cumulative* noise added to the ball perpendicular to the shooting direction as a function of the travelled distance d along this direction. This function was learned by repeating an experiment in which a player was placed at various distances in front of the center of the goal (zero y-coordinate) and shot the ball 1000 times from each distance perpendicularly to the goal line. For each instance we recorded the y-coordinate of the point where the ball entered the goal. We empirically found that, to a good approximation, the standard deviation σ of the ball perpendicular to the shooting direction was given by a monotone increasing function

$$\sigma(d) = -1.88 * \ln(1 - d/45) \tag{1}$$

with $\ln(\cdot)$ the natural logarithm. Moreover, the Central Limit Theorem [2] indicates that the ball distribution along the goal line will be approximately Gaussian with zero mean and standard deviation $\sigma(d)$ from (1), as shown in Fig. 1(a). The scoring probability is then given by the area of the Gaussian density between the two goalposts.

When the ball is shot at an angle to the goal, the ball can travel different distances (implying different deviations) before it reaches the goal line, causing the distribution along the goal line to be non-Gaussian. The key observation is that we want to compute probability *masses* and for equal masses the particular shape of the distribution that gives rise to these masses is irrelevant. Therefore, instead of computing the distribution of the ball along the goal line analytically (by solving the constrained diffusion process equations) and then integrating to find its probability mass between the two goalposts, we compute the probability mass from the identity

$$P\{\text{goal}\} = 1 - P\{\text{not goal}\} = 1 - P\{\text{out from left}\} - P\{\text{out from right}\} \tag{2}$$

where $P\{\text{not goal}\}$ denotes the probability that the ball will miss the goal, going out from the left or the right goalpost. This probability mass is easier to compute from the tails of the Gaussian distributions corresponding to the two goalposts.

---

[1] The ball velocity vector (v_x^{t+1}, v_y^{t+1}) in cycle $t + 1$ is equal to $0.94 * (v_x^t, v_y^t) + (\tilde{r}_1, \tilde{r}_2)$ where $\tilde{r}_1$ and $\tilde{r}_2$ are random numbers uniformly distributed in [-rmax, rmax], with rmax $= 0.05 * ||(v_x^t, v_y^t)||$.

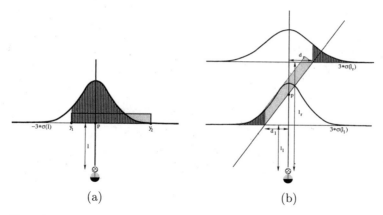

(a)   (b)

Figure 1: Two situations of shooting to the goal (light gray) together with the associated probability distributions. (a) Shooting perpendicularly. (b) Shooting at an angle.

This is shown in Figure 1(b). For example, when the ball reaches the left post, it has effectively travelled distance l_l and its distribution perpendicular to the shooting line is Gaussian with deviation $\sigma(l_l)$ from (1). The probability that the ball will go out from the left goalpost is approximately[2] equal to the shaded area on the left. Thus

$$P\{\text{out from left}\} \approx \frac{1}{\sigma(l_l)\sqrt{2\pi}} \int_{-\infty}^{-d_l} \exp\left[-\frac{y^2}{2\sigma^2(l_l)}\right] dy \qquad (3)$$

where d_l is the shortest distance from the left goalpost to the shooting line. The situation that the ball will go out from the right post is analogous. The only difference is that the ball will have to travel a larger distance, which will make its deviation larger and the corresponding Gaussian flatter. Finally, using (2) we can determine the probability that the ball will enter the goal.

2.2 Subproblem 2: Probability of Passing the Goalkeeper

The second subproblem can be stated as follows: given a shooting point in the goal, determine the probability that the goalkeeper intercepts the ball before it reaches the goal line. In our experiments we used the goalkeeper of *Robocup-2000* winner *FC Portugal*, since it appeared to be one of the best available goalkeepers. Due to lack of space we only briefly describe the proposed method, and refer for details to [1].

The main observation is that ball interception can be regarded as a two-class *classification* problem: given the player and goalkeeper position (input feature vector), predict which class (intercepting or not) is most probable. Moreover, we are interested in the *posterior* probability associated with the prediction of each class. Formalizing the problem in this way allows for direct application of a variety of methods from the field of statistical pattern recognition [3].

To collect a training set, we performed an experiment in which a player repeatedly shot the ball from a fixed position straight to the goal, while the goalkeeper was placed randomly at different positions relative to the player. A data set was formed by recording 10.000 situations, together with a boolean indicating whether the goalkeeper had

---

[2]We neglect the probability that the ball will end up to the right of the left goalpost after having travelled an 'illegal' trajectory outside the field.

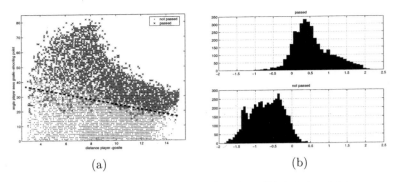

Figure 2: (a) Data set and discriminant function. (b) 1-d class histograms.

intercepted the ball or not. The relevant features for classification turned out to be (i) the absolute angle between the goalkeeper and the shooting point in the goal as seen by the player, and (ii) the distance between the player and the goalkeeper. These two values formed a two-dimensional feature vector on which the classification was based.

The recorded data set is shown in Figure 2(a) where we note that there is an almost linear discriminant function between the two classes. We determined this discriminant via regression on the class indicator boolean variable, a procedure which is known to give the optimal *Fisher's linear discriminant* [3, Ch. 3.2]. Projecting all data points perpendicularly to the discriminant line, we get a set of one-dimensional points u_i that describe, to a good approximation, the two classes. The histogram class distributions of these points are plotted in Figure 2(b). Then, we fit a univariate Gaussian function $p(u|C)$ on each class C in the overlapping region, from which we can compute the posterior probability for this class using the Bayes rule, which approximately gives the required posterior P(pass goalkeeper $|u$) as a simple sigmoid function [1].

2.3 Determining the Best Scoring Point

Having computed the probability that the ball will end up inside the goal (subproblem 1) and the probability that the goalkeeper will not intercept it (subproblem 2), the assumption of independence gives the total probability as the product of these two probabilities. This total probability is a bell-shaped function, representing the probability that the ball will enter the goal, with a valley around the position of the goalkeeper. The curve will have only two local maxima, corresponding to the left and the right starting point of the valley, which can be located with a simple hill-climbing algorithm. The global maximizer of this function is selected as the best shooting point.

References

[1] J. Kok, R. de Boer, and N. Vlassis. Towards an optimal scoring policy for simulated soccer agents. Technical Report IAS-UVA-01-06, Computer Science Institute, University of Amsterdam, The Netherlands, Oct. 2001.

[2] A. Papoulis. *Probability, Random Variables, and Stochastic Processes.* McGraw-Hill, 3rd edition, 1991.

[3] B. D. Ripley. *Pattern Recognition and Neural Networks.* Cambridge University Press, Cambridge, U.K., 1996.

Intelligent Autonomous Systems 7
M. Gini et al. (Eds.)
IOS Press, 2002

A Reactive Path-Planner using Multilayered Cellular Automata

Fabio M. Marchese

Dipartimento di Informatica, Sistemistica e Comunicazione

Università degli Studi di Milano - Bicocca

Via Bicocca degli Arcimboldi 8, I-20126, Milano, Italy

Marchese@DISCo.UniMIB.IT

Abstract. In this paper we present a Reactive Path-Planning Algorithm for a non-holonomic mobile robot. The robot we considered has to move using smoothed trajectories, without stopping and turning in place, and with a minimum steering radius. We have studied an algorithm based on a directional (anisotropic) propagation of repulsive and attracting potential values on a Multilayered Cellular Automata Model. The algorithm finds all the optimal collision-free trajectories following the minimum valley of a potential hypersurface embedded in a 4D space, built respecting the imposed constraints. Our approach turn out to be distributed and incremental: whenever changing the initial or the final pose, or the obstacles distribution, the automata start evolving towards a new global steady state, looking for a new set of solutions. Because it reacts to obstacles distribution changes, it can be also used in unknown or dynamical environments in combination with a world modeler. The path-planning algorithm is applicable on a wide class of vehicles with different kinematics, selected by changing a set of parameters.

1 Introduction

In this paper we describe a very fast, safe and complete path-planning method for robot based on Multilayered Cellular Automata. The task we want to tackle is to build a robot navigation system that works in real-time, while interacting with the environment and reacting as fast as possible to its dynamical events. Many authors have proposed different solutions during the last twenty years. Since 1979, Lozano-Pérez and Wesley first proposed a path-planning algorithm among polyhedral obstacles based on the visibility graph [8], then extended to the Configuration Space approach [9], based on a geometrical description of the environment. The path-planners working on these models generate very precise optimal trajectories and can solve really difficult problems, especially in cluttered worlds, also taking into account non-holonomic constraints, but they are very time consuming, too. On our opinion, if we want to interact with a real dynamical world, we must constantly sense the world and immediately re-plan accordingly to the new information. Some authors have developed approaches that are more similar to our purposes using the Artificial Potential Fields Methods. In the eighties, Khatib [5] first proposed this method for the real-time collision avoidance problem using a potential function in a continuous space. The worst problem affecting this approach was the presence of local minima in which the robot gets stalled. Kim and Khosla solved the problem using harmonic functions [6]. Jahanbin and Fallside first introduced a wave propagation algorithm in the Configuration Space on discrete maps [4]. They propagated a wave of distance values from the goal into all the free space (*Distance Transform*). In [2], the authors used the Numerical Potential Field Technique on the Configuration Space to build the workspace skeleton (a generalized Voronoi Diagram). Zelinsky extended the *Distance Transform* to the *Path Transform* [11]. Tzionas et al. in [10] described a path-planning algorithm for a diamond-shaped holonomic robot in a static known environment based on CA. They used the maximum clearance criterion (passing far away from the obstacles), moving the robot along the edges of a Voronoi Diagram built by the CA. In this paper, we have used CA as formalism for merging a Grid Model of the world (Occupancy Grid) with the Configuration Space of the robot and Numerical (Artificial) Potential Field Methods, with the aim to give a simple and fast solution for the path-planning problem for a non-holonomic mobile robot. This method uses a directional (anisotropic) propagation of distance values between automata to build a potential hypersurface embedded in 4D space. Using a constrained version of the descending gradient is possible to find out all the admissible, equivalent and shortest trajectories (for a given metric) that connect two configurations in the C-Space.

2 Problem Statements

A wide variety of world models can be used to describe the interaction between an autonomous agent and its environment. One of the most important is the Configuration Space [7, 9]. Let us consider a simple rigid body $\Re$ with a generic shape, a finite extension and an orientation (a preferential direction of movement). $\Re$ moves in a workspace $\mathcal{W}$ embedded in a physical n-dimensional space and has p DOF. The Configuration Space $\mathcal{C}$ (*C-Space*) of a robot is the set of all its configurations. If the robot can translate and rotate, the *C-Space* is a 3D manifold $\mathbb{R}^2 \times \mathbb{S}^1$ ($\mathbb{S}^1 = \mathbf{SO(2)}$ is the unit circle). If we make the assumption that $\mathcal{W} \subseteq \mathbb{R}^2$, the workspace can be modelled as a 2D Bitmap $\mathcal{GW}$ (*Workspace Bitmap*), represented by the application $BM : \mathcal{W} \to \{0, 1\}$, where 0s represent empty cells. In the same way, it is possible to represent the *C-Space* as a n-D Bitmap $\mathcal{GW}$ (*C-Space Bitmap*). The *C-Potential* is a function $\mathbf{U(q)}$ defined over the *C-Space* that drives the robot through the sequence of configuration points to reach the goal position and orientation [2]. Let's introduce the following assumptions on the environment in which the robot navigates: 1) the workspace is well known a priori; 2) space topology is planar with a closed boundary. There is a further assumption concerning the robot: it has a lower bound on the steering radius (non-holonomic vehicle). This assumption determines important restrictions to the type of trajectories that the robot can follow.

Cellular Automata are automata defined on a Cellular Space $\mathbb{Z}^n$ with transition functions invariant for translation [3]: $\mathbf{f_c}(\cdot) = \mathbf{f}(\cdot), \forall \mathbf{c} \in \mathbb{Z}^\mathbf{n}, \mathbf{f}(\cdot) : \mathbf{Q}^{|\mathbf{A_0}|} \to \mathbf{Q}$, where $\mathbf{c}$ is the coordinate vector identifying a cell, $\mathbf{Q}$ is the set of states of an automaton and $\mathbf{A_0}$ is the set of arcs outgoing from a cell to the neighbors. The mapping between the Robot Path-Planning Problem and CA is quite simple: every cell of the *C-Space Bitmap* $\mathcal{GW}$ is an automaton of a CA. The state of every cell contributes to build the *C-Potential* $\mathbf{U(q)}$ through a diffusion mechanism between neighbors. The initialization values of CA is essentially the *Workspace-Space Bitmap*. The trajectories are found following the minimum valley of the surface $\mathbf{U(q)}$. In this work we use a simple extension of the CA model: we associate a vector of attributes (state vector) to every cell. Each state vector depends on the state vectors of the cells in the neighborhood. There is a second interpretation: this is a Multilayered Cellular Automaton [1], where each layer corresponds to a subset of the state vector. Each subset is evaluated in a single layer and depends on the same attribute of the neighbor cells in the same layer and depends also on the states of the corresponding cell and its neighbors in other layers. In the following section, we describe each layer and the transition functions implemented in its cells.

3 Layers Definitions

3.1 Boundary and Obstacles Layer

The Boundary and Obstacles layer represents the *Workspace Bitmap* $\mathcal{GW}$ and it is used to map the obstacles and the boundary of the environment. In Regular Decomposition world models the robot is often represented as a point moving from one cell to a neighbor one. To take into account its real extension, the well-known technique of enlarging the obstacles by a quantity $n = Int(R/l) + 2$ (R maximum robot radius from the kinematics center, l size of a cell) has been used [8]. We have defined different types of workspace zones: free zones where it can move fast (wide space), safe zones where it moves carefully (cluttered space), and interdicted zones where it cannot enter (around the obstacles). The attribute evaluated in this layer (*Obstacles Repulsion Attribute*) generates a potential surface, the gradient of which is a repulsive force that keeps the robot away from the obstacles (some examples in [2, 8]). The potential function is

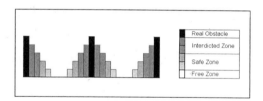

Figure 1: Cross-section of the Repulsive Potential Surface (see Fig. 2a) in a two-rooms environment after Obstacles Enhancement phase

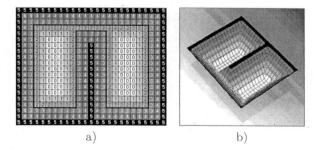

a)   b)

Figure 2: Repulsive Potential generated both in the interdiction zone and safe zones (black thick line represents the border between the two zones)

constructed using an attribute formally defined as $Rep_c(t) \in [0, \ldots, k]$, where c is a cell of the cellular space. The automata evolve, using the following transition function:

$$Rep_c(t+1) = \begin{cases} k & if\ Rep_c(t) = k \\ \max_{\forall i \in I_c} \{Rep_i(t) - 1, 0\} & otherwise \end{cases} \quad \forall t > 0, \forall c \in \mathcal{GW}$$

where I_c is the set of neighbor cells of the cell c. In this case I_c is an 8-connected neighborhood (Moore neighborhood). The setup of this layer is the original *Workspace Bitmap* $\mathcal{GW}$ (Occupancy Grid). In Fig. 2a-b is shown an example of expansion covering also the safe zones. The planner will try to find trajectories that keep the robot in the free space and only in some circumstances enter in the safe zone (e.g. when passing through the "door").

3.2   Goal Position Layer and Starting Position Layer

The *Goal Position Layer* specifies the goal pose of the robot. The attribute in each cell is a boolean: $GPos_c \in \{True, False\}^8$. Each cell c contains a vector of eight boolean variables, one for each possible outgoing direction. All the attributes of the cells are normally set to *False*; only the final orientation in the goal cells is set to *True*. It represents the desired goal configuration $q_g = (x_g, y_g, \theta_g)$, where the robot orientation θ has been sampled in eight values $\theta \in D = \{N, NE, E, SE, S, SW, W, NW\}$. Similarly, the *Starting Position Layer* represents the starting pose of the robot. It is initialized with the starting configuration $q_i = (x_i, y_i, \theta_i)$. Both layers are statical and do not evolve during the computation.

3.3   Attraction_to_Goal Layer

This is the core of the entire Path-Planning Algorithm. The *Attraction_to_Goal Layer* is defined as: $Attr_c(d_{out}) \in \mathbb{N}^8$. It is a vector of eight values corresponding to the eight defined directions. It represents the integer distance of the cell c from the Goal cell if the robot moves to the $d_{out} \in D$ direction along a collision free path. It is a digitalized representation of the C-*Potential* function $\mathbf{U}(\mathbf{q})$ defined on the C-*Space Bitmap*. To evaluate the path length, we have defined a metric on the space, assigning a cost (weight) to each robot movement: (*forward, forward_diagonal, direction_change, stop. rotation, backward, backward_diagonal*). Because of the robot non-holonomic constraints, not every movement can be done in every robot pose. We have introduced a subset of admissible moving directions $D'(c, d) \subseteq D$ depending on the robot position (cell c) and orientation d which satisfy the maximum curvature constraint. Given the configuration (c, d), the robot can arrive to a set of configurations $(\gamma, \delta), \forall \gamma \in I'_c(d), \forall \delta \in D'(\gamma, d)$, where $I'_c(d) \subseteq I_c$ is the subset of neighbor cells reachable from the configuration (c, d), and $D'(\gamma, d) \subseteq D$ is the subset of admissible orientations at the destination cell γ compatible with the leaving direction d. The Repulsive Potential also contributes to the path length calculation. The transition function is defined as follows:

$$Attr_c(d, t+1) = \begin{cases} 1 & if\ c = goal\_cell\ and\ d = final\_direction \\ Rep_c(t) + minAttr_c(d, t) & if\ Attr_c(d, t) \neq Rep_c(t) + minAttr_c(d, t) \\ Attr_c(d, t) & otherwise \end{cases}$$

$$\forall t > 0, \forall c \in GW, \forall d \in D$$

where: $minAttr_c(d, t) = min_{\substack{\forall \gamma \in I'_c(d) \\ \forall \delta \in D'(\gamma, d)}} \{Attr_\gamma(\delta, t) + Cost(\gamma, \delta)\}$. The planner, using the criterion of the minimum cost, will prefer trajectories passing only in free zones, instead of trajectories entering in the safety zone. If we change the movement costs, we can consider robots with different types of kinematics. For example, the metric $(2, 3, 1, 0, High, High, High)$ emulates a car-like kinematics moving only forward (Dubin's car), while the metric $(2, 3, 1, 0, High, 2, 3)$ emulates a common car-like kinematics moving also backward (Reed's and Shepp's car). Two main properties has be demonstrated: the termination of the propagation and the absence of local minima. The

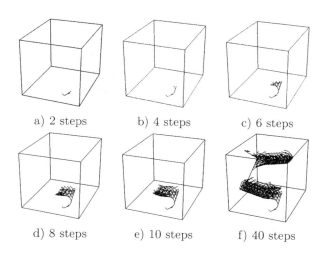

| a) 2 steps | b) 4 steps | c) 6 steps |

| d) 8 steps | e) 10 steps | f) 40 steps |

Figure 3: Attraction Potentials Skeletons at different evolution steps

graphical representation of the potential surface is quite difficult. To be able to infer its aspect, we must reduce the number of dimensions projecting from the 4D space to a 3D space. The result is a representation of a skeleton laying on the hypersurface which gives an approximate idea of its topology. The sketches in Fig. 3 show the first evolution steps of the skeleton growing up from the goal cell in the environment of Fig. 1 for a Dubin's car. In the last picture there is the final results after 40 steps.

3.4 Paths_Extraction Layer

This layer determines all the shortest paths that connect the starting point to the goal point on the base of potential hypersurface $U(q)$ computed in the previous layer. The method used extends the descent gradient methods described in [4, 11]. Because the robot has a constraint on the steering radius, not all the movement directions are admissible and the gradient method has to reflect this condition. Hence, we have to use a constrained negated gradient, i.e. the gradient vector is evaluated only for the admissible directions set. Formally it is defined as (*Path_Extraction Layer*): $Path_c(d_{in}, d_{out}) \in N^8 \times N^8$. It is a look-up table of 64 values: for each of the 8 incoming directions, it is indicated the cost to reach to goal if the robot would go out the cell following one of the 8 directions. In Fig. 4 is shown the solution found for the two rooms problem for a car-like robot moving only forward. The gray levels depend on the intensity of the repulsive force. Because of the steering constraint, the robot has to navigate into the second "room" to invert the initial direction, and pass for a second time, but with a different orientation, in the starting cell (next to the "door"). This is a remarkable result of this algorithm: it can find out more complex solutions than using conventional Artificial Potential Fields Methods [2, 11], which allow only one passage per cell. The resulting trajectories are subsets of the attraction skeleton (Fig. 4.a) i.e.

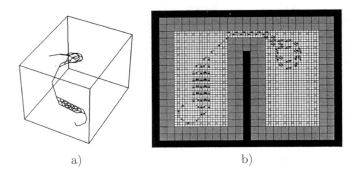

a)   b)

Figure 4: Attractive Potentials along the trajectories found: a) 3D projection, b) 2D projection

only those parts of the skeleton which connect the starting cell to the goal cell belong to the final trajectories.

4 Algorithm Properties

This algorithm has the main advantage to be computable in an asynchronous way: there is no specific evaluation order of the cells in each layer, and every layer can start to update without waiting that the other layers have reached a stationary point. A second important property is related to the consistency of the solution found. For a given environment (i.e. obstacles distribution) with a given starting and goal cell, the solution found, if it exists, is the set of all optimal paths for the given metric. The CA evolution can be seen as a motion from one point to another point of the global state space until an optimal solution is reached. This is a convergence point for the given problem or a steady global state. If we make some perturbations, such as changing the environment (adding, deleting or moving one or more obstacles) or changing the goal cell, then the point becomes unstable and the CA starts to evolve again towards a new steady state, finding a new set of optimal trajectories. We call this property *Incremental Updating*. The CA spontaneously evolves to a new steady state from the previous one, without to be reset and reinitialize. The perturbations can be done also during the CA evolution, therefore realizing a *Reactive Path-Planner*: a path-planner that reacts to external changes.

5 Some Experimental Results

In this section, we illustrate a reduced set of experimental results that we obtained using the algorithm previously described.

5.1 Testing different metrics

In Fig. 5 and Fig. 6 are shown the solutions of a path-planning problem in a more complex environment. In the first example the robot can move only forward; in the

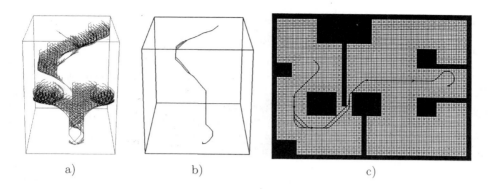

Figure 5: An example of a more complex world with metric $(2, 3, 1, H, H, H, H)$: a) Attraction Potentials Skeleton, b) Paths Skeleton, c) Trajectories

second, the robot can also move backward and rotating in the same place. The results show that a robot, also rotating in the same place, has a wider set of alternative trajectories. The performance tests, carried out with an Intel Pentium III 450 MHz PC, gives the following results: 92.45 ms for Fig. 5 and 68.24 ms for Fig. 6. These mean times are evaluated over 1,000 experiments for each problem and include the initialization phase of each test.

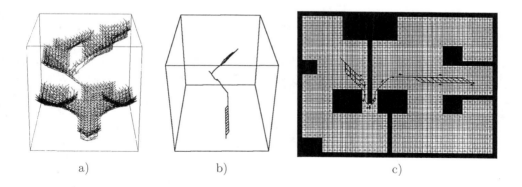

Figure 6: An example of a more complex world with metric $(2, 3, 1, 1, 2, 2, 3)$: a) Attraction Potentials Skeleton, b) Paths Skeleton, c) Trajectories

5.2 Complexity and Worst Cases

The complexity of this algorithm is strictly related to the obstacles distribution. It is quite impossible to evaluate it because of the huge number of possible obstacles distributions. We can only estimate a good upper-bound. In the worst cases, the longest paths cover nearly $N/2$ cells, and require about $2\frac{N}{2}N$ cells updates to be computed. Thus, the upper-bound of the complexity is $O(N^2)$, where N is the total number of cells of the *Workspace Bitmap* $\mathcal{GW}$. With an interdiction zone not null, the result is even better, because of a lower number of free cells.

6   Conclusions

In this paper we have described how we applied the Cellular Automata approach in the Path-Planning Problem for a non-holonomic mobile robot moving in a known dynamical environment. Some results and properties of this work can be highlighted: 1) CA formalism well correspond to one of the model often used to represent the environment (Occupancy Grid) for a fast sensorial mapping (module not described in this paper); 2) it is simple to implement this algorithm directly on a SIMD machine; 3) the algorithm is flexible and can be fitted to different types of kinematics changing the set of weights; 4) it generates collision-free and safe trajectories, also with cluttered obstacles distributions, allowing to pass more than once in the same position; 5) trajectories are smoothed respecting the imposed kinematics constraints; 6) it is a Reactive Path-Planner: it permits an incremental updating every time a modification of the environment is detected by sensors, or the goal changes (e.g. in following problem).

References

[1] Bandini S., Mauri G., Multilayered cellular automata, *Theoretical Computer Science*, **217** (1999) 99-113

[2] Barraquand J., Langlois B., Latombe J. C., Numerical Potential Field Techniques for Robot Path Planning, *IEEE Trans. on Systems, Man and Cybernetics*, Vol. **22**, No. 2 (March/April 1992) 224-241

[3] Goles E., Martinez S., Neural and Automata Networks: dynamical behavior and applications, Kluwer Academic Publishers (1990)

[4] Jahanbin M. R., Fallside F., Path Planning Using a Wave Simulation Technique in the Configuration Space in Gero J. S., Artificial Intelligence in Engineering: Robotics and Processes, Computational Mechanics Publications (Southampton 1988)

[5] Kathib O. Real-time Obstacle Avoidance for Manipulator and Mobile Robots, *Int. Conf. on Robotics and Automation* (1985)

[6] Kim J. O., Khosla P. K., Obstacle Avoidance Using Harmonic Potential Functions, *IEEE Trans. on Robotics and Automation*, Vol. **8**, No. 3 (June 1992) 338-349

[7] Latombe J. C., Robot Motion Planning, (Kluwer Academic Publishers, Boston, MA, 1991)

[8] Lozano-Pérez T., Wesley M. A., An Algorithm for Planning Collision-Free Paths Among Polyhedral Obstacles, *Comm. of the ACM*, Vol. **22**, No. 10 (October 1979) 560-570

[9] Lozano-Pérez T., Spatial Planning: A Configuration Space Approach, *IEEE Trans. on Computers*, Vol. C-32, No. 2 (February 1983) 108-120

[10] Tzionas P. G., Thanailakis A., Tsalides P. G., Collision-Free Path Planning for a Diamond-Shaped Robot Using Two-Dimensional Cellular Automata, *IEEE Trans. on Robotics and Automation*, Vol. **13**, No. 2 (1997) 237-250

[11] Zelinsky A., Using Path Transforms to Guide the Search for Findpath in 2D, *Int. J. of Robotics Research*, Vol. **13**, No. 4 (August 1994) 315-325

Intelligent Autonomous Systems 7
M. Gini et al. (Eds.)
IOS Press, 2002

3D MOTION RECOVERY WHILE ZOOMING USING ACTIVE CONTOURS

Elisa Martínez Marroquín
Enginyeria La Salle
Universitat Ramon Llull
Pge. Bonanova, 8
08022 Barcelona. Spain
elisa@salleurl.edu

Carme Torras Genís
Institut de Robòtica i Informàtica Industrial
CSIC-UPC
Llorens i Artigas, 4-6
08028 Barcelona. Spain
ctorras@iri.upc.es

Abstract. This paper considers the problem of 3D motion recovery from a sequence of monocular images while zooming. Unlike the common trend based on point matches, the proposed method relies on the deformation of an active contour fitted to a reference object. We derive the relation between the contour deformation and the 3D motion components, assuming time-varying focal length and principal point. This relation allows us to present a method to extract the rotation matrix and the scaled translation along the optical axis.

1 Introduction

The ability to zoom provides an image definition that eases a range of visual tasks common in robot vision, such as structure recovery or recognition. However, camera zooming invalidates most of the current solutions to computer vision problems (e.g., tracking or calibration), which assume constant intrinsic camera parameters, and therefore demands new approaches [6, 5]. Zooming does not only change the focal length but also the principal point, due to optical and mechanical misalignments in the lens system of the camera [4, 17]. The rest of intrinsic camera parameters (e.g., pixel size and aspect ratio) remain constant for long periods of time [16] and may be assumed known.

The process of calibration with the aid of a calibration pattern [18, 19] is inapplicable in real time or in cases where the camera optical parameters undergo frequent changes. Different approaches have recently emerged for autocalibrating the camera assuming time-varying internal parameters [17, 14, 1]. They are based only on point matches. The present work is based on an active contour and aims to recover the 3D motion parameters.

It is known that the 3D structure and motion can be recovered from a sequence of images [7, 13]. This requires a measure of the visual motion on the image plane and a model that relates this motion to the real 3D motion. The bottleneck when trying to bring this into practice is the computation of visual motion, which requires at least a set of feature matches between frames. Moreover, common methods for feature matching perform particularly poorly when zooming. Noting that the cumulative research on active contours [3, 8, 2] provides an efficient tracking of objects, this work has been motivated by the idea of building an algorithm for 3D motion recovery upon an active contour tracker.

Previous works by the authors highlight the feasibility of recovering 3D structure and motion from the analysis of an active contour fitted to a reference object. This is shown for different degrees of camera calibration [11, 10] and for uncalibrated cameras with constant intrinsic parameters [12, 9]. Here we extend the analysis to the case of time-varying internal calibration parameters due to zooming.

The paper is organized as follows. Section 2 relates the deformation of a contour to the 3D motion components and the internal calibration parameters. Then, Section 3 describes the process followed to recover the 3D motion components. Section 4 shows two examples of the experiments conducted to test the method. Finally, we draw some conclusions in Section 5.

2   Projection of 3D contour on the image plane

An active contour is fitted to the contour $\mathbf{D}(s)$ of a reference object (i.e. the target), which is marked on-line by the operator and may have any shape. It is automatically tracked along the sequence, and its corresponding shape vector is updated at each frame [12]. The shape vector provides a direct measure of image deformation that, as we will show, permits deriving the 3D relative motion between the camera and the target.

Keeping the attention on the small region used as reference, allows one to assume a simplified camera model for the tracked region, no matter if this model does not fit the rest of the image. Using a weak-perspective camera model, the projection $\mathbf{d_0}(s)$ (hereafter, the template) of the 3D contour $\mathbf{D}(s)$ in the initial frame is

$$\mathbf{d_0}(s) = \frac{f^{(0)}}{Z_0} \begin{bmatrix} K_u & 0 \\ 0 & K_v \end{bmatrix} \begin{bmatrix} X_0(s) \\ Y_0(s) \end{bmatrix} + \begin{bmatrix} u_0^{(0)} \\ v_0^{(0)} \end{bmatrix}, \tag{1}$$

where $f^{(i)}, u_0^{(i)}, v_0^{(i)}$ are the focal length and principal point for frame i, K_u, K_v denote the pixel size and Z_0 is the distance from the camera to the target at the reference frame.

Assuming rigid motion between frames, the projection of the 3D curve in frame i is

$$\mathbf{d}(s) = \frac{f^{(i)}}{Z_0 + T_z} \begin{bmatrix} K_u & 0 \\ 0 & K_v \end{bmatrix} \left(\begin{bmatrix} R_{11} & R_{12} \\ R_{21} & R_{22} \end{bmatrix} \begin{bmatrix} X_0(s) \\ Y_0(s) \end{bmatrix} + \begin{bmatrix} T_x \\ T_y \end{bmatrix} \right) + \begin{bmatrix} u_0^{(i)} \\ v_0^{(i)} \end{bmatrix}, \tag{2}$$

where R_{ij} are the elements of the rotation matrix and T_i are the elements of the translation vector. Combining equations (1) and (2),

$$\mathbf{d}(s) - \begin{bmatrix} u_0^{(i)} \\ v_0^{(i)} \end{bmatrix} =$$

$$= \frac{f^{(i)}}{f^{(0)}} \frac{Z_0}{Z_0 + T_z} \begin{bmatrix} K_u & 0 \\ 0 & K_v \end{bmatrix} \begin{bmatrix} R_{11} & R_{12} \\ R_{21} & R_{22} \end{bmatrix} \begin{bmatrix} \frac{1}{K_u} & 0 \\ 0 & \frac{1}{K_v} \end{bmatrix} \left(\mathbf{d_0}(s) - \begin{bmatrix} u_0^{(0)} \\ v_0^{(0)} \end{bmatrix} \right) +$$

$$+ \frac{f^{(i)}}{Z_0 + T_z} \begin{bmatrix} K_u & 0 \\ 0 & K_v \end{bmatrix} \begin{bmatrix} T_x \\ T_y \end{bmatrix}.$$

The above equation can be rewritten as

$$\mathbf{d}(s) - \begin{bmatrix} u_0^{(i)} \\ v_0^{(i)} \end{bmatrix} = \mathbf{L} \left(\mathbf{d_0}(s) - \begin{bmatrix} u_0^{(0)} \\ v_0^{(0)} \end{bmatrix} \right) + \mathbf{p},$$

where

$$\mathbf{L} = \frac{f^{(i)}}{f^{(0)}} \frac{Z_0}{Z_0 + T_z} \begin{bmatrix} K_u & 0 \\ 0 & K_v \end{bmatrix} \begin{bmatrix} R_{11} & R_{12} \\ R_{21} & R_{22} \end{bmatrix} \begin{bmatrix} \frac{1}{K_u} & 0 \\ 0 & \frac{1}{K_v} \end{bmatrix} =$$

$$= \frac{f^{(i)}}{f^{(0)}} \frac{Z_0}{Z_0 + T_z} \begin{bmatrix} R_{11} & R_{12}\frac{K_u}{K_v} \\ R_{21}\frac{K_v}{K_u} & R_{22} \end{bmatrix}$$

and

$$\mathbf{p} = \frac{f^{(i)}}{f^{(0)}} \frac{1}{Z_0 + T_z} \begin{bmatrix} K_u & 0 \\ 0 & K_v \end{bmatrix} \begin{bmatrix} T_x \\ T_y \end{bmatrix}. \tag{3}$$

The difference between the curve at a particular instant and the template is

$$\mathbf{d}(s) - \begin{bmatrix} u_0^{(i)} \\ v_0^{(i)} \end{bmatrix} - \mathbf{d_0}(s) + \begin{bmatrix} u_0^{(0)} \\ v_0^{(0)} \end{bmatrix} = (\mathbf{L} - \mathbf{I}) \left(\mathbf{d_0}(s) - \begin{bmatrix} u_0^{(0)} \\ v_0^{(0)} \end{bmatrix} \right) + \mathbf{p}, \tag{4}$$

where $\mathbf{I}$ is the 2×2 identity matrix.

Without loss of generality, the center of the template is assumed to be equal to the principal point, then equation (4) can be rewritten in terms of $\mathbf{d'_0}(s)$ and $\mathbf{d'}(s)$, that is, the projected contours referred to the template's centroid, as

$$\mathbf{d'}(s) - \mathbf{d'_0}(s) = (\mathbf{L} - \mathbf{I})\mathbf{d'_0}(s) + \mathbf{p} - \mathbf{\Delta u}, \tag{5}$$

where $\mathbf{\Delta u} \triangleq \begin{bmatrix} u_0^{(i)} - u_0^{(0)} \\ v_0^{(i)} - v_0^{(0)} \end{bmatrix}$. Equation (5) shows that the changes in the contour at each frame correspond to affine deformations of the template.

The affine parameters are $\mathbf{L}$ and $\mathbf{r} \triangleq \mathbf{p} - \mathbf{\Delta u}$, and are recovered from the shape of the contour, as follows. An active contour tracker is used to estimate the contour shape at each frame. This tracker is based on a Kalman filter and provides estimates of the contour's shape vector, which contains the affine parameters that relate the current shape of the contour with the template's shape (see [12] for details).

Assuming a constant aspect ratio $\mathcal{A} = \frac{K_u}{K_v}$, the $\mathbf{L}$ matrix can be rewritten as

$$\mathbf{L} = \frac{f^{(i)}}{f^{(0)}} \frac{Z_0}{Z_0 + T_z} \begin{bmatrix} R_{11} & R_{12}\mathcal{A} \\ R_{21}\frac{1}{\mathcal{A}} & R_{22} \end{bmatrix},$$

Then, taking $\mathcal{A} = 1$, a simplified matrix $\mathbf{L_s}$ can be computed

$$\mathbf{L_s} = \frac{f^{(i)}}{f^{(0)}} \frac{Z_0}{Z_0 + T_z} \begin{bmatrix} R_{11} & R_{12} \\ R_{21} & R_{22} \end{bmatrix}. \tag{6}$$

3 Extraction of 3D motion parameters

In this section we derive the relation between the affine parameters described above and the 3D motion components: 3D rotation $\mathbf{R}$ and 3D translation $\mathbf{T}$. The rotation matrix can be written in terms of the Euler angles,

$$\mathbf{R} = \mathbf{R_z}(\phi)\mathbf{R_x}(\theta)\mathbf{R_z}(\psi) \tag{7}$$

where $\mathbf{R_z}(\psi)$ and $\mathbf{R_z}(\phi)$ are rotation matrices about the Z axis and $\mathbf{R_x}(\theta)$ is a rotation matrix about the X axis.

Using the Euler notation to represent the rotation matrix, equation (6) can be rewritten as

$$\mathbf{L_s} = \frac{f^{(i)}}{f^{(0)}} \frac{Z_0}{Z_0+T_z} \mathbf{R_z}|_2(\phi)\mathbf{R_x}|_2(\theta)\mathbf{R_z}|_2(\psi) =$$
$$= \frac{f^{(i)}}{f^{(0)}} \frac{Z_0}{Z_0+T_z} \mathbf{R_z}|_2(\phi) \begin{bmatrix} 1 & 0 \\ 0 & cos\theta \end{bmatrix} \mathbf{R_z}|_2(\psi), \qquad (8)$$

where $\mathbf{R}|_2$ denotes the 2×2 submatrix of $\mathbf{R}$. Then,

$$\mathbf{L_s L_s}^T = \mathbf{R_z}|_2(\phi) \begin{bmatrix} \left(\frac{f^{(i)}}{f^{(0)}} \frac{Z_0}{T_z+Z_0}\right)^2 & 0 \\ 0 & \left(\frac{f^{(i)}}{f^{(0)}} \frac{Z_0}{T_z+Z_0}\right)^2 cos^2\theta \end{bmatrix} \mathbf{R_z}|_2^{-1}(\phi).$$

This last equation shows that θ can be computed from the ratio of eigenvalues of $\mathbf{L_s L_s}^T$, namely (λ_1, λ_2),

$$cos\theta = \sqrt{\frac{\lambda_2}{\lambda_1}},$$

where $\lambda_1 = \left(\frac{f^{(i)}}{f^{(0)}} \frac{Z_0}{Z_0+T_z}\right)^2$ is the largest eigenvalue. The angle ϕ can be extracted from the eigenvectors of $\mathbf{L_s L_s}^T$. The eigenvector $\mathbf{v_1}$ with largest eigenvalue equals the first column of $\mathbf{R_z}|_2(\phi)$,

$$\mathbf{v_1} = \begin{bmatrix} cos\phi \\ sin\phi \end{bmatrix}.$$

At this stage, isolating $\mathbf{R_z}|_2(\psi)$ in equation (8),

$$\mathbf{R_z}|_2(\psi) = \frac{f^{(0)}}{f^{(i)}}(1 + \frac{T_z}{Z_0}) \begin{bmatrix} 1 & 0 \\ 0 & \frac{1}{cos\theta} \end{bmatrix} \mathbf{R_z}|_2(-\phi)\mathbf{L_s},$$

and observing that

$$\frac{f^{(0)}}{f^{(i)}} \left(1 + \frac{T_z}{Z_0}\right) = \frac{1}{\sqrt{\lambda_1}}, \qquad (9)$$

we can find $sin\psi$ and then ψ. Once the angles ψ, θ, ϕ are known, the rotation matrix $\mathbf{R}$ can be computed as in equation (7).

¿From equation (9) the scaled depth is recovered as

$$1 + \frac{T_z}{Z_0} = \frac{1}{\sqrt{\lambda_1}} \frac{f^{(i)}}{f^{(0)}}.$$

This recovered depth depends on the relation between the focal lengths in consecutive frames. In robot vision applications one may assume that the robot controls the zooming factor. Hence, the relation between focal lengths at different time instants may be assumed known even when the exact focal length is unknown.

From equations (3) and (9),

$$\frac{f^{(0)}}{Z_0} \begin{bmatrix} T_x \\ T_y \end{bmatrix} = \frac{\mathbf{r} + \Delta\mathbf{u}}{\sqrt{\lambda_1}} \begin{bmatrix} \frac{1}{K_u} & 0 \\ 0 & \frac{1}{K_v} \end{bmatrix}.$$

Thus, we observe that the recovered scaled translation depends on the difference between the principal points in consecutive frames.

4   Experimental examples

Previously to incorporating the technique to the visual system of the robot ARGOS [15], for which it has been developed, we have performed some experiments in a more controlled setting. Two examples of the experiments conducted to test the method in the laboratory are presented. Figure 1 shows an example of the results of the 3D motion estimation while zooming. An active contour has been fitted to the target (i.e. the square). A virtual object is drawn in the middle of the image with the estimated 3D rotation and translation along Z between the camera and the target. As expected, the estimated 3D rotation is invariant to zooming, while the estimated translation along Z changes proportionally to the zoom factor.

Figure 2 shows the results obtained for different 3D motions. Again a virtual object is drawn in the middle of the screen following the motions of the target. We verify that the proposed method provides qualitatively correct results.

5   Concluding remarks

We have analysed how the deformation of an active contour can be used to extract the 3D motion components while zooming. The basis of the method draws on ideas from previously published papers by the authors [10, 9], and fills the remaining hole in the analysis of the deformation of an active contour for different assumptions about the intrinsic camera parameters.

The theoretical deduction along with the experimental results show that the 3D rotation matrix can be reliably recovered while zooming. However, as one could expect, the scaled depth is distorted by the camera zoom. On the other hand, the change in the principal point due to the zoom affects the recovery of the other two components of the 3D translation vector. The experiments have been conducted with a monocular camera, hence the results keep the ambiguities common in this case. However, the deduction may be easily extended to a stereo rig.

Acknowledgments

This research has been partially supported by the research grant "Navegación autónoma de robots guiados por objetivos visuales" CICYT DPI2000-1352-C02-01 of the Spanish Science and Technology Council.

References

[1] L. Agapito, R. Hartley, and E. Hayman. Linear calibration of a rotating and zooming camera. In *Proc. of the Computer Vision and Pattern Recognition Conference*, 1999.

[2] A. Blake and M. Isard. *Active contours*. Springer, 1998.

[3] A. Blake, M.A. Isard, and D. Reynard. Learning to track the visual motion of contours. *J. Artificial Intelligence*, 78:101–134, 1995.

[4] J.L. Crowley, P. Bobet, and C. Schmidt. Maintaining stereo calibration by tracking image points. In *Proc. Conf. Computer Vision and Pattern Recognition*, pages 483–488, 1993.

A.

B.

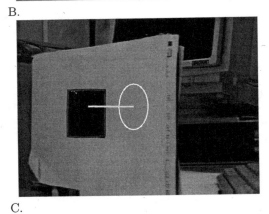

C.

Figure 1: *3D motion recovery while zooming. The estimation of the rotation is invariant to camera zooming.*

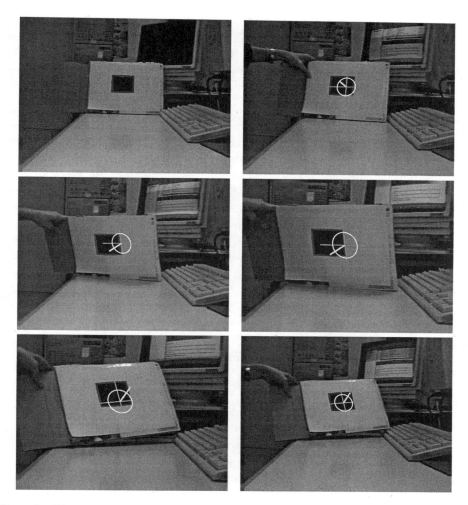

Figure 2: *3D motion recovery with an uncalibrated camera while zooming. Six samples of a video sequence taken by a static camera observing a moving target. The first image is the initial view, which is taken as the template. The subsequent images show the recovered motion after different target motions.*

[5] E. Hayman, T. Thrhallsson, and D. Murray. Zoom-invariant tracking using points and lines in affine views an application of the affine multifocal tensors. In *Proc. 7th Int. Conf. on Computer Vision*, September 1999.

[6] A. Heyden and K. Astrom. Euclidean reconstruction from image sequences with varying and unknown focal length and principal point. In *Proc. Conf. Computer Vision and Pattern Recognition*, 1997.

[7] B.K.P. Horn. *Robot vision*. MIT Press, 1986.

[8] M.A. Isard and A. Blake. Visual tracking by stochastic propagation of conditional density. In *Proc. 4th European Conf. Computer Vision*, pages 343–356, Cambridge, England, Apr 1996.

[9] E. Martínez and C. Torras. Depth map from the combination of matched points with active contours. In *Proc. of the IEEE International Conference on Intelligent Vehicles.*, pages 332–338, Dearborn, Michigan, USA, October, 2000.

[10] E. Martínez and C. Torras. Epipolar geometry from the deformation of an active contour. In *Proc. International Conference on Pattern Recognition.*, pages 534–537, Barcelona, Spain. September, 2000.

[11] E. Martínez and C. Torras. Integration of appearance and geometric methods for the analysis of monocular sequences. In *Proc. IST/SPIE 12th Annual Symp. on Electronic Imaging*, pages 62–70, San Jose. California. USA. January, 2000.

[12] E. Martínez and C. Torras. Qualitative vision for the guidance of legged robots in unstructured environments. *Pattern Recognition.*, 34(8):1585–1600, August 2001.

[13] D.W. Murray and B.F. Buxton. *Experiments in the machine interpretation of visual motion*. MIT Press, 1990.

[14] M. Pollefeys, R. Koch, and L. Van Gool. Self-calibration and metric reconstruction in spite of varying and unknown internal camera parameters. In *Proc. 6th Int. Conf. on Computer Vision, Bombay, India*, pages 90–95, January 1998.

[15] Argos Robot. *www-iri.upc.es/people/porta/robots/argos/index.html*.

[16] S. Soatto and P. Perona. Recursive estimation of camera motion from uncalibrated image sequences. In *Proc. 1st IEEE International Conference on Image Processing (ICIP)*, pages II-58–62, 1994.

[17] P. Sturm. Self calibration of a moving camera by pre–calibration. In *Proc. Conf. Computer Vision and Pattern Recognition*, 1996.

[18] R.Y. Tsai. A versatile camera calibration technique for high-accuracy 3d machine vision metrology using off-the-shelf tv cameras and lenses. *IEEE Journal of Robotics and Automation*, RA-3(4):323–344, 1987.

[19] J. Weng, P. Cohen, and M. Herniou. Camera calibration with distortion models and accuracy evaluation. *IEEE Trans. on Pattern Analysis and Machine Intelligence*, 14(10):965–980, 1992.

Intelligent Autonomous Systems 7
M. Gini et al. (Eds.)
IOS Press, 2002

SHAPE: Simulated Heterogeneous Agent Planing of an Environment

[1]Gerard McKee, [1]James Bull and [2]Paul Schenker
[1]Department of Computer Science, The University of Reading, UK
[2]The Jet Propulsion Laboratory, Pasadena, USA.

ABSTRACT

In this paper we present a multi-agent model for "shaping" an environment. The model comprises a heterogeneous pool of agents comprising three types of agent, namely roller, bulldozer and digger agents. The shaping task involves the planing, or smoothing, of an extended unstructured landscape. The paper describes the architecture of the agent system, the contribution that each agent makes to the task, and the embodiment of the task goal across the agents. A simulation environment was created to investigate the multi-agent architecture experimentally. The simulation of the environment terrain and the affect that each agent exerts on the terrain is described. Experimental studies employing the simulation are presented. These demonstrate the effectiveness of the agent model for the given task and the relative role of the agents in completing the task. The results are important for the development of models for site preparation in remote outposts, such as those planned for Mars, but are also relevant to the design and engineering of robot-based systems for environment structuring in general.

1. Introduction

It is a characteristic of human agency within an environment that the environment acquires a structure suited to human needs and patterns of behaviour. Typical structuring includes the construction of homes, office buildings, industrial plants, roads, motorways, railways, airports and shipping terminals. The transportation infrastructure here supports, in turn, a variety of vehicles including cars, lorries, trains, airplanes, and varied water-based modes of transport respectively. The construction of these systems requires some form of environment modification, whether for the infrastructure itself or the vehicles that employ the infrastructure, or indeed the power plants that keep them functioning. As humans now reach out towards the colonisation of space, the preparation of remote landscapes for science missions, and ultimately human habitation, offers a significant challenge to human ingenuity. The main challenge is that of deploying within the remote environment a surrogate active agency that will prepare the environment for these science and human needs. In the near term this agency will most likely be provided by robotic systems [1]. For example, research is currently underway investigating cooperative robot systems for the deployment of a solar photo-voltaic (PV) tent array on the Martian surface [2]. The construction of such a "power station" is an important precursor for extended robotic and human presence on Mars.

In this paper we report on a preliminary study aimed at investigating the requirements for robotic agents to actively manipulate an environment into a target structure. In its most general form the process of modification or "forming" can be viewed as the imposition of structure upon an unstructured environment. The general question is what agency (single or multiple robotic system and/or human agents) will achieve the task most effectively [3]. We model this transformation as comprising two stages, namely site preparation and subsequent construction of an artefact (a road, a building, a power station, etc.) on the prepared site. This is not to say that these two stages are separate in general, but for our purpose they can be taken as such. This two-stage model is also not to be considered as exclusive of all site preparation tasks. For example, preparing land for farming may include the introduction of an extensive network of drainage systems into the existing landscape, or the construction of borders surrounding the site – as in land reclamation from the sea.

We focus in the studies reported here on site preparation. We assume that this preparation involves levelling, or "planing", the environment. Such tasks may comprise a number of elements, including "clearing" the land of debris such as rocks, boulders, trees and shrubs. It may also include digging up material and relocating it elsewhere in the site, exporting it to locations off-site, or even importing new material onto the site. We consider the task explored in the studies reported here as a form of "shaping". It comprises the relocation of material within the site, but excludes importing or exporting of material. We also assume that the surface is homogeneous in form, varying only in density from one location to another. In these studies also we assume agents with specific, but diverse, operational characteristics. These are in part inspired by a range of online robotic educational scenarios under development at the University of Reading [4]. Finally, one of the key challenges of employing multi-agent systems is how to embody the task goals within the system. We describe a simple control architecture that distributes the task goals across the set of agents in a way that minimises their interaction.

The remainder of the paper is organised as follows. The following section describes the overall architecture of the agent system that was investigated in these studies. Three aspects are discussed in detail, namely the individual agent types, the global interaction between the agents, and the representation of the task within the agent population. Section 3 describes the experimental environment developed to investigate the agent system. It outlines the types of environments that can be created within the simulation and the representation of the impact of each agent type on its environment. Section 4 presents three sets of experimental results that illustrate the affect of the agent population on a target environment. The results show that effective smoothing, or planing, behaviour can indeed be achieved. Finally, section 5 provides a summary and conclusions.

2. Agent Architecture

We assume, to begin with, that the shaping of a terrain will require a set of "workers" and "tools". A physically self-contained robotic system may comprise both the worker and tool(s) as an integrated whole. We also assume that the "agent pool" that is to execute the shaping task will comprise a heterogeneous set of agent types. In general, each agent type will be suitable for taking on different tasks, for which they will possess tailored actuators and end effectors. The operational capabilities of each agent type, however, may not be entirely distinct. The studies presented in this paper assume three types of agent, namely roller, bulldozer and digger agents (Figures 1 & 2). The first of these, the roller agents, act locally to compact surface material. The primary action is compaction, but there is also a certain amount of lateral displacement of material, and surface smearing if the material is in small mounds. The bulldozer type agent makes a more significant impact on its environment, pushing surface material ahead and leaving a smooth, narrow, track behind. It skims material off the surface and deposits it at the end of its track. Hence, it has a greater capability to relocate surface material than the roller agent. The third agent type, the digger agent, digs up surface material from one area and deposits it in another area. In fact, its own action, digging, has an unstructuring effect locally. However, the digger agent can carry surface material for extended distances and, therefore, has a still longer range than the bulldozer agent.

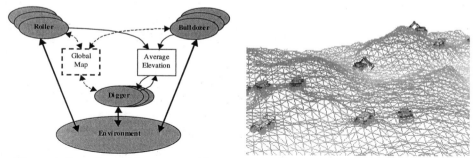

Figure 1. Overview of the agent architecture   Figure 2. A simulated environment

The fact that all three agents in one way or another relocate surface material reflects the "shaping" character of the task. Each agent has a different "reach", namely local, intermediate and long-range, reflecting the general requirements for local, intermediate and long-range relocation of surface material in order to achieve an extended, smooth surface. In order to understand the embodiment of the task within the agents it is important to understand the relative decision-making that is required of the agents within this context. The first two agents, the roller and bulldozer agents need simply to seek out terrain that suits their operational characteristics and to work it. In the simulation studies reported here this is as far as their behaviour goes. The digger agent, however, has a significant question to ask, implicit in the selection of the terrain to work on: When should it dig up material and, given that it is carrying a load, when should it deposit that load?

Qualitatively, the digger agent should dig up material from raised areas and deposit it in sunken areas, where these are measured relative to the target smoothed surface. However, it has to decide when the surface it is traversing changes from a raised area, suitable for digging, to a sunken area, suitable for depositing. This is particularly vexing since it will be difficult to estimate the level of the target surface prior to task commencement, and even during the progression of the task; it is only when the task is completed that an accurate measure of the final surface level will become apparent. The mechanism proposed here for addressing this decision is to assume that each agent in the population carries with it a sensor that can measure its local elevation. Each agent broadcasts its elevation to all of the other agents so that each can compute the average elevation of the site (Figure 1). The digger agents use this measure to decide when to dig up material and when to deposit it. A random element is incorporated in the decision in order to avoid sharply rising ridges and rising peaks at the decision boundary. In summary, therefore, the task goal is not explicitly defined anywhere, but is embodied in the behaviour of the agents to different degrees. The strongest representation is in the digger agents.

In addition to each agent type being tailored to a specific range of operations, and sharing elevation information globally, each agent must determine if the terrain it is currently traversing is suitable for its operational capabilities. For this purpose we assume that all three agent types can inspect the local terrain and classify it into one of a set of four classes. Three of these classes reflect the terrain suitable for either the agent itself or one of the other two agent types. The fourth category reflects an environment unsuitable to any agent type, but may not necessarily represent a "smooth surface" category. However, as the shaping continues this category does indeed come to represent, implicitly, the target smooth terrain. The categories are nominally labelled *roller*, *bulldozer*, *digger* and *unknown*. The agents take a small sample of the terrain elevation in five areas in their immediate vicinity, namely forward, left, right, and to the rear, and also the surface immediately below the agent. We assume that the former could be based on elevation maps from stereo camera systems [5] and the latter could be based on tilt and/or range sensors on the underbelly of the agent. The samples are treated as separate, independent measurements, for the purpose of decision making on the part of the agent. Each sample is fed to a Kohonen neural net classifier which classifies the terrain into one of the four classes. The network can be trained beforehand to recognise the terrain suitable for each category.

If at least one of the samples is suitable for the agent, then the agent can either begin working the surface if the sample corresponds to its current location, or it can drive towards the sample location. If none is suitable, the agent enters a wander behaviour, which means it moves in random directions through the environment. In these cases the agent also broadcasts the sample classification to a global map that is shared by all agents (Figure 1). This means that if the local terrain is not suitable for the agent, it can check the global map for suitable terrain and move towards it. It is only if the latter is also not the case that the agent will then enter the wander behaviour. The global map incorporates within the agent architecture the ability for an agent to identify and pass on information about terrain it has observed and that is suitable for other classes of agents. Thus, while operational capabilities are assumed different across the agents, perceptual capabilities are similar.

3. Experimental Environment

A simulation environment was specifically developed for the purpose of conducting the experiments reported here. The environment, called "SHAPE", comprises interfaces for creating terrain of various types, for defining agent pools – involving setting numbers for each agent type, for creating and training the Kohonen network, and for running the simulation. It also incorporates a primitive physics model. This section focuses on the modelling of the terrain and the modelling of the affect of each agent type on the terrain.

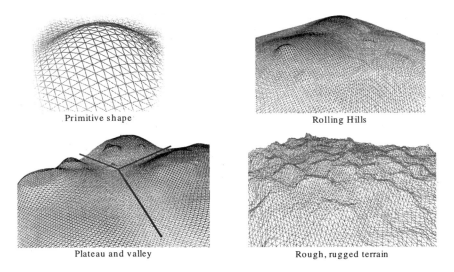

Primitive shape   Rolling Hills

Plateau and valley   Rough, rugged terrain

Figure 3. Terrain Generation

The types of terrain that can be created in SHAPE range from rolling hills, through valleys and plateaus, to rugged landscapes (Figure 3). The terrain surface is represented by a 2D array made up of surface elements, each of which holds the location in 3D space of that element and the density of terrain at that point. To generate terrain, seven variables are required from the user: bumpiness, dippiness, maximum elevation, bump width, displacement, noise, and altitude at which plateaus are formed. Bumpiness and dipiness correspond to the percentage of all surface elements that will be the central point of a surface feature (bump or dip). Width is the maximum width that any feature can be. Displacement is the distance by which the entire surface is raised above the y = 0 plane – it is useful for preventing dips from bottoming out. Noise is the amount of surface elevation added to each surface element after all features have been generated. It is uniformly distributed over the entire terrain. Each bump or dip in the terrain is generated first as a small surface array consisting only of elevation data. The elevation of each element in this array is determined by a Gaussian function (Figure 3, top left). To generate an entire surface, a series of features are added at random locations to an initially flat environment, each feature being randomly generated within the bounds specified by the user. Plateaus are formed if the surface rises above the user defined plateau level.

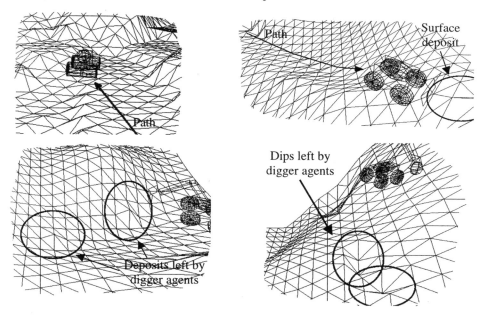

Figure 4. Environmental affect of roller, bulldozer and digger agents

When an agent exerts an affect on the environment it may result in a deformation of the surface and/or a deviation in the density of the terrain. A roller agent is capable of levelling the terrain it is situated on (Figure 4, top-left). This is simulated by taking the mean height of all the terrain under the agent and setting each surface element under it to this mean height. The roller agent, therefore, is good for levelling out areas of high frequency, low amplitude surface disturbance. A bulldozer agent works in a similar way to a roller agent, but also removes some material from the terrain it flattens, determined by the mean density of the terrain it encounters. The mean height of all surface elements covered during its traverse is reduced, taking into account the density of the underlying surface. The terrain is flattened as per the roller agent. The bulldozer moves the material it skims off the terrain to a pile at the end of the track (generated by the Gaussian feature function; Figure 4, top-right). It is suited to general surface flattening. The digger agent, as discussed previously, references the mean surface height generated from the altitudes broadcast by all agents. If the altitude of a digger agent is higher than this mean altitude (plus a random variance) and its scoop is empty it will take a scoop of material from the terrain (Figure 4, bottom-right). If the agents' location is lower than this mean (minus a random variance) and its scoop is full it will drop its load at that location (Figure 4, bottom-left). The effect that the scoop has on the environment is simulated using the Gaussian feature function to generate either a feature of positive height (a lump) or a feature of negative height (a dip).

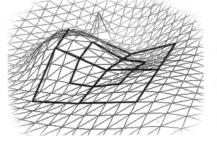

Finally, Figure 5 illustrates the five samples taken of the local environment by each agent. Each sample is a square area and overlaps with other samples as shown. The size of the samples are determined by the parameters selected by the user for the Kohonen network.

Figure 5 (right). Five samples taken about the agent's current location

4. Results and Analysis

In this section we report on three simulation experiments that address general characteristics of the agent architecture. The results presented are qualitative, based on observations of the simulation environment following a number of cycles. A cycle corresponds to an iteration of the algorithm and does not reflect any particular time period. However, the physics model employed in the simulation does reflect the basic mechanics of motion. All three experiments were conducted using terrain generated from the same set of parameters, and generated anew for each experiment. The terrain can be described as a mixture of a rolling landscape with local rugged surface features.

The first experiment represented a form of control. It employed a population of 30 agents, comprising 10 of each agent type. The Kohonen network was left untrained, meaning that the network would misclassify terrain features and therefore the target agent would be unable to make a significant impact on that terrain, and may even leave it with less structure than before. The initial form of the environment and its form following 10,000 and 20,000 cycles are shown in Figure 6 – this pattern of presentation is repeated for the other experiments. The experiment, as expected, resulted in a poorly flattened surface at the end of the run (20,000 cycles). The final terrain has a ridged structure, dotted with sharp peaks around the edges.

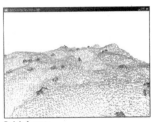

Initial state | 10,000 cycles | 20,000 cycles

Figure 6. Experiment 1 – untrained Kohonen network

The second experiment employed the same specification for the terrain and the agent population as the first, but incorporated a trained neural network map. The network takes a sample grid of size 4 by 4 and has an output layer of 6 by 6 neurons. It was trained to recognise low magnitude, high frequency bumps as suitable for the roller agent, high magnitude bumps or dips as suitable for the digger agent and peaks or small bumps as suitable for a bulldozer. The result after 20,000 cycles (Figure 7) is a qualitatively flat terrain with some small projections located on the edge of the environment.

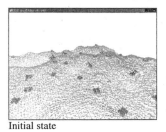 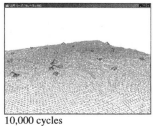 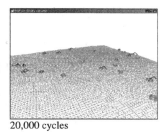
Initial state   10,000 cycles   20,000 cycles

Figure 7. Experiment 2 – trained 6x6 Kohonen network

The small projections observed during the second experiment suggested that the digger agents were unloading surface material where the surface dipped very slightly. A further experiment was therefore undertaken to explore the effect of culling the digger agents midway through the simulation run. This should ensure that a roughly flat terrain has been formed such that bulldozers and rollers might finish the job of smoothing. The initial population was set again to 30 agents, but this time just 5 each of the roller and bulldozer agents, and 20 of the digger agents. The same trained neural network map and terrain specification were employed as for the second experiment. When the digger cull took place, the number of roller and bulldozer agents were increased from 5 to 15 each to maintain the number of agents in the environment. The results for the experimental run, shown in Figure 8, were insightful. The surface remaining after 20,000 cycles was indeed smooth, but was also slightly bowed up in the middle. This reflected the fact that the rollers and bulldozers alone could not shift the mound that was left at the midway point – when the digger agents were culled; the mound therefore remained and was simply smoothed over. These results illustrate the importance of the long-range affects of the digger agents.

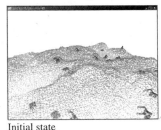 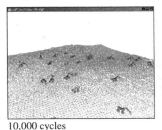 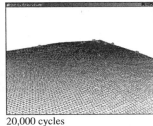
Initial state   10,000 cycles   20,000 cycles

Figure 8. Experiment 3 – digger cull

In conclusion, the combination of a population of diverse agent types and a trained classifier network provides the best results for the set of experiments reported above. The results also highlight the key role of the digger agents in relocating surface material over extended distances in the environment and the embodiment within their behaviour of the global requirements of the task.

5. Summary and Conclusions

In this paper we have presented a multi-agent model for robotic shaping of an unstructured environment. The task we have investigated can be considered characteristic of site preparation. The studies presented involved a heterogeneous pool of agents offering a range of local and long-range operational capabilities. The task itself was represented in the agents to different degrees. A number of simulation experiments showed that shaping of an extended environment could in principle be achieved. These studies are preliminary to an extended, systematic investigation of robotic work crews for environment shaping. A goal of this research will be to develop rigorous models of the relationship between the task structure and the agent work crew that embodies the task. These models will take into account the diverse behavioural characteristics of the agent population [6], the different mechanical structures and manipulation capabilities of the agents, and the interaction (implicit or explicit) between the agents, to assess the suitability of a work crew to the task.

We mention here four specific areas of further development for this research. The first is the development of more realistic models of the environment. Enriched models should allow for more varied geology including rocks, boulders and embedded structures such as outcrops and trees. Simulations, however, have inherent limitations. The development of real robot work crews based on these simulation studies is an important practical goal of the research. A second area for further development is the agent population. The current model employs three agent types based on existing terrestrial landscaping vehicles. There is considerable scope for evolving robotic agents more suited not only to the current terrain but also to other types of terrain as well. A key challenge here is identifying optimisation techniques [3] that take into account models of the terrain and the operational criteria that must be considered if such agent populations are to be dispatched to remote sites.

A third area for further development is agent behaviour. In the current model, for example, bulldozer agents do not direct their scooping trajectory such that the terminating point corresponds to a dip. Such a behaviour would be more productive. In general, there could be considerably more purpose in the behaviour of the individual agents. A fourth area of further development is the investigation of agent architectures for the second stage of environment structuring, that of building on the prepared site [1]. The key challenge is to develop methods that start from a target structure description to define a robotic agent architecture that will effect the construction. Evolutionary techniques could play an important role here.

References

[1] Robot Colonies, Special Issue of Autonomous Robots (Eds. R. C. Arkin and G. A. Bekey), Vol. 4, No. 5, 1997.
[2] P. S. Schenker, T. L. Huntsberger, P, Pirjanian, A. Trebi-Ollennu, H. Das, S. S. Joshi, H. Aghazarian, A. G. Ganino, B. A. Kennedy and M. S. Garrett, Robot work crews for planetary outposts: close cooperation and coordination of multiple mobile robots, *SPIE Proceedings on Sensor Fusion and Decentralized Control in Robotic Systems III*, Vol. 4196, pp. 210-220, 2000.
[3] L. E. Parker, Y. Guo and D. Jung, Cooperative robot teams applied to the site preparation task, in Proceedings of the 10th International Conference on Advanced Robotics, (ICAR 2001), pp. 71-77, 2001.
[4] G. T. McKee and K. Phillips, TORUS: Toys operated remotely for understanding science, *SPIE Proceeding Vol. 4195, Mobile Robots XV and Telemanipulator and Telepresence Technologies VII*, Nov. 2000.
[5] T. L. Huntsberger, E. T. Baumgartner, H. Aghazarian, Y. Cheng, P. S. Schenker, P. C. Leger, K. D. Iagnemma, and S. Dubowsky, Sensor-fused autonomous guidance of a mobile robot and applications to Mars sample return operations, in *Proceedings of the SPIE Conference on Sensor Fusion and Decentralized Control in Robotics Systems II*, Vol. 3839, pp. 2-8, 1999.
[6] T. Balch, Hierarchic Social Entropy: An Information Theoretic Measure of Robot Group Diversity, *Autonomous Robots*, Vol. 8, No. 3, July 2000.

Intelligent Autonomous Systems 7
M. Gini et al. (Eds.)
IOS Press, 2002

Resource Scheduling and Load Balancing in Distributed Robotic Control Systems

Colin McMillen, Kristen Stubbs, Paul E. Rybski, Sascha A. Stoeter,
Maria Gini, Nikolaos Papanikolopoulos
Center for Distributed Robotics
Department of Computer Science and Engineering, University of Minnesota, U.S.A.
{mcmillen,kstubbs,rybski,stoeter,gini,npapas}@cs.umn.edu

Abstract. This paper describes the latest advances made to a software architecture designed to control multiple miniature robots. As the robots themselves have very limited computational capabilities, a distributed control system is needed to coordinate tasks among a large number of robots. Two of the major challenges facing such a system are the scheduling of access to system resources and the distribution of work across multiple workstations. This paper discusses solutions to these problems in the context of a distributed surveillance task.

1 Introduction

There are many advantages in using a distributed control system to manage the operation of a group of robots. Such a system could control a wide variety of heterogeneous devices and do so over great physical distances. This system could be very modular, supporting many kinds of devices. However, designing and implementing a distributed control system such that it works effectively and efficiently across varying hardware configurations and in spite of differing computational demands is a non-trivial task.

For instance, one challenge is to ensure that various system components have access to the resources (such as communications channels, computational units, robot chassis, or sensor inputs) that they need in order to accomplish their tasks. Not all resources can accommodate many simultaneous access requests, and many resources can only handle a single request at a time. Another challenge is to determine how to spread the demand for resources over the entire system. This helps to ensure that the computational and resource load is balanced, and that no one component is taxed too greatly.

A distributed control system has been developed for managing a group of small, mobile robots which have extremely limited on-board sensing and computing capabilities. These robots, called Scouts, are completely reliant upon a proxy processing scheme for their operation. In this scheme, the "bodies" of the robots are physically separate from the "brains" and are connected only by wireless data links. Sensor information from the Scouts is also broadcast over a wireless data link. Any workstation within range of a particular Scout robot is capable of receiving and interpreting its transmissions. Because the control software for the Scouts runs on one or many external workstations, it has access to the resources of other participating workstations on the Internet. This means that the workstation processing the sensor input does not necessarily have to be the same as the one that is transmitting motion commands back to the Scout robots.

The communication channels that the Scouts use to send and receive information are very limited in power and available throughput. As a result, access to these channels must be explicitly scheduled so that the demand for them can be met while maintaining the integrity of the system's operation. The Scout control architecture has been developed to take these factors into account. Additionally, because the architecture can make use of multiple networked workstations, it attempts to improve the entire system's performance by balancing the computational load across these computers.

This paper provides a general overview of the Scouts' control architecture and describes the architecture's resource management and load balancing capabilities. Experimental results which illustrate the utility of these features are presented.

2 Related Research

A software architecture designed for controlling groups of robots must allow for distributed operations, handle resource allocation, and support real-time operations with graceful degradation. A number of architectures have been proposed, many of them described in [5]. Our architecture has some similarities with CAMPOUT [7], a distributed hybrid-architecture based on behaviors. The major difference is that we focus on resource allocation and dynamic scheduling, while CAMPOUT is mostly designed for behavior fusion. We rely on CORBA [4] as the underlying technology for distributed processing, while in CAMPOUT each robot runs an instance of the architecture and uses sockets for communication with other robots. Our architecture also has some similarities with ALLIANCE [6], which provides distributed and fault-tolerant control for teams of homogeneous robots. The issues that our architecture addresses are more general, allowing for control of heterogeneous robot teams as well as not putting any restrictions on the methodology for the robot control (deliberative or reactive).

Resource allocation and dynamic scheduling are essential to ensure robust execution. Our work focuses on dynamic allocation of resources at execution time, as opposed to analyzing resource requests off-line, as in [1, 3], and modifying the plans when requests cannot be satisfied. Our approach is specially suited to unpredictable environments, where resources have to be allocated in a dynamic way that cannot be predicted in advance. We rely on the wide body of algorithms that exists in the area of real-time scheduling [9] and load balancing [2]. Our architecture provides support for distribution of resources across robots, use of shared resources, and seamless integration of autonomous and human-supervised control [10].

3 Distributed Robotic Control

The Scout robot is a cylindrical, two-wheeled robot that is 40 mm in diameter and 110 mm in length. Designed for portability and the ability to access hard-to-reach areas, Scouts have potential applications in the fields of urban surveillance, search-and-rescue scenarios, and other situations requiring a team of small, autonomous, maneuverable robots. One of the primary missions that these robots are designed for is a distributed reconnaissance and surveillance task where larger robots called Rangers are used to transport and deploy the Scouts into areas where they can gather further information. In any particular mission, there could be up to ten Scouts assigned to a single Ranger. More detail about the Ranger/Scout hardware, the reconnaissance and surveillance missions they are designed to do, and some experimental validation of these missions is beyond the scope of this paper and instead can be found in [8].

To facilitate the operation of team-based autonomous behaviors, a software architecture has been developed. This software architecture consists of several components.

Many of these components are **Resource Controllers** (RCs), which control access to various system resources, such as the Scout robots, radio and video transmission frequencies, and various framegrabber cards. Some RCs, such as those that manage the radio frequencies, are *sharable*, which indicates that multiple system components may be granted concurrent access to these resources. Other RCs, such as those that manage the framegrabber cards, are *non-sharable*, which indicates that only one system component may access them at a time. Other components, called **Behaviors**, utilize the system RCs to perform autonomous tasks. **Behaviors** can be joined together in a hierarchical structure to perform arbitrarily complex tasks. To simplify the writing of **Behaviors**, an abstraction called the **Aggregate Resource Controller** (ARC) has been developed. ARCs encapsulate **Behaviors'** requests for simultaneous access to groups of RCs.

One of the more complex autonomous behaviors the system supports is **Place Sensor Net**, which is the establishment of a sensor network of Scout robots (Figure 1). This is the deployment phase of the mission, where a Ranger sends a number of Scouts into an area so they can better view the environment. When the **Place Sensor Net** behavior is activated, all Scouts involved first search for a dark location and servo toward it, attempting to hide. After each Scout is hidden, it turns toward the light and watches for motion. This task is relatively demanding of the software architecture, making it an ideal example of how the architecture schedules access to resources and how it shares the workload among multiple machines.

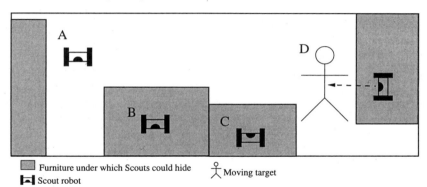

Furniture under which Scouts could hide
Scout robot
Moving target

A: This Scout has started in an open part of a room surrounded by furniture. It starts by searching for a dark place in which to hide.

B: This Scout has moved to a dark area (the shadow underneath a piece of furniture) and hidden in it.

C: This Scout has turned toward light (in this case, the open area of the room).

D: This Scout is watching for motion, alerting the user to the presence of any moving targets within its line of sight.

Figure 1: Example of the **Place Sensor Net** behavior.

4   Resource Controller Manager

The RESOURCE CONTROLLER MANAGER (RCM) is the core service responsible for scheduling access to all system resources. A behavior requests simultaneous access to a group of RCs by creating an ARC and having the ARC request a schedule from the RCM. Each ARC has several parameters, set by the behavior, which are of interest to the RCM. The most important parameter is needed RCs, a list of all the requested RCs. For each sharable RC in this list, the behavior must provide one or more bandwidth parameters, which quantify (in an RC-specific way) how the behavior intends to utilize the resource. Other ARC parameters set by the behavior include a priority and a minimum runtime. The latter parameter specifies the minimum amount of time that the behavior must control all requested RCs in order to accomplish any useful work. This parameter is necessary in our system because some components, such as the Scouts' video cameras, can take several seconds to activate, during which the behaviors can take no useful action.

Whenever the RCM computes a proposed schedule, it must query all sharable RCs that will be in use. These queries determine whether the sharable RCs can handle the proposed sets of ARCs, given the ARCs' bandwidth parameters. This is done because the RCM cannot be expected to know the implementation details of every sharable RC. The sharable RCs must use the bandwidth parameters provided to determine whether they can handle the requested load. If a sharable RC determines that it cannot handle the load, it rejects the proposed schedule and the RCM is forced to come up with a different schedule proposal. In this way, the domain-specific knowledge of a sharable RC can be used as a veto power, influencing the RCM's scheduling decisions.

A UML sequence diagram showing the behavior of the RCM and related components during a typical schedule request is shown in Figure 2. This diagram shows a Place Sensor Net behavior creating arc0, an ARC which requests a schedule from the RCM. The RCM computes a proposed schedule and sends it off to all the sharable RCs on the system. All sharable RCs accept the schedule, so the RCM notifies all scheduled ARCs (arc0 and any others that may have already been running) that they now have access to their needed RCs. The Place Sensor Net behavior can then use these resources to complete its task. When the ARC's minimum runtime has expired or the RCM has determined that the ARC should be preempted, the ARC is notified that its resources are no longer available. In turn, the ARC notifies its behavior that the resources are no longer available, and the behavior stops processing.

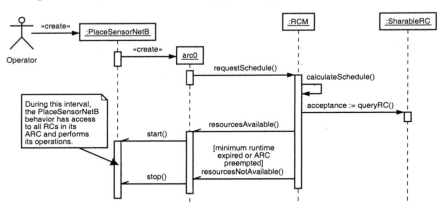

Figure 2: UML sequence diagram showing a typical RCM schedule request.

Two different strategies for computing the schedule have been tried. In the first, an optimal schedule is produced such that the number of RCs in use at any given time is maximized. The problem with this approach is that, for large numbers of

RCs, computing the optimal schedule is exponentially complex (in terms of time). The computational complexity for this approach is dominated by the calculation of all possible sets of RCs.

The second scheduling strategy reduces total runtime by choosing a schedule that can be computed in polynomial time. This approach considers several factors when calculating a schedule. If there is resource contention between multiple ARCs, their priority values are first examined, with ARCs of higher priority having the chance to run first. If the priorities of multiple ARCs are the same, the RCM looks at their minimum runtime parameters. The RCM schedules ARCs with lower minimum runtime values first. If the minimum runtimes are also identical, the RCM arbitrates the tie by giving priority to the first ARC that requested scheduling. ARCs which do not immediately get the chance to run are queued; the RCM examines this queue every second[1] to determine if any queued ARCs can be started. If there is a queue of ARCs waiting for resources, a new ARC making a resource request must also contend with the ARCs in the queue. An additional feature of this scheduling algorithm is its ability to preempt currently-running ARCs (which may not have been running for their requested minimum runtime) if an ARC of higher priority requests a schedule. Any preempted ARCs that have not completed their minimum runtime enter the queue of waiting ARCs until their needed resources are free again. In this case, the preempted ARCs will be allowed another full slot of minimum runtime. The runtime complexity of this scheduling algorithm is dominated by the number of ARCs making RC requests and the number of total possible RCs each ARC can request.

Preliminary testing illustrates significant performance difference between the two RCM scheduling algorithms. The two versions of the scheduler were run in a controlled experiment in which ten RCs were available to the system. Twenty behaviors were written to simultaneously contend for these resources; each behavior requested exclusive access to up to five RCs. Over ten trials, the first version of the scheduling algorithm took a mean of 39.3 seconds to calculate a schedule, while the second version calculated a schedule in a mean time of 0.59 seconds. The second algorithm is preferred because it is clearly more responsive to access control requests.

To demonstrate the operation of this scheduling algorithm, a version of the Place Sensor Net behavior was run on an implementation of the RCM. In this scenario, there are four resources available to the system. Two of these resources correspond to Scouts, while two correspond to the framegrabber cards used to process video. (To simplify this demonstration, no sharable RCs were used.) The Place Sensor Net behavior creates a total of four ARCs to encapsulate resource requests; all of these ARCs request a schedule within the first two seconds. After eleven seconds have elapsed, a human operator takes manual control of the system, requiring two Scouts and a framegrabber card. The human controller has priority over all autonomous behaviors, so the ARC created on the operator's behalf has a very high priority.

Table 1 summarizes the resource requests made by the various ARCs during the scenario. The ARCs are ordered by the time each requested the RCM for a schedule. Table 2 shows how the scheduler allocated the resources to this set of ARCs.

Initially, arc0 and arc1 request a schedule from the RCM. However, they contend for FGrab1, the first Framegrabber RC. Since their priority and minimum runtime parameters are identical, the RCM chooses to schedule arc0 first, because arc0 was the first to initiate a schedule request. At time $t = 1$, arc2 makes a schedule request, but

---

[1]The time quantum of this architecture is one second. This value was chosen because behaviors and other components do not make requests faster than once a second.

Table 1: Summary of resource requests made during the Place Sensor Net scenario.

ARC	Sched. Req. Time	priority	min. runtime	needed RCs
arc0	$t = 0$ sec.	1	5 sec.	Scout33, FGrab1
arc1	$t = 0$ sec.	1	5 sec.	Scout35, FGrab1
arc2	$t = 1$ sec.	1	4 sec.	Scout33, FGrab2
arc3	$t = 2$ sec.	1	3 sec.	Scout33, FGrab2
arc4	$t = 11$ sec.	10	3 sec.	Scout33, Scout35, FGrab1

Table 2: Allocation of resources during the Place Sensor Net scenario.

Time	0	1	2	3	4	5	6	7	8	9	10	11	12	13	14	15	16	17
Scout33	0	0	0	0	0	3	3	3	2	2	2	4	4	4	2	2	2	2
Scout35	-	-	-	-	-	1	1	1	1	1	-	4	4	4	-	-	-	-
FGrab1	0	0	0	0	0	1	1	1	1	1	-	4	4	4	-	-	-	-
FGrab2	-	-	-	-	-	3	3	3	2	2	2	-	-	-	2	2	2	2

is blocked because arc0 is using Scout33. At $t = 2$, arc3 makes a schedule request. This ARC has a shorter minimum runtime than arc2, but requires the same resources, which are still in use by arc0. Hence, it is also blocked. At $t = 5$, arc0's minimum runtime has expired, so it is released from the schedule, and the RCM now has to allocate the new set of free RCs among the waiting ARCs. Since arc1, arc2, and arc3 are all of the same priority, arc3 gets its resources allocated first, because it has the shortest minimum runtime. The free RCs remaining are exactly those needed by arc1, so it is also scheduled to run at this time. After three seconds have elapsed, arc3's resources are given to arc2. arc2 continues until $t = 11$, when arc4 requests control of many resources on behalf of a human operator. This ARC preempts arc2, since it has a higher priority. arc2 is placed back into the queue of waiting ARCs until $t = 14$, when arc4 finishes and its RCs are freed. arc2 then runs for its full four-second minimum runtime before it is finished. After $t = 17$, arc2's resources are freed, and the system returns to an idle state.

5   Load Balancing

Load balancing is another important part of the architecture. Without load balancing, the placement of components across computers will likely be done in a suboptimal way. If many components are started on the same workstation, or if the selected workstation has limited computational power, the overall performance of the system will decrease.

Load balancing is done by the LOAD BALANCER and the LOAD REPORTER, two core services dedicated specifically to this task. When a new component (such as a behavior, an ARC, or an RC) is to be started, several conditions are analyzed to determine the workstation with the least amount of load. The component is then started on that machine. Each machine capable of starting software components hosts a LOAD REPORTER, which periodically sends relevant load information to the central LOAD BALANCER service. When the COMPONENT PLACER needs to start a new component, it queries the LOAD BALANCER for the best host on which to start the component, given the list of possible hosts. The LOAD BALANCER returns the name of the best host based on a set of specified conditions, such as current load average, processor speed, and available memory. The COMPONENT PLACER may choose which metric it wishes the LOAD BALANCER to use based on runtime conditions.

The current load balancing system has both advantages and disadvantages; these can best be seen by examining two simulated **Place Sensor Net** scenarios (Figure 3), which illustrate how the LOAD BALANCER operates in different situations. In the first situation (Figure 3(a)), one **Place Sensor Net** is started using all of the available Scouts simultaneously. Since none of the computers in the network have a very high load, all of the components are started on one machine. Once the behavior begins to run, it is obvious that system resources are not being used effectively. Without process migration, components are confined to the machine on which they were started, regardless of how its load changes. The current system does not compensate well for this, but it does perform load balancing well for components which are started at staggered intervals. The second situation, shown by Figure 3(b), illustrates the starting of three different **Place Sensor Net** behaviors. These sensor nets are placed in three different areas which require varying amounts of time for the three groups of Scouts to reach, so the **Place Sensor Net** behaviors are not started at the same time. As the graphs indicate, the LOAD BALANCER starts up components on each of the available machines because it takes into account which machines have heavier loads each time the COMPONENT PLACER needs to start a new component. The addition of the LOAD REPORTER and LOAD BALANCER to the core services helps to ensure that new software components are distributed appropriately and efficiently over the set of available hosts.

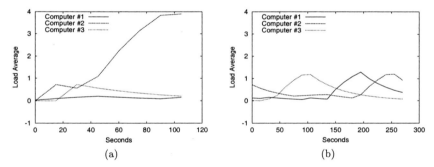

Figure 3: Load average on three computers with (a) simultaneous behavior start times and (b) staggered behavior start times.

6 Summary and Future Work

Details of the scheduling and load-balancing capabilities of a distributed robotic control architecture have been presented. The scheduling algorithm used by the RESOURCE CONTROLLER MANAGER does not necessarily guarantee the generation of an optimal schedule of RCs, but it is capable of generating a reasonable schedule within an acceptable amount of time. The LOAD BALANCER service tries to make sure that the computational load is spread equally among all workstations in the system.

One area of future research is process migration. The current load balancing system decides only where components should be started, but it cannot move them from one machine to another. The next step would be to allow components to be moved from one computer to another while minimizing the interruption of their operation and preserving their internal state. Because this is a distributed system running over connections that may have high latency, there is currently no way to guarantee real-time performance of robotic behaviors. All behaviors are written with this in mind and must explicitly adapt if the rate of sensor data reception diminishes. More advanced load balancing capabilities would help to decrease this potential performance bottleneck.

Another area for future work involves determining which data is most useful to provide to the LOAD BALANCER so that it can make better decisions about the best host on which to start a component. This may also involve allowing the core services to detect at runtime which pieces of information should be used to make this decision.

Acknowledgements

Material based upon work supported by the Defense Advanced Research Projects Agency, Microsystems Technology Office, ARPA Order No. G155, Program Code No. 8H20, issued by DARPA/CMD under Contract #MDA972-98-C-0008.

References

[1] E. M. Atkins, T. F. Abdelzaher, K. G. Shin, and E. H. Durfee. Planning and resource allocation for hard real-time, fault-tolerant plan execution. *Autonomous Agents and Multi-Agent Systems*, 4(1/2), Mar. 2001.

[2] G. Cybenko. Dynamic load balancing for distributed memory multiprocessors. *Journal of Parallel Distributed Computing*, 7(2):279–301, 1989.

[3] E. H. Durfee. Distributed continual planning for unmanned ground vehicle teams. *AI Magazine*, 20(4):55–61, 1999.

[4] O. M. Group. *The Common Object Request Broker: Architecture and Specification*. Object Management Group, 1998.

[5] D. Kortenkamp, R. P. Bonasso, and R. Murphy. *Artificial Intelligence and Mobile Robots*. AAAI Press/MIT Press, 1998.

[6] L. E. Parker. ALLIANCE: An architecture for fault tolerant multi-robot cooperation. *IEEE Trans. on Robotics and Automation*, 14(2), 1998.

[7] P. Pirjanian, T. Huntsberger, A. Trebi-Ollennu, H. Aghazarian, H. Das, S. Joshi, and P. Schenker. Campout: a control architecture for multirobot planetary outposts. In *Proc. SPIE Conf. Sensor Fusion and Decentralized Control in Robotic Systems III*, Nov. 2000.

[8] P. E. Rybski, N. Papanikolopoulos, S. A. Stoeter, D. G. Krantz, K. B. Yesin, M. Gini, R. Voyles, D. F. Hougen, B. Nelson, and M. D. Erickson. Enlisting rangers and scouts for reconnaissance and surveillance. *IEEE Robotics and Automation Magazine*, 7(4):14–24, Dec. 2000.

[9] J. Stankovic, M. Spuri, K. Ramamritham, and G. Buttazzo. *Deadline Scheduling For Real-Time Systems: EDF and Related Algorithms*. Kluwer Academic Publishers, Boston, 1998.

[10] S. A. Stoeter, P. E. Rybski, M. D. Erickson, M. Gini, D. F. Hougen, D. G. Krantz, N. Papanikolopoulos, and M. Wyman. A robot team for exploration and surveillance: Design and architecture. In *Proc. of the Int'l Conf. on Intelligent Autonomous Systems*, pages 767–774, Venice, Italy, July 2000.

Intelligent Autonomous Systems 7
M. Gini et al. (Eds.)
IOS Press, 2002

Cooperation between Omnidirectional Vision Agents and Perspective Vision Agents for Mobile Robots

Emanuele Menegatti, Enrico Pagello†
Intelligent Autonomous Systems Laboratory
Department of Informatics and Electronics
University of Padua, Italy
†*also with*: Institute LADSEB of CNR
Padua, Italy
{emg,epv}@dei.unipd.it

Abstract. Multiple robot systems in which every robot is equipped with a vision sensor are more and more frequent. Most of these systems simply distribute the sensors in the environment, but they do not create a real Cooperative Distributed Vision System. In this paper we propose an approach to realize a Cooperative Distributed Vision System within a team of heterogeneous mobile robots. We present two research streams which we are working on, along with theoretical and practical insights.

1 Introduction

In this paper, we present our researches on Distributed Vision for mobile robots. The most promising applications of Distributed Vision probably are surveillance and monitoring systems. Most of the previous works dealt only with static vision systems, we think that the introduction of mobile robots can improve the performances and robustness of the system. As an example, think of an industrial site controlled with a video monitoring system composed only of cameras with fixed locations (even pan-tilt-zoom cameras). If an alarm or a meaningful event happens outside the field of view of any of the cameras, the system cannot *"see"* this event. This is because the system has a *predetermined field of view*. If we mount some cameras on mobile robots, the system can send a robot to inspect the new location of interest. The system is more flexible and we realize what we call a *dynamic field of view*.

Nowadays, the relatively low cost of the vision sensors allows to equip every member of a multi-robot team with a vision sensor.

Introducing mobile robots fitted with cameras distributes the sensors in the environment, but this is not enough, a set of cameras scattered in the environment needs to communicate over a network in order to became a unique Distributed System. In the following we will prefer the term *Vision Agent* (VA) instead of *"vision system"*. The term VA (Vision Agent) emphasizes that the vision system is not just one of the several sensors of a single robot, but that it interacts with the other vision systems to create an intelligent distributed system.

2 Previous Works

Our work has been inspired by the work of Ishiguro [2]. He proposed an infrastructure called *Perceptual Information Infrastructure* (PII). In his paper, he proposed an imple-

Figure 1: **(Left)** A close view of the vision system of Nelson. On the left, the perspective camera. In the middle, pointed up-ward the omnidirectional camera. **(Right)** A close view of two of our robots. Note the different vision systems.

mentation of the PII composed by static VAs, i.e. fixed cameras with a certain amount of computational power. This realisation of the *PII* was called Distributed Vision System. The assumption that every VA is static simplifies the problem and allows the use of very simple vision algorithms, but implies that the whole system is not scalable to include Mobile VAs.

A parallel but independent work is the one of Matsuyama. He explicitly introduced mobile robots in the theoretical frame of its Cooperative Vision System. In the experiments presented in [3], he used active cameras mounted on a special tripod in order to have a fix view point. This allowed the use of a simple vision algorithm, not scalable to mobile robots.

As far as we know, no attempt has been tried to realize a DVS with truly mobile robots running robot vision algorithms.

3   The aim of our work

Our aim is to introduce a real Mobile Vision Agent in the DVS architecture, i.e. to apply the ideas and the concepts of Distributed Vision to a mobile robot equipped with a camera. We are on the way to create a Distributed Vision System within a team of heterogeneous robots fitted with heterogeneous vision sensors. The redundancy of observers (and observations) is a key issue for system robustness.

4   Implementation

4.1   Two VAs mounted on the same robot

The first step is to realise a Cooperative behavior between two heterogeneous vision agents embodied in the same robot: an omnidirectional and a perspective vision system. The omnidirectional vision sensor is a catadioptric system composed of an omnidirectional mirror and standard colour camera looking at it, Fig. 3 (Left). The omnidirectional mirror is a multi-part mirror with a custom profile we designed[1], Fig. 3 (Right).

The omnidirectional camera is mounted on the top of the robot and offers a complete view of the surroundings of the robot [6][5]. The perspective camera is mounted in the

---

[1]For details on the procedure we used to design the custom profile of the mirror, please refer to [4]

front of the robot and offers a more accurate view of objects in front of it. These two cameras mimic the relationship between the peripheral vision and the foveal vision in humans. The peripheral vision gives a general, and less accurate, information on what is going on around the observer. The foveal vision determines the focus of attention and provides more accurate information on a narrow field of view. So, the omnidirectional vision is used to monitors the surroundings of the robot for detecting the occurrence of particular events. Once one of these events occurs, the Omnidirectional VA (OVA) sends a message to the Perspective Vision Agent (PVA). If the PVA is not already focused on a task, it will move the robot in order to put the event in the field of view of the perspective camera. This approach was suggested by our previous researches presented in [1].

Experiments on such a system are running and they will provide more insights on the cooperation of two heterogeneous VAs.

Figure 2: The mirror with the custom profile we designed

4.2 Coordination of several VAs mounted on different robots

Another stream of research is the creation of a Cooperative Distributed Vision System for a team of robots depicted in Fig. 3. The first step is to implement the idea of the Cooperative Object Tracking Protocol proposed by Matsuyama [3]. In the work of Matsuyama the central notion is the concept of *agency*. An **agency**, in the definition of Matsuyama, is *the group of the VAs that sees the objects to be tracked and keeps an history of the tracking.* This group is neither fixed nor static. In fact, a VA exits the agency, when it is not able to see the tracked object anymore, and a new VA can joint the agency as soon as the tracked object comes in its field of view. The described algorithm has been realised by Matsuyama with his fixed view point cameras. As mentioned before, in such a system there is not a truly mobile agent and the system is not scalable to Mobile VA.

Our novel approach is to implement the Cooperative Object Tracking Protocol within a team of mobile robots. This requires a totally new vision approach, in fact, the point of view of the VA is continuously changing while the robot moves. So, the changes in the image are due not only to the changes in the world (as in the Matsuyama testbed), but also to change of position of the VA. Moreover, we have to introduce a measure of uncertainty in the estimation of the positions of the tracked objects, because the location of the Vision Agents is not exactly known anymore.

To explain these issues, let us see a simple example. In the cooperative tracking of an object, if a VA sees the tracked object, it sends a message to the agency's master. This

Figure 3: Our team of heterogeneous robots

checks if the latter has seen the correct object. Errors can result either because the robot sees another object resembling the tracked one or because it is not properly localized and so it reports a fallacious position for the object. To cope with the uncertainty in the objects position, every VA transmits to the master the calculated object position with the confidence of this estimation. The master dispatches to the other robot a position calculated as an *average* of the different position estimations, weighted by the confidences reported by every VA.

The master role is crucial for the correct functioning of the agency. The master role cannot be statically assigned. If the object is continuously moving, the first robot that sees it will not have the best observational position for long. So, the master role must pass from robot to robot. The processes of swapping the master role is critical, if the master role is passed to a robot that sees an *incorrect object* the whole agency will fail in the *tracking* task. The simplest solution could be to pass the master role to the robot with the highest confidence on the object's position. This means to shift the problem to identify a reliable confidence function. This makes sense. In fact, if a robot is correctly localized and correctly calculates the relative distance of the object, it will have strong weight in the calculation of the object position; given this, it can reliably take the role of master.

4.2.1   The confidence function

The confidence function ψ_{abs} associated to the reliability of the estimation of the absolute object position is a combination of several factors. In fact, the position of the object is calculated by a vectorial sum of the relative distance of the object from the robot and the absolute position of the robot. So, the confidence of the estimation of the absolute position of the object is the sum of the confidences function associated to the self-localisation, ψ_{sl}, and of the confidence function associated to the estimation of the relative position of the object with respect to the robot, ψ_{rel}. In our experiments, the self-localisation process uses the vision of landmarks. The process is run only by time to time and if the landmarks are visible. Between two of these processes the position is calculated with the odometers. This means that the localisation information degrades with time. The confidence function associated with the self-localisation is the result of the following contribution: type of vision system (perspective, omnidirectional, etc.), absolute error made from the vision system in the calculation of the landmarks position (estimated a priori) and time passed after the last self-localisation process. The relative position of the object with respect to the robot is calculated as in [4] and the confidence function depends on: type of vision system and distance from the robot.

At the moment the exact definition of the confidence function is under testing. The experiments will tell us how much every contribution should weight in the final function.

5 Conclusions and Acknowledgments

In this paper we presented two of the research streams we are following to implement a Cooperative Distributed Vision System. We proposed to realise the DVS with heterogeneous mobile Vision Agents. We suggested a way to fuse the information coming from two heterogeneous VAs mounted on the same robot. Regarding the problems introduced by the mobile Vision Agents, we suggested a way to cope with the uncertainty introduced in the localisation of the objects of interest.

At the time of writing experiments are running on such a systems providing theoretical and practical insight.

We wish to thanks the student of the ART-PD and Artisti Veneti Robocup teams who built the robots. This research has been partially supported by: the Italian Ministry for the Education and Research (MURST), the Italian National Council of Research (CNR) and by the Parallel Computing Project of the Italian Energy Agency (ENEA).

References

[1] S. Carpin, C. Ferrari, E. Pagello, and P. Patuelli. Bridging deliberation and reactivity in cooperative multi-robot systems through map focus. In M.Hannebauer, J. Wendler, and E. Pagello, editors, *Balancing Reactivity and Social Deliberation in Multi-Agent Systems,*, LNCS. Springer, 2001.

[2] H. Ishiguro. Distributed vision system: A perceptual information infrastructure for robot navigation. In *Proceedings of the Int. Joint Conf. on Artificial Intelligence (IJCAI97)*, pages 36–43, 1997.

[3] T. Matsuyama. Cooperative distributed vision: Dynamic integration of visual perception, action, and communication. In W. Burgard, T. Christaller, and A. B. Cremers, editors, *Proc. of the Annual German Conf. on Advances in Artificial Intelligence (KI-99)*, volume 1701 of *LNAI*, pages 75–88, Berlin, Sept. 1999. Springer.

[4] E. Menegatti, F. Nori, E. Pagello, C. Pellizzari, and D. Spagnoli. Designing an omnidirectional vision system for a goalkeeper robot. In A. Birk, S. Coradeschi, and P. Lima, editors, *Proceeding of RoboCup 2001 Int. Symposium (to appear in RoboCup-2001: Robot Soccer World Cup V.)*. Springer, 2001.

[5] E. Menegatti, M. Wright, and E. Pagello. A new omnidirectional vision sensor for the spatial semantic hierarchy. In *IEEE/ASME Int. Conf. on Advanced Intelligent Mechatronics (AIM '01)*, pages 93–98, July 2001.

[6] Y. Yagi. Omni directional sensing and its applications. *IEICE TRANS. INF. & SYST.*, VOL. E82-D(NO. 3):pp. 568–579, MARCH 1999.

Intelligent Autonomous Systems 7
M. Gini et al. (Eds.)
IOS Press, 2002

Synthesis of Supervision Policies for Robust Sensory-Motor Behaviors

Benoit Morisset and Malik Ghallab
LAAS - CNRS, Toulouse, France
bmorisse@laas.fr, malik@laas.fr

Abstract. We are proposing here an approach and a system, called ROBELS,that enables a designer to specify and build a supervision system which learns from experience very robust ways of performing a given task such as "navigate to". The designer specifies a collection of Hierarchical Tasks Networks (HTN) that are complex plans, called *modalities*, whose primitives are sensory-motor functions. Each modality is a possible way of combining some of these functions to achieve the desired task. The relationship between supervision states and the appropriate modality for pursuing a task is learned through experience as a Markov Decision Process (MDP) which provides a general policy for the task. This MDP is independent of the environment; it characterizes the robot abilities for that task.

1 Introduction

In robotics today, representations and techniques available for task planning are mostly effective at the abstract level of mission planning. Primitives for these plans are tasks such as "navigate to location l", "retrieve and pick-up object o". These tasks are far from being *primitive* sensory-motor functions. Their design is very complex.It is not much helped out by task planning techniques - which may explain the weak interest of the robotics community in task planning. Our purpose here is exactly the design of such tasks in a robust, generic way. We do claim that task planning can be helpful for this design. Certainly not as a collection of plug-and-play planners. But planning representations and techniques are useful for specifying alternative complex plans achieving a task. They are useful for learning a domain independent policy that chooses, in each supervision state, the best such a plan for pursuing the task. The robots we are experimenting with are autonomous mobile platforms in structured environments. They are equipped with several sensors - sonar, laser, vision - and actuators, eventually with a robot arm. Our robot architecture has a functional level, for the sensory-motor functions, a control level, called executive, and a decision level for supervision and planning [1]. The architecture relies on several development tools, in particular Genom [2] for the specification and integration of sensory-motor modules, and Propice [3], which is a PRS-like environment for programming the supervision system. The functional level of our robots is fairly rich. It has several modules for the same function, e.g., for localization, for map building and updating, or for motion planning and control. This is needed since, in addition to possible failures, no single method or sensor has a universal coverage. Each has its weak points and drawbacks. Robustness requires a diversity of means for achieving a sensory-motor function. Robustness also requires an ability to combine consistently these functional means and to chose among them the most appropriate one for the current context. The ambition of ROBELS, the system proposed here, is to address this last requirement. ROBELS enables a designer to specify and build a supervision system which learns from experience very robust ways of performing a task such as "navigate to" (this is the task illustrated here on which we have extensively experimented). The designer specifies a collection of Hierarchical Tasks Networks (HTN) [4, 5, 6] that are complex plans, called *modalities*, whose primitives are sensory-motor functions. Each modality is a possible way of combining some of these functions to achieve the desired task. A modality has a rich context-dependent control structure. The relationship between supervision states and the appropriate modality for pursuing a task is far from being obvious. In ROBELS this relationship is learned through experience as a Markov Decision Process (MDP) [7] which provides a general policy for achieving the task. This MDP characterizes the robot abilities for that task; it is independent of the environment although it may get improved through learning if the robot is moved to another environment.

To summarize, primitives for ROBELS are redundant low-level sensory-motor *functions* of a robot, which are precisely modeled with their advantages and weak points. In our architectures these functions are formalized as Genom modules [2]. Some of them are briefly introduced in section 3. Since the task at hand here is navigation, our representations of space are mostly relevant; they are introduced in section 2. ROBELS proposes a representation for specifying a set of HTN *modalities*, described in section 4, for combining and controlling consistently a subset of functions in order to perform a task. Section 5 details the representation of the state of the supervision system on which an MDP-based graph is learned providing a *policy* which chooses the appropriate modality for pursuing a *task*. The last section presents ongoing experimental results and discusses the approach with respect to the state of the art and to future work.

2 Environment representation

Metric map. Most functions described here and the ROBELS supervision system itself rely on a model of the environment learned and maintained by the robot. The basic model is a 2-D map of obstacle edges acquired by the laser. A preliminary learning phase has to take place before autonomous navigation using ROBELS modalities and before starting the learning phase of a supervision policy.

Topological graph. A labeled topological graph of the environment is further added to the map by hand specification in current implementation. Cells are polygon that partition the metric map. Each cell is characterized by its name and a *color*. We distinguished 7 different colors or navigation properties: Corridor, Corridor_With_Posters, Large_Door, Narrow_Door, Confined_Area, Open_Area, Open_Area_With_Camera. Edges of the topological graph are labeled by an estimate of the transition length from one cell to the next and by a heuristic estimate of how easy is such a transition.

3 Sensory-Motor Functions

In our architecture, sensory-motor functions are defined as a set of Genom modules [2]. A module may integrate several functions each corresponding to a specific query to the module. A report is sent back by a module once the query has been executed, indicating to the controller either the end of a nominal execution, or giving it additional information for non nominal cases. Some of the sensory-motor functions of the robot, needed later, are introduced below, indicating for each one their main non-nominal reports and the associated control.

Segment-based localization. This Simultaneous Map Building and Localization procedure uses an Extended Kalman Filter to match the local perception with the previously built model [8]. It relies on a 2-D map of obstacle edges incrementally built by the robot from laser range data. It offers two modes, a re-localization mode, used when the robot is lost and a continuous position updating mode. *Advantages:* this function works in a tracking mode with a frequency of 1 Hz. It is generally reliable and robust to partial occlusions. *Weaknesses:* laser occlusion gives unreliable data. Moreover, in long corridors the laser obtains no data along the corridor axis. *Main associated control:* a report warns that the imprecision of the robot position has exceeded the allowed threshold. The robot stops, turns on the spot to find a non-ambiguous corner and activates the re-localization mode.

Localization on landmarks. This function uses monocular, grey level vision to detect known landmarks that are quadrangular, planar objects, e.g. doors or wall posters. It derives from the perceptual data an accurate estimation of the robot position [9].
Advantages: this function gives a very accurate estimate of the robot position (about one centimeter). *Weaknesses:* landmarks are available and visible in few areas of the environment. *Main associated control:* a report of a potentially visible landmark indicates that the robot enters an area of visibility of a landmark. The robot searches it using the pan-tilt mount. A failure report notifies that the landmark was not identified. Eventually, the robot retries from a second predefined position.

Absolute localization. A set of fixed calibrated cameras that cover a particular area is used to recognize and localize the robot [10].
Advantages: very few occlusions occur in the vertical axis. The localization is very accurate and robust. *Weaknesses:* this function works only when the robot is within the covered area. *Main associated control:* a report of a recognition failure can be due to a robot position outside of the covered area or to other reasons such as insufficient light, occluded recognition pattern, etc.

Path planner. This function plans a feasible path for holonomous as well as non-holonomous robots in a metrical model towards some goal. [11]. The space model used is a discretized bitmap derived from the learned map.
Advantages: this path planner is fairly generic and robust. *Weaknesses:* The planned path doesn't take into account environment change that occurs during navigation. *Main associated control:* a report

may warn that a goal is too close to an obstacle. The planner computes a corrected goal position and a new path is computed. Another warning report is issued if the initial robot position is too close to an obstacle. The robot has to be moved away from the obstacles by a reactive motion function before a new path is computed.

Elastic Band for Plan Execution. This function updates and maintains dynamically a flexible trajectory. External forces are associated to obstacles perceived by the laser. These forces are applied to all configurations in order to dynamically update the path away from obstacles [12, 13].
Advantages: a band is a series of robot configurations. Its format makes it very easy to couple to a planed path giving a very robust method for long range navigation. *Weaknesses:* a mobile obstacle can block the band against a static obstacle. The band may fall into local minima. The dynamic deformation is a costly step; this may limit the reactivity in certain cluttered, dynamic environments. *Main associated control:* a report may warn that the band execution is blocked by a temporary perceived obstacle that cannot be avoided (e.g. a closed door, an obstacle in a corridor). If the band relies on a planed path, the new obstacle is added to the map. A new trajectory taking account the unexpected obstacle is computed, and a new elastic band is executed.

Reactive Obstacle Avoidance. This function provides a reactive motion capability. It works in two steps. First, it extracts from sensory data a description of free regions. It selects one of them and computes a motion command to that region [14].
Advantages: this method offers reactive motion capability that remains efficient in very cluttered space. *Weaknesses:* like all the reactive methods, it may fall into local minima. Without a reference path this method is not appropriate to perform long range navigation. *Main associated control:* a failure report is generated when the reactive execution is blocked.

4   Task Modalities

A navigation task (Goto x y θ) given by a user or by a mission planning step requires an integrated use of several functions among those presented earlier. Each consistent combination of these functions gives a new behavior, a new way to perform the task, with its specific characteristics that make it more appropriate for some contexts or environments. Each such combination is a particular plan called a *modality*.We have specified and implemented 4 different and complementary modalities for the navigation task. Let us exemplify one such modality before giving the detail of the HTN representation for modalities and the associated control system.

4.1   Example of a modality

Modality $M1$ uses 3 functions: the path planner, the elastic band for the dynamic motion execution, and the laser-based localization. When $M1$ is chosen to carry out a navigation, the laser-based localization is initialized. The robot position is maintained with a frequency of 1 Hz. A path is computed and is carried out by the elastic band function. Stopping the modality interrupts the band execution and the localization loop. Suspending the modality stops only the band execution. The path, the band, the localization loop are maintained. A suspended modality can be resumed by restarting the execution of the current elastic band.

4.2   Representation for Modalities

Modalities are represented as Hierarchical Task Networks [6]. The HTN formalism is adapted to our needs because of its expressiveness and its flexible control structure [4],[5]. HTNs offer a middle ground between programming and automated planning, allowing the designer to express control structure when available, as in our case. Our HTNs for modalities are for the moment specified manually, but the synthesis of all modalities from generic specifications seems a feasible objective because of this representation.

Our HTNs are And/Or trees. An internal node is a task or a subtask that can be pursued in different context-dependent ways, which are the Or-connectors. Each such Or-connector is a possible decomposition of the task into a conjunction of subtasks. There are two types of And-connectors with sequential and parallel branches. Branches linked by a sequential And-connector are traversed sequentially in a depth-first manner. Branches linked by a parallel And-connector are traversed in parallel, in a breadth-first way. The leaves of the tree are primitive actions, each corresponding to a unique query addressed to a sensory-motor function. Thus, a root task is dynamically decomposed, according to the context, into a set of primitive actions organized as concurrent or sequential subsets.

Execution starts as soon as the decomposition process reaches a leaf even if the entire decomposition process of the tree is not complete.

A primitive action can be *blocking* or *non-blocking*. In blocking mode, the control flow waits until the end of this action is reported before starting the next action in the sequence flow. In non-blocking mode actions in a sequence are triggered sequentially without waiting for a feedback. A blocking primitive action is considered ended after a report has been issued by the function and after that report has been processed by the control system. The report from non-blocking primitive action may occur and be processed after an unpredictable delay.

In our case, every modality tree starts with 6 Or-connectors labeled start, stop, suspend, resume, succeed and fail. The start connector represents the nominal modality execution, the stop connector the way to stop it and to restore the neutral state, characterized by the lack of any function execution. Furthermore, the environment model modified by the modality execution recovers its standard form. The suspend and resume connectors are triggered by the control system described below. The suspend connector allows to stop the execution by freezing the state of the functional level. The resume connector restarts the modality execution from such a frozen state. The fail (resp. succeed) connector is followed when the modality execution reaches a failure (resp. a success) end. These connectors are used to restore the neutral state and allow certain executions required in these specific cases.

4.3 Control and resource sharing

The feedback from sensory-motor functions to modalities has to be controlled as well as the resource sharing of parallel activities. The control system catches and react appropriately to reports emitted by functions. We found it convenient to also specify the control as HTNs. Reports from functions play the same role in the control system as tasks in modalities. A report of some type activates its own dedicated control HTN in a reactive way. A control tree represents a temporary behavior and cannot be interrupted . Each non nominal report points out a non nominal function execution. The aim of the corresponding control tree is to recover to a nominal modality execution. Some non nominal reports can be non recoverable failures. In these cases, the corresponding control sends a "fail" message to the modality pursuing this function. Nominal reports may notify the success of the global task. In this case, the "success" alternative of the modality is activated.

Resources to be managed are either physical non-sharable resources (the motors, the cameras, the pan-tilt mount) or logical resources (the environment model that can be temporally modified). The execution of a set of concurrent non-blocking actions can imply the simultaneous execution of different functions. Because of that, several reports may appear at the same time, and induce the simultaneous activation of several control activities. To manage this a potential resource conflict, a resource manager organizes the resource sharing according to priorities:

• Each non-sharable resource is semaphorized. The request for a resource takes into account the priority level of each consumer. This priority level is specified by the designer.

• The execution of a control HTN is not interruptible. If another HTN requires a resource already in use by a control execution, the message activating this HTN (either modality or control) is added to a spooler according to its priority level.

• A control HTN has a priority higher than those of start and suspend connectors but lower than those of stop and fail connectors.

When a non-nominal report is issued, a control HTN starts its execution. It requests the resource it needs. If this resource is already in use by a start connector of a modality, the manager sends to this modality a suspend message, and leaves a resume message for the modality in the spooler according to its priority. The suspend alternative is executed freeing the resource, enabling the control HTN to be executed. If the control execution succeeds, waiting messages are removed and executed until the spooler becomes empty. If the control execution fails, the resume message is removed from the spooler and the fail alternative is executed for the modality.

4.4 Other navigation modalities

We present here briefly three other modalities that have different execution conditions and rely on different functions.

Modality M2 uses the reactive obstacle avoidance function. The robot position is given by the odometer which accumulates inaccuracy. This inaccuracy can be locally reset by the visual localization function when the robot goes by a known landmark. Reactive navigation between landmarks allows to cross a corridor without an accurate knowledge of the robot position. Typically this M2 modality can be used in long corridors. The growing inaccuracy can make it difficult to find out the next landmark.

The search method allows some inaccuracy on the robot position by moving the cameras but this inaccuracy cannot exceed one meter. For this reason landmarks should not to be too far apart with respect to the required updating of odometry estimate. Furthermore, the reactive navigation of M2 may fall into a local minima.

Modality M3   relies on the reactive obstacle avoidance function. The robot position is given by the absolute localization system. The start connector starts the reactive motion and activates periodically the position processing. When the laser is occluded and when no landmark is visible, this last localization method is very efficient if the robot is within the area covered by fixed cameras

Modality M4   relies on the robot position given by the laser-based localization. The start connector starts the reactive motion and activates the localization loop. The localization function is not limited to a particular area. The reactive motion is not attached to a predefined trajectory. For these two reasons, modality M4 offers an efficient alternative in narrow environments like offices, and in cluttered spaces. It can be preferred to modality M1 to avoid unreliable re-planning steps if the band is blocked by a cluttered environment. Navigation is only reactive, hence with a local minima problem. The weakness of the laser localization in the long corridors is also a drawback for M4.

5   Supervision policies

5.1   Control attributes

The supervision system has to choose a modality for pursuing a task which is most appropriate to current supervision state. In order to do this, components of this state, or supervision attributes, have to reflect control information of sensory-motor functions. Let us list briefly these supervision attributes.
• **Cluttering of the environment.** This is an important information to establish the execution conditions of motion and localization functions. This attribute is defined as a weighted sum of the distances to nearest obstacles perceived by the laser, with a dominant weight along the robot motion axis.
• **Precision of the position estimate.** The quality of the position estimate is computed from co-variance matrix maintained by each localization function.
• **Confidence in the position estimate.** The inaccuracy is not sufficient to qualify the localization. Each localization function supplies a confidence estimate about the last processed position.
• **Properties of current area.** When the robot position estimate falls within some labeled cell of the topological graph, corresponding attributes are taken into account. These are the color label (i.e., Corridor, Corridor_With_Posters, Large_Door, Narrow_Door, Confined_Area, Open_Area, Open_Area_With_Camera).
• **Modality in use.** This information is essential to assess the supervision state and possible transitions between modalities.

5.2   Supervision graph

In order to work out with a discrete state space, continuous supervision attributes are discretized over few significant intervals. This state space defines a supervision graph, noted SG. Nodes and arcs of SG are learned from the robot experiences. Arcs are labeled by the modality applied to state transition; they are weighted according to their observed control conditions.

Note that the same transition label m_i can be given to different arcs issued from a state s. This is due to the inherent non determinism of the system. Unpredictable external events may modify the environment state, e.g. someone passing by may change the value of the cluttering attribute, or the localization inaccuracy attribute. Therefore the execution of the same modality from a state may lead to different adjacent states. A probability distribution, issued from observed statistics is associated to the set of identically labeled transitions.

5.3   Search in the supervision graph

The Supervision graph SG is formally a Markov Decision Process. As an MDP, SG could be used reactively on the basis of some universal policy which selects for each state the next best modality. The policy usually optimizes a general utility criterion that abstract away the current navigation goal. We are proposing here another more precise approach that takes into account explicitly the goal,

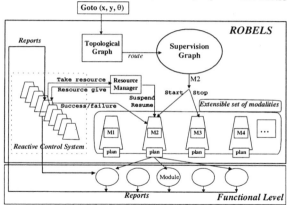

Figure 1: Robels

transposed into SG. It relies a look-ahead mechanism which is based on a search for a path in SG that reflects a topological route to the navigation goal.

Giving a navigation task (Goto x y θ) a search in the topological graph is achieved for an optimal route to the goal, taking into account estimated costs of edges between cells. This route is characterized by the corresponding sequence of cell labels, mainly their colors. Now, a path in SG defines also a sequence of colors, those of traversed states. By definition, such a path in SG corresponds to a topological route if it defines a sequence of colors which can match that of the topological route. Among all paths corresponding to the topological route to the goal, it is simple to search for an optimal one with respect to a criterion that combines the cost of each transition and the probability of occurrence.

Having found a path in SG from current state to the goal, the first modality is executed. When the robot reaches a new state, the search for the new modality is repeated until the control reports a success or a failure.

A sequence of randomly generated navigation goals is given to the robot. During its motion, new supervision states are met and new transitions are recorded or updated. Several modality selection strategies can be applied for learning SG.

The supervision state is updated with a high frequency, 10 Hz in our implantation, i.e. 15cm at the highest robot speed. The cost of a transition is the average of all the durations recorded for each execution of this transition. The fact that a modality required a control step during its execution, together with the outcome of that control are taken into account to weight the evaluation of a transition cost. The probabilities of occurrence are updated each time a transition is used. When a transition leads to a failure situation the corresponding arc points at a failure state.

Several strategies are defined to generate SG:

• **Persistent with no supervision graph.** When a modality is chosen for a given task, this modality is kept until either it succeeds or a fatal failure is notified. In this case, a new modality is chosen randomly and is executed according to the same principle. SG is not searched to choose a modality. This strategy is used to expand SG.

• **Persistent with supervision graph.** Each time the state changes, SG is searched to choose a modality. If no transition exists from the current state, no modality is returned, then the persistent principle described above applies. If a modality is returned, this modality is applied. This strategy uses a graph already developed.

• **Random with no supervision graph.** The aim of this strategy is to evaluate heterogeneous transition classes. Basically, each time a new supervision state is reached, a new modality is randomly chosen. We further added some random change. Each time a new state is reached, we randomly determine with some given probability if a new modality has to be applied. This strategy is especially used to expand a graph.

• **Random with supervision graph.** Each time a new state is reached, SG is searched for a modality. If no modality is returned, a new modality is randomly applied. Otherwise the returned modality is performed. This strategy is interesting to use with a graph already developed. The transitions added will be heterogeneous.

6 Discussion

This paper addressed the issue of robust supervision for task executing in an autonomous robot. We believe to have brought here two contributions. We have introduced a representation based on

HTNs which enables to combine various sensory-motor functions in order to specify complex behaviors as a modalities that have very flexible control structures. We have developed in Propice, a PRS-like environment for programming reactive controllers, 4 such modalities devoted to navigation tasks. In the current state of our implementation, these 4 HTN modalities have been fully implemented on the Diligent Robot [15, 16]. They have been extensively tested. Two of these modalities (M1 and M2) have been demonstrated together [9] using a set of selection rules specified manually. The development of these rules triggered our work on learning a robust supervision system. We are achieving the implementation of the MDP-based supervision graph with the search method coupled to the topological graph and with the learning procedures.

We started generating the SG graph. Its reasonable size of about 1000 states and few thousands transitions enables to update the supervision state. We are not yet able to report on results of the complete integrated ROBELS system, with experiments and measures of robustness. We plan to test how portable is the learned supervision graph by bringing the robot in a new environment and by studying its behavior and the evolution of SG. Another significant test will be the incremental addition of a new modality.

This is certainly not the first contribution that relies on a planning formalism and on plan-based control in order to program an autonomous robot. For example, the *Structured Reactive Controllers* [17] are close to our concerns and have been demonstrated effectively on the Rhino mobile robot. The efforts for extending and adapting the Golog language [18] to programming autonomous robots offer another interesting example from a quite different perspective, that of the Situation Calculus formalism [19]. The "societal agent theory" of [20] offers also another interesting approach for specifying and combining sequentially, concurrently or in a cooperative mode several agent-based behaviors; the CDL language used for specifying the agent interactions is similar to our Propice programming environment. Let us mention also the "Dual dynamics" approach of [21] that permit the flexible interaction and supervision of several behaviors. These are typical examples of a rich state of the art on possible architectures for designing autonomous robots (see [1] for a more comprehensive survey). Our particular and to our knowledge original contribution for the specification of behaviors or modalities relies in the formal use of HTNs with a dynamic expansion and traversal of task networks. The approach is effective because of the position of this representation between programming and automated planning. It seems indeed feasible to synthesize modalities from generic specifications. Furthermore, we have stressed here the need for, and we have studied the consistent use of several redundant HTN modalities to achieve robustness.

The second contribution of this work is an original approach for learning from the robot experiences an MDP-based supervision graph which enables to choose dynamically a modality appropriate to the current context for pursuing the task.

Here also the use of MDPs for supervision and control of robot navigation tasks is not new. Several authors expressed directly Markov states as space cells of a navigation grid and addressed navigation through MDP algorithms, e.g. value iteration [22, 23, 24]. Learning systems have been developed in this framework. For example, XFRMLEARN extends these approaches further with a knowledge-based learning mechanism that adds subplans from experience to improve navigation performances [25]. Other approaches considered learning at very specific levels, e.g., to improve path planning capabilities [26]. Our approach stands at a more abstract and generic level. It addresses another purpose: acquiring autonomously the relationship from the set of supervision states to that of redundant modalities. We have proposed a convenient supervision space. We have also introduced a new and effective search mechanism that projects a topological route into the supervision graph. The learning of this graph relies on simple and effective techniques. But the learned MDP is independent of a particular environment and characterizes the robot capabilities.

In addition to future work directions mentioned above, an important test of ROBELS will be the extension of the set of tasks, e.g., adding manipulation tasks such as "open a door". This significant development will require the integration of new manipulation functions, the design of redundant behaviors for these tasks and their associated control, and the extension of the supervision state. We believe ROBELS to be generic enough to support and permit such developments.

References

[1] R. Alami, R. Chatila, S. Fleury, M. Ghallab, and F. Ingrand. An Architecture for Autonomy. *International Journal of Robotics Research*, 17(4):315–337, April 1998.

[2] S. Fleury, M. Herrb, and R. Chatila. Genom: a tool for the specification and the implementation of operating modules in a distributed robot architecture. In *IROS, Grenoble, France*, volume 2, pages 842–848, September 1997.

[3] F. F. Ingrand, R. Chatila, R. Alami, and F. Robert. PRS: A High Level Supervision and Control Language for Autonomous Mobile Robots. In *IEEE ICRA, St Paul, (USA)*, 1996.

[4] K. Erol, J. Hendler, and D.S. Nau. HTN planning: Complexity and expressivity. In *AAAI*, 1994.

[5] K. Erol, J. Hendler, and D.S. Nau. Complexity results for hierarchical task-network planning. In *Annals of Mathematics and AI*, pages 18:69–93, 1996.

[6] E. Sacerdoti. A structure for plans and behavior. In *American Elsevier Publishing*, 1977.

[7] D.J. White. Markov decision processes. In *John Wiley & Sons*, 1993.

[8] P. Moutarlier and R. G. Chatila. Stochastic Multisensory Data Fusion for Mobile Robot Location and Environment Modelling. In *Proc. International Symposium on Robotics Research, Tokyo*, 1989.

[9] J.B. Hayet F. Lerasle M. Devy. Planar landmarks to localize a mobile robot. In *SIRS'2000, Berkshire, England*, pages 163–169, July 2000.

[10] S. Fleury, T. Baron, and M. Herrb. Monocular localization of a mobile robot. In *IAS-3, Pittsburgh, USA*, 1994.

[11] J. P. Laumond, P. E. Jacobs, M. Taix, and R. M. Murray. A motion planner for nonholonomic mobile robots. *IEEE Transactions on Robotics and Automation*, 10(5):577–593, 1994.

[12] S. Quinlan and O. Khatib. Towards real-time execution of motion tasks. In R. Chatila and G. Hirzinger, editors, *Experimental Robotics 2*. Springer Verlag, 1992.

[13] M. Khatib. *Contrôle du mouvement d'un robot mobile par retour sensoriel*. PhD thesis, Université Paul Sabatier, Toulouse, December 1996.

[14] J. Minguez L. Montano. Nearness diagram navigation (ND): a new real time collision avoidance approach. In *IROS Takamatsu, Japan*, pages 2094–2100, 2000.

[15] R. Alami, R. Chatila, S. Fleury, M. Herrb, F. Ingrand, M. Khatib, B. Morisset, P. Moutarlier, and T. Simon. Around the lab in 40 days... In *IEEE ICRA, San Francisco, (USA)*, 2000.

[16] R. Alami, I. Belousov, S. Fleury, M. Herrb, F. Ingrand, J. Minguez, and B. Morisset. Diligent:towards a human-friendly navigation system. In *IROS'2000,Takamatsu, Japan*, 2000.

[17] M. Beetz. Structured reactive controllers - a computational model of everyday activity. In *3rd Int. Conf. on Autonomous Agents*, pages 228–235, 1999.

[18] Levesque H. et al. Golog: a logic programming language for dynamic domains. In *J. of Logic Programming*, pages 31:59–84, 1997.

[19] Reiter R. Natural actions, concurrency and continuous time in the situation calculus. In *KR*, pages 2–13, 1996.

[20] D.C. MacKenzie R.C. Arkin J.M. Cameron. Multiagent mission specification and execution. In *Autonomous Robots, 4(1):29 V52*, 1997.

[21] J. Hertzberg H. Jaeger Ph. Morignot U.R. Zimmer. A framework for plan execution in behavior-based robots. In *ISIC-98 Gaithersburg MD*, pages 8–13.

[22] Thrun S. et al. Map-learning and high speed navigation in rhino. in ai-based mobile robots: case studies of successful robot systems. In *(Eds.), MIT Press,*, 1998.

[23] Kaelbling L. et al. Acting under uncertainty: discrete bayesian models for mobile-robot navigation. In *IEEE/RSJ Int. Conf. on Intelligent Robots and Systems*, 1996.

[24] T. Dean and M. Wellman. Planning and control. In *Morgan Kaufmann*, 1991.

[25] M. Beetz and T. Belker. Environment and task adaptation for robotics agents. In *ECAI*, 2000.

[26] K. Z. Haigh and M. Veloso. Learningsituation-dependent costs: Improving planning from probabilistic robot execution. In *In 2nd Int. Conf.on Autonomous Agents*, 1998.

Intelligent Autonomous Systems 7
M. Gini et al. (Eds.)
IOS Press, 2002

Memory registers optimization in stochastic functional self-organized sorting performed by a team of autonomous mobile agents

Sorinel Adrian Oprisan [1]
Department of Psychology, University of New Orleans
New Orleans, LA 70148

Abstract. The self-organized activity of a team of autonomous mobile agents is modeled based on their capabilities to perform a random walk, recognize and move different "objects". The emergent behavior is a distributed sorting and clustering based entirely on local information processing. Theoretically derived optimum memory weighting function is based on *intermediate steady-states* assumption. Computational optimization of the angular second-moment agrees with theoretical results.

1 Introduction

The reasons for growing interest in swarm intelligence applications are twofold. Some tasks may be inherently too complex or impossible for a single robot to perform. On the other hand, a swarm of simple robots may also be more flexible without the need to reprogram the robots, and more reliable and fault-tolerant because one or several robots may fail without acting task completion. In addition, self-organization and decentralization made possible via interactions taking place through the environment, point to the possibility of significantly reducing communications between robots. Most contributions to date in swarm intelligent robotics present statistical analyses of the results [3] which, however, do not allow experimenters to identify the system parameters which have influenced the performance. Furthermore, the stochastic nature of the swarm intelligence approach usually leads to high variability in performance, as clearly shown in [14]. Swarm intelligence models can, at least partially, describe the collective activities of social insects, include the formation of trail networks and foraging patterns in many ant species [5, 7, 9], rhythmical patterns of activity in ants (Leptothorax) [4], thermoregulation in clusters of bees [22], the piling of dead bodies by ants (Pheidole) [6], larval sorting by ants (Leptothorax) [6], or the dynamics of colony development in wasps (Polistes) [13]. Self-organization has also been applied to the modeling of the social organization, including hierarchical differentiation [20, 12], division of labor [18, 20], and age (or temporal) polyethism [21]

In our study, the robots' mission is to search and collect "food-items" in a foraging area and sort them in disjoint piles. A theoretical expression for the time-dependent memory radius, r, for the functional self-organization process [1, 10, 15, 16, 17] was derived. The optimization procedure is based on intermediate steady states assumption and its validity was numerically tested using the angular second-moment feature as aggregation degree [11].

2   The mechanism of stochastic functional self-organization based on memory-weighted model

Our model is based on the following assumptions: 1. The environment is a two-dimensional periodic lattice with $N_x \times N_y$ sites. The periodic lattice (torus) was considered in order to eliminate the finite size effects.

2. The lattice sites are occupied by *objects* denoted by letters a, b, c and so forth. A free site is occupied by a ϕ-type object.

3. The agents, called *robot-like-ant (RLA)*, move randomly through the lattice. Anytime the RLAs transport an object. The carried object might be of ϕ-type and, therefore, the robot move freely.

4. When a robot moves to a given site it must decide if there are conditions to put down the carrying object and to pick up the existing one. The swapping condition writes

$$f_\alpha \geq f_\beta, \tag{1}$$

where f_α is the weighted frequency of the carried α-type object and f_β is the corresponding weighted frequency of the encountered β-type object. Walking through the lattice, every RLA records in its *memory register* the object-types encountered. By analyzing its content, the RLA extracts information about the spatial arrangement of the objects. In the current implementation, a binary string with the following structure characterizes every object-type:

$$s_{\alpha,\tau} : u_{\alpha,1} u_{\alpha,2} \dots u_{\alpha,\tau}, \tag{2}$$

where

$$u_{\alpha,i} = \begin{cases} 1 & \text{if an } \alpha\text{-type object was encountered at } i\text{-th step}, \\ 0 & \text{otherwise}. \end{cases} \tag{3}$$

The following conservation rules take place $\sum_{i=1}^{\tau} u_{\alpha,i} = n_\alpha$, for any $\alpha = \overline{1, T}$, where n_α is the total number of α-type object encountered and $\sum_{\alpha=1}^{T} u_{\alpha,i} = 1$, for any $i = \overline{1, \tau}$, where T is the total number of distinct object types.

An important memory register model, proposed by Deneubourg [6], consists in a shift register of fix length with equal weight. As the time passes, the whole record is shifted one place, the older (less significant) record is removed, and on the first (most significant) place enters the new record. Another important memory register model, proposed by Oprisan [1, 10, 15, 16], uses a first order recurrence to define the actual state of the CA. The main advantage is its computational efficiency and long (temporal) correlation. The computational efficiency reflects in a very economic memory allocation: only two real variables must be stored - the previous state of CA and the state associated with the newly encountered object-type. Long (temporal) correlation means that the actual state depends on the whole history of the system's states. Actually, it was proved (see [15, 16, 17]) that there is an intrinsic limit of the temporal correlation imposed by the machine accuracy. However, the temporal length of correlation for the model proposed is orders of magnitude bigger than for the fix number of shift register.

In our weighted memory register, every object-type is characterized, at any instant τ, by a weighted frequency:

$$f_\alpha(\tau) = \frac{\sum_{i=1}^{\tau} w(i) u_{\alpha,i}}{\sum_{i=1}^{\tau} w(i)}, \tag{4}$$

where $w(i)$ is an appropriate weighting function. The weighting function is

$$w(i) = \frac{1}{r^{i-1}}, \qquad (5)$$

indicating that for $r \gg 1$ the contribution of the τth step (with $\tau \gg 1$) to present decision is quite insignificant (*short-type memory*). This choice simulates the long-term memory effect. The limit case $r = 1$ corresponds to an infinite and equally weighted record. When $r < 1$ the contribution of older steps become significant [1, 10]. This type of memory function, that enhances the effect of past history, was suggested as a possible microscopic mechanism in a model of metastasis [16, 1, 17]. The above definition of the memory register leads to a first order recursive definition of the weighted memory associated to every object-type [16].

Numerical simulations have shown that the memory radius, (r), should depend on the cluster dimension (i.e. aggregation stage) in order to optimize the computational effort [15, 16]. Thus, clustering process occurs for any $r > 1$ but the speed with which the system reaches his sorted steady state depends sensibly on r: an abruptly decreasing memory radius drives the system into equilibrium state consisting in many small clusters (nonequilibrium state). The present study provides a proper memory radius-time relationship in order to allow sorting and avoid local equilibrium states.

3 The results

3.1 Theoretically derived relationship between memory radius and aggregation stage. The intermediate steady-states hypothesis

The swapping process that take place between nth and $n+1$th computational time lead to

$$f_\alpha^n < f_\beta^n, \quad f_\alpha^{n+1} > f_\beta^{n+1}. \qquad (6)$$

For simplicity, only two object-types, a, b, were considered. Based on (3), it results $u_{\beta,i}^n = 1 - u_{\alpha,i}^n$, and

$$f_\beta^n = \sum_{i=0}^{\tau} \frac{u_{\beta,i}^n}{r^i} = \sum_{i=0}^{\tau} \frac{1}{r^i} - f_\alpha^n. \qquad (7)$$

Substituting above relationship into (6a), it gives $f_\alpha^n < \frac{1}{2} \sum_{i=0}^{\tau} \frac{1}{r^i}$. An "1" bit on the most recent entry of the α-type object binary string forces swapping after $n + 1$ time step. Using (4), the swapping condition writes $f_\alpha^n > r \left(\frac{1}{2} \sum_{i=0}^{\tau} \frac{1}{r^i} - 1 \right)$, where the following recurrent relationship was used $f_\alpha^{n+1} = 1 + \frac{1}{r} f_\alpha^n$. Summarizing, in the limit of large memory $\tau \to \infty$, the swapping conditions (6) is:

$$\frac{r(2-r)}{2(r-1)} < f_\alpha^n < \frac{r}{2(r-1)}. \qquad (8)$$

The aggregation process takes place progressively: starting with a random distribution, first only two-objects clusters appear, then only three-objects clusters and so forth. This *intermediate steady-states assumption* help us to establish a quantitative relationship between the memory radius, (r), and the aggregation stage. For example, after nth computational step, a two-objects cluster appears if, based on (8), the memory radius satisfies $r \in (2^{1/2}, 2)$. For three-objects clusters the memory radius must satisfies $r \in$

$(r^{1/3}, r^{1/2})$. Generally, it is straightforward that transition from p-objects clusters steady state to $(p+1)$-objects clusters requires $r \in (2^{1/(p+1)}, 2^{1/p})$.

To determine how long it will take, in conventional iteration steps, to realize a complete two-objects clusters steady state and then a three-objects one and so on, a mean field view-point was adopted. Let us denote by N_α the number of α-type objects in the lattice and by $\sqrt{N}$ the linear dimension of the square lattice. The mean distance, λ, between the same object-types, for a uniform distribution of the objects, is given by conservation condition $N_\alpha = \left(1 + \frac{\sqrt{N}}{\lambda}\right)^2$, which implies $\lambda = \frac{\sqrt{N}}{2} \frac{1+\sqrt{4N_\alpha-3}}{N_\alpha-1}$. During the aggregation process the mean-free distance between the objects increases due to decreasing of the clusters' number. We propose as a conventional iteration time step needed to switch between p and $p + 1$-objects steady clusters a quantity proportional with the sum of the corresponding mean-free distance between the clusters. For example $\lambda^{1\to2} = \frac{\sqrt{N}}{2} \sum_{i=\frac{N_\alpha}{2}}^{N_\alpha} \frac{1}{\sqrt{4i-1}-1}$, where the superscript $1 \to 2$ design the initial and, respectively, final aggregation steady states. The proportionality constant depends on the successive identical steps required to visit the same cluster in order to transport them to the nearest neighbor ones. Therefore, taking into account this successive (minimal) number of repetitive visits the above relation can be generalized as follows

$$\lambda^{k\to k+1} = (2k + 1)\frac{\sqrt{N}}{2} \sum_{i=\frac{N_\alpha}{k+1}}^{\frac{N_\alpha}{k}} \frac{1}{\sqrt{4i-1}-1}. \tag{9}$$

Replacing above sum by $\sum_{i=\frac{N_\alpha}{k+1}}^{\frac{N_\alpha}{k}} \frac{1}{\sqrt{4i}}$, valid for $N_\alpha \gg 1$, on the bases of Euler formula

$\sum_{i=a}^{b} f(i) \approx \int_a^b f(x)dx + \frac{1}{2}(f(a) + f(b)) + \frac{1}{12}(f'(b) - f'(a)) + \ldots$, it results

$$\lambda^{k\to k+1} \approx (2k + 1)\frac{\sqrt{N}}{2} \sum_{i=\frac{N_\alpha}{k+1}}^{\frac{N_\alpha}{k}} \frac{1}{\sqrt{4i}} \approx 2N(2k + 1)\frac{\sqrt{k+1} - \sqrt{k}}{\sqrt{k(k+1)}},$$

which simplifies to

$$k \approx \frac{2N}{\lambda}.$$

On the other hand, we previously found that to built clusters with k objects (in the k-objects steady state sorting stage) the memory radius, (r), must lie in the range $(2^{1/(k+1)}, 2^{1/k})$. Therefore, the last two relations give

$$r \propto \exp\frac{\lambda \ln 2}{2N}. \tag{10}$$

This is the theoretically derived optimized relationship between the memory radius, r, and the computational time step N.

3.2   Numerical results and the annealing rule

The functional self-organization algorithm above described (see [1, 10, 15, 16] for details) requires a well-defined measure of the aggregation stage. Our choice is a texture analysis using features, which considers that texture-context information is contained in the

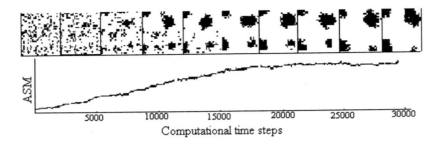

Figure 1: Different snapshots are uniformly sampled with 500 time-steps and shown on the top panel. The behavior of the contrast feature against the iteration time-step is plotted (lower panel). Here the environment is a rectangular 100×100 lattice, with 10 % black objects' concentration, 40 RLAs and a constant memory radius $r = 1.1$. As the system approaches its final steady state, the contrast feature saturates. The minimum and maximum values were analytically evaluated and, therefore, can be used to measure the distance between the current aggregation stage and the steady state.

overall spatial relationship between its gray tones [11]. Let $p(i,j)$ denote the normalized matrix of relative frequencies with which two cells, separated by distance d, occur on the image, one with gray tone i and the other with gray tone j. A relevant feature is *the angular second-moment (ASM)*, which represents a measure of homogeneity of the image. In a homogeneous image this features has a great values and decrease if the texture become less homogeneous (texture with different clusters):

$$ASM = \sum_{i=1}^{N_g} \sum_{j=1}^{N_g} p(i,j)^2, \qquad (11)$$

where N_g is the number of gray levels present in the image under investigation. Numerical simulations demonstrate that the above-defined feature is sensitive to aggregation stage and offer a quantitative meaning of this fuzzy concept (see Figure 1). Based on the above-defined global measure, we performed extensive numerical simulations to find the optimal time-dependence of the memory radius, (r), in order to validate the theoretical derived relationship (10) and its background hypothesis - the intermediate steady states assumption. We found that at the very beginning of the numerical simulations there is a quasi-linear relationship between the slope of the feature and the time step (see Figure 2).

Analyzing Figure 2, we conclude that it is advantageous, in order to reduce the computational effort, to start the numerical simulations with a high value for the memory radius. A high value of the memory radius means a very abrupt decrease of the feature and a rapidly slowing-down of the algorithm. If numerical simulation continues along this path, the system needs a long computation to reach its final steady state. On the other hand, an initial low memory radius determines a slow change of the features but the end of linear region is more closely to its final steady state. Our optimization procedure tries to combine the high speed of the features decreases for initial high memory radius with the lowest quasi steady state at the end of linear region for low memory radius. To determine the limits of the linear domain, which means that for a particular value of the memory radius the algorithm enters the slowing down regime, we performed numerical simulations until the linear correlation coefficient maximizes.

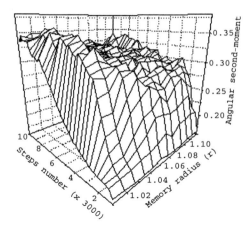

Figure 2: The 3D plot of the contrast feature against the time-step and memory radius r is shown. With a constant memory radius, a linear dependence of the feature on the time step, at the very beginning of the numerical simulations, can be observed. Decreases of the memory radius determine a decrease of the slope of the feature. On the other hand, the surfaces are highly fractured and, therefore, an appropriate time-dependent memory radius relationship must be chosen in order to avoid trapping the simulation in a local minimum.

At that point the final stage of aggregation was reached, for a particular value of the memory radius (r). We recorded the slope of the contrast feature for that specific memory radius value and the simulations were started over for another value of r. Figure 3A summarize the computationally derived optimal relationship between the slope of the contrast feature and the memory radius in order to ensure a minimum computational time. Once the relationship between microscopic control parameter (memory radius r) and the macroscopic measure of aggregation stage (the contrast feature) was established we get practical instrument to optimize the aggregation process. The computational procedure is as follows. We monitored the slope of the contrast feature and change the memory radius according to computationally derived relationship (see Figure 3A).

The plot of the optimally controlled memory radius against the iteration step (see Figure 3B) shows that the interpolation curve (continuous line) agrees with our theoretically derived time-dependent memory radius (see equation (10)).

4   Discussion and conclusions

Previous studies suggested that a realistic approach on the problem of local decision in the aggregation process performed by a team of mobile agents is the first order recurrent memory function [1, 10, 15, 16]. The present study analyses the dynamical aspects of the memory feature, particularly, its correlation length or *memory radius*. Based on *intermediate steady state* hypothesis we derived a theoretical time-dependent memory radius that leads to a minimum aggregation time. The intuitive idea behind our approach is that each two-, three-, four-, etc. clusters are metastable and the inherent stochastic behavior of the robot-like-ants (RLA) is the mechanism that drives the systems from intermediate (metastable) steady states to a final (stable) steady

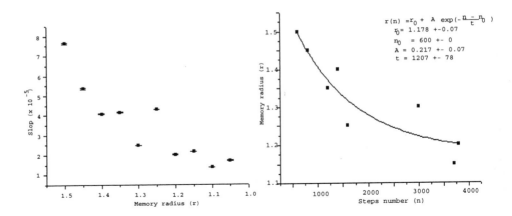

Figure 3: The plot of the steady slope of contrast feature against the memory radius r (A). The environment is a rectangular 100×100 lattice, with 10 % black objects' concentration and 40 RLAs. The chosen slope of the features maximizes the linear correlation coefficient and, therefore, is an indication that the steady state was reached. The plot of the memory radius (r) against the time step (n) when the sorting steady state was reached (B). The best fit is an exponential decaying function with the indicated parameters.

state. Once we found a time-dependent memory radius, which we thought to be the optimum, the next step was to check our finding using numerical simulation.

On the other hand, using a swarm of robots inspired from social insects behavior has some drawbacks. For example, stagnation is one: because of the lack of a global knowledge, a group of robots may find itself in a deadlock, where it cannot make any progress. Another problem is to determine how these so-called "simple" robots should be programmed to perform user-designed tasks. The pathways to solutions are usually not predefined but emergent, and solving a problem amounts to finding a trajectory for the system and its environment so that the states of both the system and the environment constitute the solution to the problem: although appealing, this formulation does not lend itself to easy programming. Until now, we implicitly assumed that all robots were identical units: the situation becomes more complicated when the robots have different characteristics, respond to different stimuli, or respond differently to the same stimuli, and so forth; if the body of theory that roboticists can use for homogeneous groups of robots is limited, there is virtually no theoretical guideline for the emergent design and control of heterogeneous swarms.

References

[1] D. Amarie, S.A. Oprisan and M. Ignat, Random walk systems behavior based on record function, *Physics Letters A*, **254** (1999) 112-118.

[2] R. Beckers, O.E. Holland and J.L. Deneubourg, From Local Actions to Global Tasks: Stigmergy and Collective Robotics, Proceedings of the Fourth International Workshop on the Synthesis and Simulation of Living Systems Artificial Life IV (1994): 181-189.

[3] B. J. Cole, Short-term activity cycles in ants: generation of periodicity by worker interaction, *The American Naturalist*, **137** (1991) 244-259.

[4] J.-L. Deneubourg and S. Goss, Collective patterns and decision making, *Ethology, Ecology and Evolution*, **1** (1989) 295-311.

[5] J.-L. Deneubourg, S. Gross, N. Franks, A. Sandova-Franks, C. Detrian, L. Chretien, The dynamics of collective sorting : Robot-like ant and ant-like robot, *From Animals to Animats: Proceedings of the First International Conference on Simulation of Adaptive Behavior* , edited by J. A. Meyer and S. W. Wilson (MIT Press, pp. 356-365, 1991).

[6] L. Edelstein-Keshet, J. Watmough. and G. B. Ermentrout, Trail following in ants: individual properties determine population behaviour, *Behavioral Ecology and Sociobiology*, **36** (1995) 119-133.

[7] N. R. Franks, The blind leading the blind in army ant raid patterns testing a model of self-organization, *Journal of Insect Behaviour*, **4** (1991) 583-607.

[8] C. V. Giuraniuc and S.A. Oprisan, Short range and long range coupling in stochastic functional self organization, *Physics Letters A*, **259** (1999) 334-338.

[9] Haralick R., K. Shanmugan, I. Distein, *IEEE Transactions on System Man and Cybernetics*, **3** (1973) 610-621.

[10] P. Hogeweg and B. Hesper, Socioinformatic processes: MIRROR modelling methodology, *Journal of Theoretical Biology*, **113** (1985) 311-330.

[11] I. Karsai, Z. Penzes. and J. W. Wenzel, Comb building in social wasps: self-organization and stigmergic script, *Behavioral Ecology and Sociobiology*, **39** (1996) 97-105.

[12] Melhuish C., O. Holland, and S. Hoddell, Collective sorting and segregation in robots with minimal sensing(1998): Preprint.

[13] S. A. Oprisan , V. Holban and B. Moldoveanu, Functional self-organization performing wide-sense stochastic processes, *Physics Letters A*, **216** (1996) 303-306.

[14] S. A. Oprisan, Convergence properties of the functional self-organization stochastic algorithm, *Journal of Physics A: Mathematical and General*, **31** (1998) 8451-8463.

[15] S.A. Oprisan, A. Ardelean and P.T. Frangopol, Self-organization and competition in the immune response to cancer invasion. A phase-orientated computational model of oncogenesis, *Bioinformatics*, **60** (2000) 1-5.

[16] S. W. Pacala, D. M. Gordon and H. C. J. Godfray, Effects of social group size on information transfer and task allocation, *Evolutionary Ecology*, **10** (1996) 127-165.

[17] G. Theraulaz, J. Gervet. and S. Semenoff-Tian-Chansky, Social regulation of foraging activities in Polistes dominulus Christ: a systemic approach to behavioural organization, *Behaviour*, **116** (1991) 292-320.

[18] C. Tofts and N. R. Franks, Foraging for work: how tasks allocate workers, *Animal Behaviour*, **48** (1994) 470-472.

[19] J. Watmough, and S. Camazine, Thermoregulation of honeybee clusters, *Journal of Theoretical Biology*, **176** (1995) 391-402.

Intelligent Autonomous Systems 7
M. Gini et al. (Eds.)
IOS Press, 2002

An Architecture for Robot Learning

Williams Paquier, Raja Chatila
LAAS/CNRS,
7 avenue du Colonel Roche, F-31077 Toulouse Cedex 04, France
wpaquier@laas.fr, raja@laas.fr

Abstract.   We propose a new approach to develop a learning robotic system. Learning starts from simple basic representations and actions. The robot is driven by the realization of goals that will need more complex and adapted representations and actions which are synthesized and selected according to their capacity of better achieving the goals. The system is based on sets of pulsed neural networks.

1 Introduction

There is a very wide body of literature on learning robots which encompasses skill learning and acquisition, learning by showing, learning environment structures, reinforcement learning, etc.[2, 1] In most approaches, there are predefined actions and representations, and predefined combination rules, which makes the system too much rigid to acquire new sensory-motor capabilities if they require completely different kinds of representations and skills. We propose a different approach in which robots acquire new capabilities by building new kinds of representations as needed, starting from a very elementary predefined set, and by synthesising new actions (or skills) also from a very limited elementary set as a starting vocabulary. Learning and adaptation rely on an exploration process that enables to build and reinforce the representations and actions that are most rewarding in achieving the goals. In the next section, we present the global system architecture, starting by introducing the basic system component, pulsed neural networks and their properties (2.1). Then we present the learning mechanism (section3) and how it enables to build adequate representations and actions.

2 System Architecture

2.1 Pulsed Neural Networks

The system relies on the properties of pulsed neural networks (PNN). Neurons are based on a discrete integrate and fire model [3] and each neuron receives pulse sequences (or trains) on its inputs and releases pulse sequences on its outputs. Each incoming pulse increases the neuron potential with its synaptic weight value. When this potential overpasses the neuron's threshold, a pulse is released. At each discharge, the potential is subject to an exponential leak which lowers it. The global system is composed of sub-systems (section 2.2) each of which is a set of connected *maps*. Maps are 2D groupings of neurons (with 8 neighbours each in the map).

The neuron dynamics is a simple integration of its afferent connections. Contrary to multi-layer perceptrons this system is totally asynchronous and information is propagated by the pulse state of neurons. For a given neuron, it is computationally cheaper to update its efferent outputs when it discharges than scanning systematically its afferent inputs and waiting for a discharge to integrate them. PNN are an economical way to code information, and they share the same structural description: the function is inherited from their place in the global network. This property is also used in spiking neurons based on rank order coding [5].

2.2 The global sensory motor loops

The global structure of the learning robot system is composed of seven highly connected sub-systems $(\pi, \pi^+, \gamma, \beta, \varepsilon, \alpha^+, \alpha)$ as shown in Figure 1. Each of these systems (except ε) is a PNN map network.

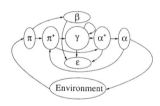

Figure 1: Global system and
sensory-motor loops.

• π (perception) is the input of the global system and is the frontier between the environment and the neural space. It is composed of maps of neurons wherein the potential values are static and dynamic "images" of external or internal stimulations. These images could be of many modalities: visual, auditive, proprioceptive, tactile, etc. π encodes these images into pulse sequences across time according to the rule: *The higher a potential, the earlier a pulse sequence is released.* The neuron have a sustained activity proportional to the potential value.

• π^+ receives inputs from π and also a copy of α outputs (α is the action part of the system). π^+ builds more complex representations and is composed of a multilayer network of maps (Figure 2) with a pyramidal structure. For a given layer, each map includes neurons that have vertical (from 2D map to 2D map - left to right in Figure 2) receptive field with excitatory connections and lateral (within a same layer) receptive fields with inhibitory connections. The pyramidal structure provides a full image perception for high-end neurons in π^+. For each map, we use shared synaptic weights so each map does the same analyses on each location of its input pulse images.

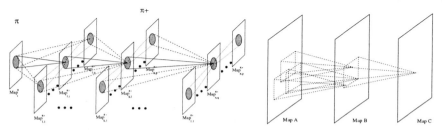

Figure 2: Left: the π and π^+ pyramidal multi layer structure. Right: detail of the structure showing how the receptive fields of neurons in π^+ grow with their distance from the input π.

• γ represents the system context. It receives inputs from π^+ and itself. This sub-system must have the properties of an associative memory. It must permit to activate new states in γ or to activate high-end neurons of α^+. It is responsible of learning new skills and determines the global system behaviour. Each neuron of γ is connected to ε.

• α^+ and α are the action sub-systems. α^+ produces sequences of actions. Its outputs are connected to α. Topologically it is composed of a multilayer map network like π^+ but does not have lateral inhibitions, but rather lateral excitations. α is the interface between neuron/pulse space and environment. It is a one-layer set of maps where each map drives a degree of freedom. For each map, neurons code elementary actions in amplitude and speed. Activation of α neurons provides a complex action which is the sum of the elementary actions which have been activated.

• β expresses the system drives (or basic instincts). It tunes ε according to π.

• ε is a system which associates each state to a set of fields or dimensions and a value in each dimension (i.e., a vector of size equal to the number of dimensions) [4]. ε and its variations represent the rewards assigned to the system states. ε is the sum of the values associated to β, γ and α^+. The β associations (dimensions and values) are fixed at system design phase. They represent the system's basic drives (values of elementary goals). The associations with γ and α^+ are learned during the system experiences. This learning process is explained in 3.2.

3   The Learning Mechanism

The learning mechanism involves the operation of all the sub-systems in parallel processes. Initially, the system is devoided of sensory-motor knowledge. It has a set of local rules (described above) and some global connectivity properties. Learning new behaviours depends on its capability to build perceptive representations of the environment, simple sequences of actions that do not depend on the environment, and on linking them under the constraint of maximizing their positive effect on the system (through ε). Initially all the sub-system connections are associated with random synaptic weights depending on their location in the system (except for inhibition which have a constant negative value).

There are two main learning processes. The first is relative to how the system learns the effects of its actions, the second, to the synaptic weights evolution. In the first process each neuron in β, γ and α^+ is connected to ε. This is a relation between an internal state and the effect it produces on the global system. ε is a global value that can be accessed by all neurons. It is used to learn sequences of actions in α^+.

The second learning rule is relative to the synaptic weight. We distinguish two classes of learning neurons: a) in π^+, the learning rule is a simple hebbian rule driven by the pulse state and can be computed just after the integration process; b) in γ and α^+, the hebbian rule and the perceived effect of action are mixed through ε.

3.1 Building Perceptive Elements

The environment is composed of complex and dynamic shapes. To represent them efficiently, the robot must extract some patterns which are categorized into classes according to some invariance criteria. π neurons potentials are composed by images of projections of these shapes. The role of π^+ is to find what is common and what is different among the features that compose these shapes. π^+ has two main inputs: the first comes from π and the second provides a copy of the actions that have just been produced in α. Each of these two links implements the same mechanism. We discuss the π-π^+ link here and the α-π^+ link in 3.4.

For each perceptual modality (visual, auditive, etc.), a layer of maps in π^+ is connected to the π input, and decomposes the images of potentials by applying groups of competitive filters systematically and simultaneously on each part of the images. Let's suppose that initially the weight kernels are randomly distributed and that they are different for all maps. Let's suppose also that the neuron thresholds of these maps are low enough to permit discharges when images are presented to the system. For a given position, one weight kernel is necessarily better adapted to the image contents (feature). According to the hebbian rule, and to the competition due to lateral inhibition, it will become more and more adapted to this feature while the kernels of the other π^+ maps in the same layer become less and less adapted. The layered architecture of π^+ permits to repeat this process across the structure so that the neurons at the top of π^+ correspond to a receptive field as large as the whole image. At this level of representation, input images have been diffracted in the sub-system, decomposed in spatial frequencies and pattern contents, and recomposed in more complex structures. This decomposition and recomposition evolves across time and converges toward a stable state. All the learned weight kernels are based on the image frequencies and feature contents. The activity of the high-end neurons of π^+ provides this distributed representation of the environment.

As top level neurons of π^+ learn to categorize projections of shapes, it is necessary to bind them in a non-projection dependent point of view[1] and for fusion of different pathways. The end of the π^+ sub-system is connected to γ which is a group of maps horizontally and vertically (long the pyramid direction) connected with associative memory properties. As shapes move during environment evolution, their projection are subject to variations. If the first γ neurons can reactivate themselves, namely by connecting their output to their input, they will remain active while the scene changes, and therefore two temporally shifted neuronal representations of the same shapes in the world will be active at the same time. The associative property will link these two neuron assemblies to each other. This associative memory property will permit to have γ assemblies of neurons selective to shapes but not view dependent.

3.2 Learning State Effect

The effect of actions and system state is measured by the values of ε. Initially (γ) and (α^+) effects are neutral $(\varepsilon = 0)$ and the only sub-system that is able to activate actions for the whole system is β. They are activated by hard-coded set of inputs, shortcuting the π^+ structure. The rule used to learn effects in (γ, α^+) is the following: *When the system has a global effect, all the active neurons have their ε component converge to the global ε values*[2]. This rule is used by the system to learn the effects of learned perceptive states by their temporal correlation with non learned and hard coded state effects. During the global system evolution, local and global effects (ε values) are learned from experience,

---

[1] this is true for the visual modality.
[2] the global ε values is simply the sum of all ε values.

first on the basis of a time correlation with a predefined state effect from β, then with learned state effects from γ. Thus the system explores and learns the effects of its own actions.

3.3 Building Sequences of Actions

The global system output is composed by the α and α^+ sub-systems. Contrary to the perceptive system, α^+ weight kernels are distributed, so each neuron is able to learn its own sequence of efferent neurons activation. At the end of the chain each neuron produces a complex sequence of α actions. The synaptic weights are the place where sequences are stored. Initially, all these weights are randomly distributed. For a given neuron in α^+ pulsing sustainly, the sequence reflects the relation between synaptic weights and the thresholds of efferent neurons. The first efferent neuron that will discharge is the one that has a synaptic weight closer to its threshold. The order of elements in the sequence is given by the decreasing order of the relation between threshold and synaptic weight. Each degree of freedom is covered by independant sets of maps that learn independantly. Initially, the system produces random actions. Then the actions reflect the state of γ (which reflects the perceptive context and internal state). The structure provides reusable elements of actions, that can be used in different sequences of actions.

3.4 How the Loop Develops New Behaviours?

From the above descriptions, we can summarize system operation as follows. The system decomposes inputs information into elementary patterns and uses these descriptions to produce more complex representations. Concurrently, internal states and perceptive context force the system to produce sequences of actions, which are initially random. One important point of the learning process is the ability to perceive not only the environment and its variations, but to associate these variations to the system's actions (the copy of the α state). The consequence of this external and internal loop is the specialisation of perceptive cells in π^+ that become sensitive to variations of the environment and to a sequence of actions at the same time. The top of this pathway between perception and action is localized in γ, which means that perception of system actions has an effect (through ε).

But how to connect the neurons responsible of the actions and those responsible of their perception? To close the loop between these two structures we use an associative network of connections between perception of actions and production of actions (the π^+-α^+ connection). Initially with very low synaptic weights, and thanks to the hebbian rule, and the fact that action neurons (α^+) remain active until the perception of action they produced, the synaptic weights increase statistically between two correlated couples of action/perception neurons. As a result, the action production neurons effects converge to the action perception effects because they remain active at the same time. The effects of actions are used by γ to change the components of actions to become more adapted to the obtention of a better effect (local and global).

4 Conclusion

We presented in this paper a global architecture for robot learning composed of perception and action sub-systems and the mechanisms that associate them together according to environment changes and to an evaluation of robot goal achievement. The system takes advantage of the properties of pulsed neural networks to efficiently code the state and its variations. Implementation is in progress.

References

[1] Rodney A. Brooks, Cynthia Breazeal, Robert Irie, Charles C. Kemp, Matthew Marjanovic, Brian Scassellati, and Matthew M. Williamson. Alternative essences of intelligence. In *Proceedings of the 15th AAAI, AAAI Press*, pages 961–968, 1998.

[2] Leslie Pack Kaelbling, Michael L. Littman, and Andrew P. Moore. Reinforcement learning: A survey. *Journal of Artificial Intelligence Research*, 4:237–285, 1996.

[3] L. Lapicque. Recherches quantitatives sur l'excitation électrique des nerfs traitée comme une polarisation. *J. Physiol. Pathol. Gen., 9:620–635*, 1907.

[4] E. L. Thorndike. *Animal Intelligence.* Hafner, Darien, CT, 1911.

[5] S. Thorpe and J. Gautrais. Rank order coding: A new coding scheme for rapid processing in neural networks. In *Computational Neuroscience: Trends in Research, J. Bower, Ed, (Plenum Press, New York)*, pages 113–118., 1998.

Intelligent Autonomous Systems 7
M. Gini et al. (Eds.)
IOS Press, 2002

Progress in Pheromone Robotics

David Payton, Regina Estkowski, Mike Howard
HRL Laboratories, LLC 3011 Malibu Canyon Road, Malibu CA 90265
{payton, regina, mhoward}@hrl.com

Abstract: In our Pheromone Robotics research, we are developing techniques for coordinating the actions of large numbers of small-scale robots to achieve useful large-scale results in surveillance, reconnaissance, hazard detection, and path finding. This paper describes progress in the areas of world-embedded computation and world-embedded displays that are enabled by our use of the biologically inspired notion of a "virtual pheromone." We describe the pheromone logic primitives that enable our robot collective to become a distributed computing mesh embedded within the environment, while simultaneously acting as a physical embodiment of the user interface.

1. Introduction

Emerging technologies in micro machining and MEMs hold the promise of creating extremely small robots. Although limited in size and power, such robots can work together in large numbers to conceivably accomplish a wide range of significant tasks [5],[7],[10],[15], including surveillance, reconnaissance, hazard detection, path finding and payload conveyance.

Coordinating and interacting with very large numbers of robots involves issues not encountered when dealing with one or a few robots [2],[8],[11]. Coordination schemes that require unique identities for each robot, explicit routing of point-to-point communication between robots, or centralized representations of the state of an entire swarm can be overwhelmed when dealing with extremely large numbers. To assure that our multi-robot system is both scaleable to large robot swarms and tolerant to individual robot failures, we focus on techniques that effectively limit each robot's interactions to its own local neighborhood.

We are inspired by techniques used by ants and termites for communication and coordination [4],[9],[17]. We implement "virtual pheromones" using simple transceivers mounted on each robot. Like their chemical counterparts, our virtual pheromones facilitate simple communication and emergent coordinated movement with only minimal on-board processing. But virtual pheromones go a step further, transforming a robot swarm into a distributed computation grid embedded in the world. This grid can be used to compute non-local information about the environment such as bottlenecks and shortest paths, in ways that are foreign to insect colonies.

We envision a scenario in which a rescue team enters an unfamiliar building after a disaster, and needs to quickly locate any survivors. The team, at the entrance to the building, releases thousands of very tiny robots, which quickly disperse into the open spaces using simple attraction/repulsion behaviors. Upon detection of a survivor, a robot emits a virtual pheromone message signaling the discovery. This message is propagated locally between robots only along unobstructed paths, producing a gradient as it is propagated. Ultimately, the message makes its way back to the entrance where rescue team members can now follow the pheromone gradient to the survivor. To do this, the robots themselves serve as a distributed display of guideposts leading the way along the shortest unobstructed path. We are pursuing this vision with 20 small robots specially designed to support our virtual pheromone type communication protocols. In an earlier paper [14] we described our approach. In this paper we briefly review key features of the approach, and then focus on recent developments.

2. Overview of the approach

There are three key features of our approach: *Virtual Pheromones* is a diffusive local-neighborhood interaction mechanism by which the robots communicate and coordinate. *World-Embedded Computation* is a technique for performing many graph-theoretic algorithms in a distributed fashion, without requiring a centralized intermediate representation. *World-Embedded Display* is a method for interfacing to a robot swarm by using the robots as annotations on the real world. These features will be described in the next three subsections.

2.1 Virtual Pheromones

Inspired by the chemical markers used by ants and termites for communication and coordination, we developed the notion of a "virtual pheromone," implemented by messages relayed from robot to robot, with specific features that enable them to travel in decaying wavefronts from the origination point. Virtual pheromones facilitate simple communication and coordination and require little on-board processing. This approach is applicable to future robots with much smaller form factors (e.g., to dust-particle size) and is scaleable to large, heterogeneous groups of robots.

   The design of virtual pheromones preserves some of the essential properties of natural pheromones that make them effective in facilitating group organization. (1) Pheromones are locally transmitted without specifying a recipient, obviating the need for unique identities that are impractical in large groups [8]. (2) Pheromone diffusion gradients provide important navigational cues and encode useful information about barriers in the environment that block pheromone propagation. (3) Pheromones decay over time, reducing obsolete or irrelevant information.

   Virtual pheromones are not faithful copies of chemical pheromones, for several practical reasons. For example, natural odors diffuse in patchy plumes whose concentration does not fall off uniformly with distance or time. But virtual pheromones are transmitted at a known intensity, and their signal strength decreases linearly with distance. Therefore, receivers can reliably estimate distances on the basis of signal strength alone. Second, if an originating source for a virtual pheromone moves, the gradient will adjust quickly without the persistence of chemical pheromones. Third, virtual pheromone messages can contain optional data beyond pheromone type, for use in distributed world-embedded computations.

2.2 World-Embedded computation

Traditional approaches to path planning and terrain analysis operate on an internal map of terrain features [12],[13], or they perform the steps of sensing, data transmission to a central point, and map generation before the data can be processed. This is especially disadvantageous when the environment is rapidly changing. Pheromone robots (or *pherobots*) require no centralized map, nor do they maintain a detailed history.

   Our approach treats pherobots as embedded processing elements in the environment. Each actively senses and computes simultaneously based only on local terrain features and pheromones passed by neighboring robots. Inter-robot communication is via line of sight signaling rather than wireless, because it results in a communications grid that embodies mobility costs in its connectivity structure.

D. Payton et al. / Progress in Pheromone Robotics

Rather than trying to overcome communication loss due to obstacles, we exploit this effect to sense the terrain and determine optimal traversal paths.

In fact, our rules for message propagation provide a distributed version of the wavefront propagation method used in Dijkstra's shortest-path algorithm [6]. Global properties such as shortest routes, blocked routes, and contingency plans are computed in a robust, distributed manner, with each member of the population of simple processors contributing a small piece of the result.

2.3 World-Embedded Display

Pherobots implement an efficient and versatile distributed computer. In some applications the results of the computation are sent back to the user via relayed messages. But the real novelty of the system is when the robot swarm acts as a

Figure 1: User's view of a pheromone gradient as seen through an augmented reality display

distributed display embedded in the environment. In effect, each robot becomes a pixel, or an annotation on the immediate environment. The robot's position within the environment provides context to interpret the meaning of the transmitted information.

As an example, consider the notional view in Figure 1. Here, the robot swarm has dispersed, and a pheromone gradient has been established. A user, seeking to follow that gradient, needs only to be able to see the local gradient vector at each robot location. One way to read the distributed display is to use an augmented reality (AR) system. AR refers to the visual presentation of information, in geometric registration with true objects seen in the environment. The combination of this world-embedded interface with our world-embedded computation means that the results of complex distributed computations can be mapped directly onto the world with no intermediate representations required.

3. Progress

The rest of the paper will relate the progress we have made toward realizing the vision described in section 2. In particular, we briefly summarize the existing robot hardware and previously reported capabilities, then detail advances in the pheromone control system and user interface. These advances include the introduction of new behavior primitives and pheromone messaging primitives.

3.1 The robot platform

Our swarm of 20 pherobots was custom designed for us by Diversified Enterprises (Figure 2). The control system was originally written in Java, using the Teambots [16] environment, which also allows us to test our algorithms on multiple simulated robots. Java on the PalmVx robot controller, using a small virtual machine called WABA, proved to be too slow. By porting to C++ and making some other speedups, we have reduced cycle time from 3 seconds to under 0.4 seconds.

Figure 2: Pherobot Swarm

Atop each robot is a set of eight radially-oriented directional InfraRed (IR) transmitters and receivers (transceivers) as shown in Figure 3. Baffles are installed between transceivers to reduce sensor overlap. Sensing is a side effect of communication; when a robot receives a message it just sent, it is considered a reflection off an obstacle. Distance to the obstacle is calculated as a function of the power of the reflected message. If there is no message to send out of a particular port, an "obstacle ping" is sent to

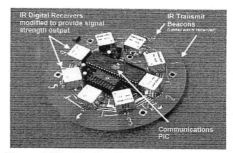

Figure 3: Transceiver for virtual pheromones

sense obstacles. Other robots sense distance and direction to a robot by detecting both the messages and the obstacle pings that it transmits. Further details on the logic for message processing are given in section 3.3.

Our choice of this particular robotic platform was a tradeoff between size and robustness. A small platform forces us to confront issues of low power and limited mobility, and indeed we initially considered a much smaller platform. However, the smaller the platform, the more time is spent struggling with practical issues that distract from theoretical issues. Our pherobots provide a reasonable compromise.

3.2 Behavior Primitives

Our robot behaviors, environmental representations, and data structures are developed to use only local interactions and information. The primary behavior primitives are based on notions of attraction and repulsion (Figure 4). Repulsion keeps robots from colliding with obstacles and other robots and urges the robot swarm to spread out and maximally cover a space. Robots can also be repelled by virtual pheromones, so that one robot can serve to block others from certain areas. Attraction keeps robots in communication range with each other and allows them to follow contours of terrain features, or to follow a pheromone message gradient to its source. These elemental behaviors may be

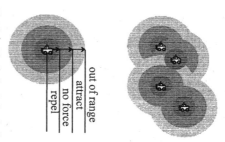

Figure 4: Attraction and repulsion zones implement gas expansion

combined to produce a variety of emergent group behaviors [1]. As a simple example, the gas expansion model of robot dispersal emulates gas particles filling a vacuum. The robot's control system generates a repulsion vector to get away from things that are too close, and an attraction vector to get closer to things that are too far away. In between is a neutral zone, to avoid attraction-repulsion oscillations. These simple behaviors allow a robot swarm to expand from a tight grouping into a maximal dispersion that maintains nearest-neighbor communications. The necessary complement to expansion is contraction; an effective approach is described by Gage [7].

While standard potential-field methods can be used to compute net attraction or repulsion by summing force vectors between a robot and its local neighbors, it is often more effective to reason about the free space surrounding a robot. To do this, we have developed a set of control system classes that partition space surrounding a robot into equally sized pie-like 2D sectors or "bins" with the robot at the center. By default, bins correspond identically to the sectors formed by the IR sensors, but they can be made smaller or larger according to the resolution needs of each behavior. In each sector, we store the range to the nearest object detected within that sector. From this simple representation, we can perform a few basic logical operations that provide us with mobility primitives that would otherwise be difficult to obtain from vector operations alone.

Foremost among these operations are those that look for contiguous blocks of free space around a robot. For example, it is easy to determine if a robot is surrounded, to find the largest open space for a robot to move into, or to locate an empty sector that is close to the robot's current heading. We can threshold on distance such that bins are considered empty so long as they contain no object within a given distance. The bins representation is particularly helpful for reasoning about whether a robot is on the edge of a swarm and can move quickly to open space, or if it is in the middle of the swarm and should remain stationary. Different types of sensed objects can be stored in separate sets of bins. This allows various logical operations to be performed between corresponding bins. For example, we maintain separate sets of bins for nearby robots and for fixed obstacles. These can then be merged to find sectors with robots but no obstacles or obstacles but no robots.

3.3 Pheromone Logic

Virtual pheromones are used both to mediate robot behavior and to facilitate world-embedded computations. We therefore use an internal representation for virtual pheromones that allows them to play a role both in motion control logic and in messaging logic. We begin with a standard 2D-vector representation, consisting of a direction field and a length field. For received pheromones, these fields convey the direction to the source, and the received signal strength. For pheromones to be transmitted, these fields convey a direction to send and a transmit intensity. With this representation alone, virtual pheromones are completely compatible with existing control system logic, and may be summed or otherwise combined with any other control vector to produce movement vectors dependent on the received transmissions of neighboring robots.

To support the pheromone-specific messaging logic needed for world-embedded computations, we extend the 2D-vector representation with additional fields. These include a field-of-view (FOV), a type code, a hop count, and a data field. The FOV is used in combination with the vector direction to specify which transmitters a robot uses to send a message. The FOV specifies the width of the beam, while the vector direction determines the center axis. The type code allows us to have multiple pheromone types to emulate the multiple chemical types found in real pheromones. The hop count is used to represent the strength of a virtual pheromone, analogous to the chemical concentration of a real pheromone. The hop count plays a fundamental role in producing virtual pheromone gradients. The data field is optional, and is available to support a variety of extensions to pheromone messaging.

There are many ways to use virtual pheromones both in robot control and in distributed computation, so it is important to be able to construct complex operations from a set of simpler building blocks. Table 1 shows some pheromone messaging primitives we have developed, sorted into three basic types of interaction between the control system and the messaging system. There are basic requests for pheromone message data, repeating processes such as continual or conditional message forwarding, and commands to send an individual pheromone message.

Pheromone Data Request Primitives	Have I sensed type X?
	Give me all of type X
	Give me the "best" of type X
	Give me the "best" of type X in each bin
Pheromone Message-Passing Primitives	Forward X when Y is true/false
	Forward X, incrementing (or decrementing) hop count by P
	Forward X, setting FOV and pheromone type
	Forward last X for K time cycles
	Forward X for K cycles then nothing for T cycles
	Forward X relative to received direction
	Forward X or Y depending on which has the max or min hop-counts.
	Forward X and Y, alternating between them.
Pheromone Sending Primitives	Initiate Sending X (once, forever, for time T)
	Send X with priority P
	Send X in directions (N S E W...)

Table 1: Pheromone Messaging Primitives

We have implemented most of the above functionality in a set of classes specifically designed for pheromone messaging. Instances of these classes may be linked together to form a logical data flow path from received messages to output messages. Along this path, operations such as changing hop-counts or changing FOV may be performed in successive pipelined stages. Each messaging class instance, in addition to performing operations on immediately available data, is also capable of retaining local state so that operations dependent on prior values of a pheromone message can be performed.

The above primitives can be combined in numerous ways to support distributed computations. Foremost among these is gradient formation, which is central to our world-embedded computation approach. In gradient formation, a robot relays received messages of a given pheromone type by selecting the message with the highest hop-count and re-transmitting it after decrementing hop-count field. If messages are received from several different directions, the message with the highest hop-count is considered the "best" message, and this is the one used for relaying, as shown in Figure 5. The net result is a pheromone gradient expressed as uniformly descending hop counts from the originating source.

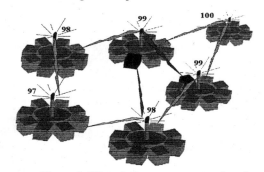

Figure 5: Virtual pheromones are relayed with a lower hop-count by each subsequent robot.

An example use of the virtual pheromone data field can be seen in our method for activating different behavior assemblages within the control system. In some cases, we want to change the control mode of just a single robot, and in other cases, we want to command an entire robot swarm to change control modes. To facilitate this, a specific pheromone type called a *command pheromone* is used. The data field in the command pheromone contains a byte to specify which behavior assemblage to run, and a bit to determine whether the pheromone is to be propagated to others. Any robot receiving a command pheromone with the propagation bit set will re-transmit that pheromone to its neighbors. A command to a single robot can thus end up switching the entire swarm to a new control mode.

3.4 User Interface

Figure 6: Augmented reality head-mounted display system with camera.

To achieve a world-embedded display as depicted in Figure 1, we incorporate an augmented reality (AR) system [3] combined with specially designed beacons mounted on each robot. We have developed a system that detects specially coded infrared signals emitted by each robot and presents information on a see-through head-mounted display (HMD) worn by the user. The infrared signals are detected by a monochrome camera that is mounted atop a Virtual-IO head-mounted display as shown in Figure 6. The camera, filtered to receive only infrared light, receives the infrared signals as blinking dots on a black background. These blinking dots are then tracked and are decoded into arrows that are displayed on the HMD. Proper alignment of the camera with the user's display allows the arrows to be

Figure 7: Coded infrared signals are decoded into arrows for overlay onto a real scene.

superimposed directly over the robot's positions, as shown in Figure 7, without any explicit knowledge of the robot or user location. The real-time tracking of the blinking dots allows the decoded arrows to appear to float over the robots even as the user's head moves.

The infrared beacons (Figure 8) on each robot are designed such that a robot can display an arrow at a

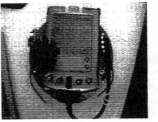

Figure 8: Augmented Reality Mast projects information from the robots to a user's head-mounted display.

desired direction regardless of the user's orientation relative to the robot. Consider, for example, a user looking at a robot from the left side and seeing an arrow pointing to the left. Another user, looking at that same robot from the right side should see an arrow pointing to the right. We achieve this desired result by having the hooded LED beacons on each robot transmit different coded messages in each of eight different directions. Depending on the user's orientation with respect to the robot, they will receive a different coded message that will result in a different arrow displayed in their HMD. Thus, the arrow will appear to be pointing in the same world direction regardless of user viewpoint.

4. Conclusion

The approach to pheromone robotics described in this paper offers robustness because it needs no explicit maps or models of the environment, and requires no explicit knowledge of robot location. Collections of robots will be able to perform complex tasks such as leading the way through a building to a hidden intruder or locating critical choke points. This is possible because the robot collective becomes a computing grid embedded within the environment, while also acting as a physical embodiment of the user interface. Over the past decades, the literature on path planning and terrain analysis has dealt primarily with algorithms operating on an internal map containing terrain features. Our approach *externalizes* the map, spreading it across a collection of simple processors, each of which determines the terrain features in its locality. The terrain processing algorithms of interest are then spread over the population of simple processors, allowing such global quantities as shortest routes, blocked routes, and contingency plans to be computed by the population.

The user interface to this distributed robot collective is itself distributed. Instead of communicating with each robot individually, the entire collective works cooperatively to provide a unified display embedded in the environment. Robots that have dispersed themselves throughout a building will be able to guide a user toward an intruder by collectively signaling the shortest path to the intruder. Using augmented reality, robots will be able to present complex displays. Users wearing a see-through head-mounted display, and a head-mounted camera that detects and tracks infrared beacons emanating from the robots will be able to see a small amount of information superimposed over each robot. Each robot, in effect, will be a pixel that paints information upon its local environment. The combination of this world-embedded interface with our world-embedded computation means that the results of complex distributed computations can be mapped directly onto the world with no intermediate representations or maps required.

5. Acknowledgements

This work is supported by the Defense Advanced Research Projects Agency under contract N66001-99-C-8514. Any opinions, findings, and conclusions or recommendations expressed in this material are those of the authors and do not necessarily reflect the views of the Defense Advanced Research Projects Agency.

6. References

[1] Arkin, R.C. 1998. *Behavior-Based Robotics*. MIT Press, Cambridge, MA.
[2] Arkin, R.C. and Bekey, G.A. (editors). 1997. *Robot Colonies*. Kluwer Academic Publishers.

[3]   Azuma, R., Hoff, B., Neely III, H., Sarfaty, R. 1999. A Motion-Stabilized Outdoor Augmented Reality System, *Proc. IEEE VR '99*, Houston, TX, pp. 252-259.

[4]   Bonabeau, E., Dorigo, M., and Theraulaz, G., 1999. *Swarm Intelligence: From Natural to Artificial Systems.* New York, Oxford University Press.

[5]   Deneubourg, J. and Goss, S. 1984. Collective Patterns and Decision-Making, *Ethology, Ecology, and Evolution,* 1:295-311.

[6]   Dijkstra, E.W. 1959. A Note on Two Problems in Connection with Graph Theory, *Numerische Mathematik,* 1:269-271.

[7]   Gage, D.W. 1992. Command and Control for Many-Robot Systems, In *Unmanned Systems Magazine,* 10(4):28-34.

[8]   Gage, D.W. 1993. How to Communicate with Zillions of Robots, In *Proc. SPIE Mobile Robots VIII,* Boston, MA, 2058:250-257.

[9]   Goss, S., Beckers, R., Deneubourg, J., Aron, S., and Pasteels, J. 1990. How Trail Laying and Trail Following Can Solve Foraging Problems, In *Behavioral Mechanisms of Food Selection,* ed. R. Hughes, Springer-Verlag, Heidelberg, Germany, pp. 661-678.

[10]   Holland, O. and Melhuish, C. 2000. Stigmergy, self-organization, and sorting in collective robotics. *Artificial Life,* 5:2.

[11]   Lewis, M.A., and Bekey, G.A. 1992. The Behavioral Self-Organization of Nanorobots Using Local Rules," In *Proc. 1992 IEEE/RSJ Int. Conf. Intelligent Robots and Systems,* Raleigh, NC.

[12]   Mitchell, J.S.B., Payton, D., and Keirsey, D. 1987. Planning and Reasoning for Autonomous Vehicle Control, *Int J. Intelligent Systems,* Vol. 2.

[13]   Payton, D.W. 1990. Internalized Plans: A Representation for Action Resources, in *Designing Autonomous Agents,* ed. Pattie Maes, MIT Press, Cambridge, Mass, pp. 89-103.

[14]   Payton, Daily, Estkowski, Howard, Lee, *"Pheromone Robots",* in Autonomous Robots, Kluwer Academic Publishers, Boston, MA, Vol. 11, No. 3.

[15]   Ünsal, C. and Bay, J. 1994. Spatial Self-Organization in Large Populations of Mobile Robots," *IEEE Int. Symp. on Intelligent Control,* pp. 249-254.

[16]   Teambots Web site: http://www.teambots.org/

[17]   Werger, B.B., and Mataric, M.J. 1996. Robotic food chains: Externalization of state and program for minimal-agent foraging, In *Proc. 4th Int. Conf. Simulation of Adaptive Behavior: From Animals to Animats 4,* MIT Press, pp. 625-6

Intelligent Autonomous Systems 7
M. Gini et al. (Eds.)
IOS Press, 2002

Autominder: A Planning, Monitoring, and Reminding Assistive Agent

Martha E. Pollack*   Colleen E. McCarthy[†]   Ioannis Tsamardinos[‡]
Sailesh Ramakrishnan*   Laura Brown*   Steve Carrion*   Dirk Colbry*
Cheryl Orosz*   Bart Peintner*

*Artificial Intelligence Laboratory
University of Michigan
pollackm@umich.edu

[†]Department of Computer Science   [‡]Department of Biomedical Informatics
University of Pittsburgh   Vanderbilt University

Abstract.
The percentage of elderly people in the population is increasing at
a phenomenal rate [14]. A significant challenge faced by many elderly
is a decline in cognitive functioning, particularly in memory. In this
paper, we describe Autominder, an automated agent designed to serve
as a "cognitive orthotic", assisting an elderly client in carrying out the
required activities of daily life (ADLs), by providing her with timely and
appropriate reminders. In generating these reminders, the goal is to balance three objectives: (i) maximizing the client's compliance in performing ADL's; (ii) maximizing the level of caregiver and client satisfaction
with the system; and (iii) avoiding making the client overly reliant on
the system. Towards these ends, Autominder stores and updates plans
representing a client's ADLs, tracks their execution, learns the typical
behavior of the client with regard to the execution of these plans, and
provides select reminders of the activities to be performed. Autominder
is being designed as part of the Initiative on Personal Robotic Assistants
for the Elderly [12], a project aimed at developing robotic systems to
assist elderly persons with memory impairment.

1   Introduction

The percentage of elderly people in the population is increasing at a phenomenal rate
in the United States [14], as well as in many other parts of the world. Indeed, the
number of people residing in nursing homes in the U.S. is projected to double or triple
by 2030. It has been shown that the quality of life for people remaining in their own
homes is generally better than for those who are institutionalized [17]; moreover, the
cost for institutional care can be much higher than the cost of care for a patient at
home. Unfortunately, a significant challenge faced by many elderly people is a decline
in cognitive functioning, particularly in memory. Such a decline can make it difficult
for someone to organize and regularly perform their necessary activities of daily living
(ADLs), such as taking medicine correctly, eating, drinking water, toileting, performing physical exercises (e.g., "Kegel" bladder exercises), performing routine hygiene,
engaging in recreational activities (e.g., watching television, attending a Bingo game),

and going to medical appointments.[1] This inability to adequately perform ADLs can necessitate institutionalization.

In this paper, we describe Autominder, an automated agent designed to serve as a "cognitive orthotic", assisting an elderly client in carrying out the required activities of daily life by providing her with timely and appropriate reminders. In generating these reminders, the goal is to balance three objectives: (i) maximizing the client's compliance in performing ADL's; (ii) maximizing the level of satisfaction with the system of both the client and the caregiver(s); and (iii) avoiding making the client overly reliant on the system and possibly decreasing, rather than increasing, her independence. Towards these ends, Autominder stores and updates plans representing a client's ADLs, tracks their execution, learns the typical behavior of the client with regard to the execution of these plans, and provides carefully chosen and timed reminders of the activities to be performed. Autominder relies on a number of AI techniques, including interleaved planning and execution, sophisticated temporal reasoning, and reasoning under uncertainty.

Autominder is being designed as part of the Initiative on Personal Robotic Assistants for the Elderly (Nursebot)[12], a multi-university collaborative project.[2] The initial focus of this initiative is the design of an autonomous robot, currently called Pearl, that will "live" in the home of an elderly person. Autominder is a central element of Pearl's software. Several prototype versions of Autominder have been fully implemented in Java and Lisp. Although the most recent version has not yet been installed on Pearl, an earlier version was used in an exploratory field test with elderly users in June, 2001.

In the next section, we provide a thorough overview of Autominder's architecture, and of the existing and novel AI techniques we are using. Section 3 briefly discusses the issue of the kinds of platforms–robotic or software–on which Autominder might be installed. Section 4 describes related work on cognitive orthotics, and finally, Section 5 summarizes and points to ongoing and future work on this topic.

2 Autominder Architecture

Autominder grew out of our earlier work on plan management, in particular, the Plan Management Agent (PMA), a prototype intelligent calendar tool [15]. PMA consists primarily of a plan manager, a system that stores a client's plans, updating them as the client adds, deletes, or modifies constraints on those plans, and/or executes actions in them. A central task for PMA is to ensure that there are no conflicts amongst the client's plans, instead suggesting alternative ways to resolve potential conflicts. An extension of PMA's main component now serves as the Plan Manager (PM) for Autominder. There are two additional components essential to Autominder: a Client Modeler(CM) and a Personal Cognitive Orthotic (PCO). The overall architecture is illustrated in Figure 1. What is not apparent in the figure is that the system is event-driven and all communication between components is routed through a message-handling component.

---

[1]In fact, the list of activities we are covering extends beyond the set usually included under the heading of ADLs. We should also note that in the early versions of the Autominder, we are not directly issuing reminders about medicine-taking, due to safety concerns: we want to ensure the correctness of Autominder before seeking FDA approval to include medicine reminders.

[2]The initiative includes researchers from the University of Pittsburgh School of Nursing and Department of Computer Science, Carnegie Mellon University Robotics Institute and Human-Computer Interaction Department, and the University of Michigan Department of Electrical Engineering and Computer Science.

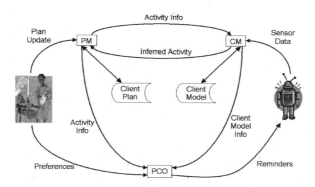

Figure 1: Autominder Architecture (simplified: Message Handler Omitted)

However, to make the flow of information clearer, we have omitted this component from the diagram, and just show the intended source and destination of each type of message.

In Autominder, the caregiver initially inputs a description of the activities the client is supposed to perform, as well as any constraints on, or preferences regarding, the time or manner of their performance. Subsequently, updates to the plan (e.g., a new doctor's appointment) can be made by a caregiver and, with certain restrictions, the client herself. Plan information flows directly to the PM, which, like PMA, checks for consistency and provides ways of resolving potential conflicts (e.g., using the toilet before leaving for the doctor's office).

Pearl, the robot on which Autominder is installed, has various sensors–camera, microphone, infrared, etc. - and it sends sensory information to the CM. Note that the pixels-to-predicates problem is solved by software outside of the Autominder: the Autominder receives reports of the form "client went to kitchen" or "toilet flush heard". The CM uses the sensor information, along with the client's plan itself, to infer whether there is an indication that a planned activity has been initiated or has ended (e.g., going to the kitchen around the normal dinner time may indicate that the client is beginning dinner). If the likelihood is high that a planned activity is being executed, the CM reports this to the PM, which can then update the client's plans by recording the time of execution and propagating any affected constraints to other activities (e.g., if the client is supposed to take medicine no less than two hours after eating, the time for medicine-taking can be made more precise upon learning that the client is having dinner). Over time, the CM also constructs a model of the client's typical plan execution patterns (e.g., that the client usually remembers to take medicine in the morning, but frequently forgets in the evening). It is important to distinguish between the *client plan*, which models the activities that the caregiver would like the client to perform and is maintained by the PM, and the *client model*, which models the system's expectations of what the client has done and will do.

The final component of Autominder is the PCO, which uses both the client plan and the client model to determine what reminders should be issued and when. Ultimately, the PCO will also make use of information provided by the caregiver or client about their preferences as to how and when the activities should be executed, consistent with the requirements of the plan.

2.1 The Plan Manager

The primary job of the Plan Manager (PM) is to maintain an up-to-date model of the plan (the ADLs) that the client should execute. Initially, a routine daily plan is submitted to the PM. This plan may then be changed in one of three ways: (i) by the addition of new activities [3]; (ii) by the modification or deletion of (constraints on) activities already in the plan; (iii) by the execution of one of the planned activities. In the first two cases, PM performs plan merging [21, 8, 20, 19]: to ensure that the change does not introduce a conflict. In the third case, it propagates the constraints affected by activity execution, as described in the example above.

To adequately represent the client plans, it is essential to support a rich set of temporal constraints: for example, we may need to express that the client should take a medication within 15 minutes of waking, and then eat breakfast between 1 and 2 hours later. We model client plans as Disjunctive Temporal Problems (DTP) [13, 18] and use an efficient algorithm for checking their consistency, which we developed in [19]. The DTP is an expressive framework for temporal reasoning problems that extends the well-known Simple Temporal Problem (STP) [5] by allowing disjunctions, and the Temporal Constraint Satisfaction Problem (TCSP) [*ibid.*] by removing restrictions on the allowable disjunctions. Formally, a DTP is defined to be a pair $<V, C>$, where

- V is a set of variables (or nodes) whose domains are the real numbers, and

- C is a set of disjunctive constraints of the form: $C_i : x_1 - y_1 \leq b_1 \vee \ldots \vee x_n - y_n \leq b_n$, such that x_i are y_i are both members of V, and b_i is a real number.

A solution to a DTP is an assignment to each variable in V such that all the constraints in C are satisfied. If a DTP has at least one solution, it is *consistent*.

Within the PM, we assign a pair of DTP variables to each activity in the client's plan: one variable represents the start time of the activity, while the other represents its end time. We can easily encode a variety of constraints, including absolute times of events, relative times of events, and event durations, and can also express ranges for each of these.

To propagate new constraints and to check for consistency, the PM uses our Epilitis system [19]. The approach to DTP solving taken in the literature has been to convert the original problem to one of selecting one disjunct, $x_j - y_j \leq b_j$, from each constraint $C_i \in C$, and then checking that the set of selected disjuncts forms a consistent STP. Checking the consistency of and finding a solution to an STP can be performed in polynomial time using shortest-path algorithms [5]. The computational complexity in DTP solving derives from the fact that there are exponentially many sets of selected disjuncts that may need to be considered; the challenge is to find ways to efficiently explore the space of disjunct combinations. This has been done by casting the disjunct selection problem as a constraint satisfaction processing (CSP) problem [18, 13] or a satisfiability (SAT) problem [1]. Epilitis combines and extends the previous approaches, in particular by adding no-good learning, and achieves a speed-up of two orders of magnitude on a range of benchmark problems [19]. For typical problems we have so far studied in the Autominder domain, performance is well within the acceptable range, typically taking less than 10 seconds.

---

[3]Note that the PM includes a library of precomputed methods for common activities, so that information need only be provided about "top-level" activities, such as going to a doctor's appointment. Lower level activities, such as arranging for transportation, will then automatically be inserted in the plan

2.2   Client Modeler

The second major component of Autominder is the Client Modeler (CM). As the Autominder client goes about her day, sensor information is sent to the CM. The CM is then responsible for two tasks: (i) inferring what planned activities the client has performed, given sensor data; and (ii) learning a model of the client's expected behavior. These tasks are synergistic, in that the client model developed is used in the inference task, while the results of the inference are used to update the model.

The client's expected behavior is represented with a new reasoning formulism called a Quantitative Temporal Dynamic Bayes Net (QTDBN). Essentially, a QTDBN combines a standard Bayes net which reasons about all temporal aspects of the client's activities, and a dynamic Bayes net (DBN) which reasons about the activities currently being executed. Together, they represent an entire day of activities. Nodes in each time slice of the DBN are random variables representing all of the following:

1. the incoming sensor data (e.g., client has moved to kitchen);

2. the actual execution of planned activities (e.g., client has started breakfast); and

3. whether a reminder for each activity has already been issued.

Initially, the model is derived from the client plan, by making two assumptions: first, that all activities in the plan will, with high probability, be executed by the client without reminders within the time range specified in the plan, and second, that the actual time of an activity can be described by a uniform probability density function over the range associated with that activity.

The CM uses sensor data and the current time to update the model. Each time sensor data arrives, the CM performs Bayesian update. If an activity execution node's probability rises above a threshold, the activity is believed to have occurred, and the CM notifies the rest of the system.

Over time, the CM should revise its model of the client's expected behavior. As suggested above, it might learn that the client usually remembers on her own to take medicine in the morning, but forgets in the evening–or it might learn that if the client eats breakfast early, she usually eats lunch early in the allowable lunch period also. By default, the CM creates its model based solely on the client plan created by the PM. For example, if the plan states that lunch must be eaten 3-4 hours after breakfast, the CM will encode that information in the probability table of the EatLunch action. Over time, the CM may learn that this relation does not hold when the client eats breakfast before 7am. The CM can then adjust the probability table to encode both the original rule and the learned exception.

2.3   Personalized Cognitive Orthotic

We have described how Autominder stores and updates the client's plan, tracks its execution, and learns the client's typical behavior patterns. We now describe the Personalized Cognitive Orthotic (PCO), the system component that decides what reminders to issue and when. The PCO identifies those activities that may require reminders based on their importance and their likelihood of being executed on time as modeled in the CM. It also determines the most effective times to issue each required reminder, taking account of the expected client behavior, and any preferences explicitly provided by the client and the caregiver.Finally, the PCO provides justifications as to why particular activities warrant a reminder.

The PCO treats the generation of a reminder plan as a satisficing problem. It is relatively easy to create a reminder plan that is minimally acceptable: it simply involves issuing a reminder at the earliest start time of every activity. However, such a plan is likely to do a poor job of satisfying the caregiver and client, and it does not attend at all to the objective of avoiding overreliance on the part of the client. Producing a higher-quality reminder plan is more difficult: not only does such a plan need to take account of whether a reminder is really necessary, but it must also take account the client's expected behavior, her preferences, and interactions amongst planned activities. The PCO handles this problem by adopting a local-search approach called Planning-by-Rewriting (PbR) [3, 2]. It begins by creating the initial reminder plan as just suggested (reminders at the earliest possible time), and then performs local search, using a set of plan-rewrite rules to generate alternative candidate reminding plans. For example, the system contains a rule that deletes reminders for activities that have low importance and that are seldom forgotten by the client. Another rule spaces out reminders for activities for the same type of action: for instance, instead of issuing eight reminders in a row to drink water, the PCO will attempt to spread these reminders out through the day. Note that if the resulting reminders would violate any constraints in the client plan, then it will not be considered further. Rules may also be domain dependent, encoding specific preferences of the client or the caregiver, e.g., finish drinking all water by 5pm if possible.

The PCO eliminates any plan that does not contain reminders for all activities that are mandatory for the safety and well-being of the client, such as doctor's appointments and dietary requirements. Beyond that, the ascribed quality of a reminder schedule will be increased if the reminder times take account of the expected and preferred times of execution; if the schedule includes a single reminder for two or more activities that may overlap temporally and that share preconditions; if potential conflicts among activities have been identified and avoided; if reminders are generally separated in time rather than clustered into a short time period; and if reminders are not included for activities that have already been initiated.

The PCO is also designed to enable the generation of justifications for reminders. Justifications are motivated by the hypothesis that client adherence to plans may be improved when the reasoning behind the existence and timing of a reminder is provided. For example, a reminder of the form "If you take your medicine now, you will not have to do it in the middle of your show," may be more compelling than the simple message "Time for medicine." In generating a justification for a reminder, PCO can make use of the underlying client plan, the preferences of the caregiver and the client, and the particular rewrite rules used in creating the current reminder plan.

3   Autominder on the Robot Platform

A reasonable question to ask is whether a mobile robot is an appropriate platform for a cognitive orthotic; competing alternatives range from hand-held devices, to traditional desktop or laptop computers, to "intelligent houses" with multiple sensors [9]. We see several potential advantages to the use of mobile robots. Handheld devices and desktop/laptop computers have impoverished sensing capabilities and little to no reminding capabilities; moreover, handheld devices may be inappropriate for the targeted class of users, who may have a tendency to misplace them. While intelligent houses can perform sophisticated sensing, they are expensive to build, and elderly people may not want to move from the homes in which they already live. Retrofitting an existing house may also be quite expensive, and once the client moves out, the sensors may no longer be useful. In contrast, an intelligent robotic assistant can "move" to the home of a new client once a previous client is done with it. Additionally, there may be independent reasons to furnish an elderly person with a mobile-robot assistant, for instance if the robot can stimulate social interaction and/or can provide physical assistance (e.g., help in getting out of chair). In that event, it would be cost-effective to piggyback a cognitive orthotic onto the mobile robot. It is worth noting, however, that the Autominder architecture could be readily used with other sorts of platforms.

4 Related Research

The literature on cognitive orthotics is relatively new, the first survey of the cognitive prosthetic field was done by [4]. Cognitive prosthetics and/or orthotics deal with a large number of varying physiological deficiencies, traumatic brain injury, stroke, neurological disease, Alzheimers, etc. Early approaches to organizing activities and providing clues were developed by Kirsch & Levine[10] and Henry & Friedman et al. [7]. The PEAT system[11] however, is the most similar system to Autominder that we are aware of, and the first to use AI techniques. PEAT is a commercial system delivered on a handheld device, which, like Autominder, is designed to provide its user with reminders about her daily activities. PEAT maintains and dynamically updates a calendar of its client's activities. Autominder differs from PEAT in a number of ways: Autominder handles client plans with complex temporal constraints, it attempts to infer its client's actions, it learns the client's typical behavior patterns, and it reasons about the quality of alternative reminder plans. The large literature on workflow systems (e.g., [6]) is also relevant to Autominder, since workflow systems are designed to guarantee that structured tasks are performed by humans in a timely manner. Discussion of some efforts to integrate AI planning technology with workflow tasks is given in [16].

5 Conclusions

We have described the architecture of Autominder, an agent that provides plan-management assistance to an elderly client. We have shown how we combine a range of AI technologies to provide cognitive orthotic capabilities, and we have argued that incorporation of real-time client data is integral to the effectiveness, autonomy, and user-friendliness of the system. In addition, we have suggested some reasons for using mobile robots as a platform for Autominder.

Prototype versions of Autominder have been implemented, integrated onto a mobile robot (Pearl), and field-tested with elderly people on an abbreviated set of activities. In the current version of the system, the CM does not yet learn client behavior over time, and the PCO does not yet handle preferences. All other mechanisms described above have been implements. In the near future we will be conducting two types of evaluations. First, we will perform more extensive field tests to determine whether the set of activities we current model are adequate for actual monitoring of elderly clients. Second, we will conduct systematic experiments in which we simulate many different executions for given client plans, and generate alternative reminder plans by varying the heuristic evaluation functions. These reminder plans will then be assessed by professional healthcare workers.

References

[1] E. Giunchiglia A. Armando, C. Castellini. Sat-based procedures for temporal reasoning. In *5th European Conference on Planning*, 1999.

[2] José Luis Ambite. *Planning by Rewriting*. PhD thesis, University of Southern California, Los Angeles, CA, 1998.

[3] José Luis Ambite and Craig Knoblock. Planning by rewriting: Efficiently generating high- quality plans. In *Proceedings of the Fourteenth National Conference on Artificial Intelligence*, Providence, RI, 1997.

[4] Elliot Cole. Cognitive prosthetics: an overview to a method of treatment. *NeuroRehabilitation*, 12:39–51, 1999.

[5] R. Dechter, I. Meiri, and J. Pearl. Temporal constraint networks. *Artificial Intelligence*, 49:61–95, 1991.

[6] Dimitrios Georgakopoulos, Mark Hornick, and Amit Sheth. An overview of workflow management: From process modeling to workflow autonomation infrastructure. *Distributed and Parallel Databases*, 3:119–153, 1995.

[7] Kimberly Henry, Mark Friedman, Shirley Szekeres, and Debra Stemmler. Clinical evaluation of prototype portable electronic memory aid. In *Proceedings of the RESNA 12th Annual Conference*, pages 254–255, June 1989.

[8] John F. Horty and Martha E. Pollack. Evaluating new options in the context of existing plans. *Artificial Intelligence*, 127(2):199–220, 2001.

[9] C. D. Kidd, R. J. Orr, G. D. Abowd, C. G. Atkeson, I. A. Essa, B. MacIntyre, E. Mynatt, T. E. Starner, and W. Newstetter. The aware home: A living laboratory for ubiquitous computing research. In *Proceedings of the Second International Workshop on Cooperative Buildings*, 1999.

[10] Ned Kirsch, Simon P. Levine, Maureen Fallon-Krueger, and Lincoln A Jaros. The microcomputer as an 'orthotic' device for patients with cognitive deficits. *Journal of Head Trauma Rehabilitation*, 2(4):77–86, 1987.

[11] Robert Levinson. Peat – the planning and execution assistant and trainer. *Journal of Head Trauma Rehabilitation*, 1997.

[12] Nursebot: Robotic assistants for the elderly. Avail at http://www.cs.cmu.edu/~nursebot.

[13] Angelo Oddi and Amedeo Cesta. Incremental forward checking for the disjunctive temporal problem. In *European Conference on Artificial Intelligence*, 2000.

[14] National Institute on Aging and United States Bureau of the Census. Aging in the united state: Past, present, and future. Avail at http://www.census.gov/ipc/prod/97agewc.pdf.

[15] Martha E. Pollack and John F. Horty. There's more to life than making plans: Plan management in dynamic environments. *AI Magazine*, 20(4):71–84, 1999.

[16] Martha E. Pollack and Ioannis Tsamardinos, and John F. Horty. Adjustable Autonomy for a Plan Management Agent. *1999 AAAI Spring Symposium on Adjustable Autonomy*, Stanford, CA, March, 1999.

[17] A. M. Rivlin. Caring for the disabled elderly: Who will pay?, 1988.

[18] Kostas Stergiou and Manolis Koubarakis. Backtracking algorithms for disjunctions of temporal constraints. In *15th National Conference on Artificial Intelligence*, 1998.

[19] Ioannis Tsamardinos. *Constraint-Based Temporal Reasoning Algorithms, with Applications to Planning*. PhD thesis, University of Pittsburgh, Pittsburgh, PA, 2001.

[20] Ioannis Tsamardinos, Martha E. Pollack, and John F. Horty. Merging plans with quantitative temporal constraints, temporally extended actions, and conditional branches. In *Proceedings of the 5th International Conference on Artificial Intelligence Planning and Scheduling*, 2000.

[21] Qiang Yang. *Intelligent Planning: A Decomposition and Abstraction Based Approach*. Springer, New York, 1997.

Intelligent Autonomous Systems 7
M. Gini et al. (Eds.)
IOS Press, 2002

Online Execution Control Checking for Autonomous Systems

Frédéric Py,Félix Ingrand
LAAS/CNRS,
7 Avenue du Colonel Roche, F-31077 Toulouse Cedex 04, France
{felix,fpy} @laas.fr

Abstract. This paper presents some recent developments of the LAAS architecture for autonomous systems. In particular, we clarify and specify the role of the Execution Control level of our architecture. This level has a fault protection role with respect to the commands issued by the decisional level, which are transmitted to the real system (through the functional level). To implement this Execution Control level, we propose an approach and a tool inspired from the model checking domain. We present a new language, used to specify the model of acceptable and required states of the system (valid contexts for requests to functional modules and resources usage). The model written in this language is then compiled in an OBDD (Ordered Binary Decision Diagram) like structure which is used online to check in real-time the constraints and the rules specified. Such model checking approach, used in a synchronous context, provides critical dependable properties. Moreover, this approach can be further used to check off line more complex temporal properties of the system.

1 Introduction

There is an increasing need for advanced autonomy in complex embedded real-time systems such as robots, satellites, or UAVs. The growing complexity of the decision capabilities of these systems raises a major problem: how to prove that the system is not going to engage in dangerous states? How to guarantee that the robot will not grab a sample with its arm, while moving (which could supposedly break the arm)? How to make sure that satellite RCS jets are not fired when the camera lens protection is off? etc. A partial response to this problem is to use a planner which will only synthesizes valid and safe plans. Yet, high level planners do not (cannot) have a complete model representing the full extend of their actions. Moreover, some of these actions are refined by the supervisor/executive, therefore the particular sequence of commands sent to the physical system is not completely controlled by the planner.

A solution to guarantee this fault protection property is to integrate a system that formally controls the validity of the commands sent to the physical system and prevents it for entering into an inconsistent state. This controller must check system consistency online during system execution without affecting the system basic functionalities, such as computation time. This checking is made in a synchronous hypothesis execution context.

The LAAS[1] architecture, presented in section 2, foresaw such mechanism in its execution control level, but for various reasons, the approach and tools proposed to fill this functionality were not used. Section 3 presents the Execution Control Level roles and requirements, with a state of the art of related works. Section 4 gives an informal description of the proposed approach, the tool and the language we use for the Execution Control Level, while section 5 presents the link between our proposed approach and the Ordered Binary Decision Diagram model. We then conclude the paper and consider future works and research directions.

2 The LAAS Architecture

The LAAS architecture [1] was originally designed for autonomous mobile robots. This architecture remains fairly general and is supported by a consistently integrated set of tools and methodology, in order to properly design, easily integrate, test and validate a complex autonomous system.

As shown on figure 1, it has three hierarchical levels, having different temporal constraints and manipulating different data representations. From the top to the bottom, the levels are:

---

[1]LAAS Architecture for Autonomous System.

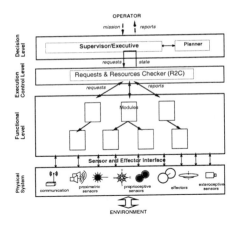

Figure 1: The LAAS Architecture.

- *A decision level:* This higher level includes the deliberative capabilities such as, but not limited to: producing task plans, recognizing situations, faults detections, etc. It embeds at least a supervisor/executive [10], which is connected to the underlying level, to which it sends requests that will ultimately initiate actions and start treatments. It is responsible for supervising plans or procedures execution while being at the same time reactive to events from the underlying level and commands from the operator. Then according to particular application it may integrate other more complex deliberation capabilities, which are called by the supervisor/executive when necessary. The temporal properties of the supervisor are such that one guarantees the reaction time of the supervisor (i.e. the time elapsed before it sees an event), but not much can be said for other decisional components.

- *An execution control level:* Just below the decisional level, the Requests and Resources Checker (R2C) checks the requests sent from above to the functional level, as well as the resources usage. It is synchronous with the underlying functional modules, in the sense that it sees all the requests sent to them, and all the reports coming back from them. It acts as a filter which allows or disallows requests to pass, according to the current state of the system (which is built online from the past requests and past replies) and according to a formal model of allowed and forbidden states of the functional system. The temporal requirements of this level are hard real-time. This is the level on which this paper focuses.

- *A functional level:* It includes all the basic built-in robot action and perception capabilities. These processing functions and control loops (image processing, motion control, ...) are encapsulated into controllable communicating modules [7]. Each module provides a number of services and treatments available through requests sent to it. Upon completion or abnormal termination, reports (with status) are sent back to the requester. Note that modules are fully controlled from the decisional level through the R2C. Modules also maintain so called "posters"; data produced by the modules, such as the current position and speed (from the locomotion module) or current trajectory (from the motion planning module) which can be seen by other modules and the levels above. The temporal requirements of the modules depend on the type of treatments they do. Modules running servo loop (which have to be ran at precise rate and interval without any lag) will have a higher temporal requirement than a motion planner, or a localization algorithm.

This architecture naturally relies on several representations, programming paradigms and processing approaches meeting the precise requirements specified for each level. We developed proper tools to meet these specifications and to implement each level of the architecture: IxTeT a temporal planner, Propice a procedural system for tasks refinement and supervision/executive, and $G^{en}{}_oM$ for the specification and integration of modules at that level. These various tools share the same namespace (i.e. the name of the modules, requests, arguments and posters).

This paper focuses on the Execution Control Level. Until recently, this level was implemented using the KHEOPS system, but for various reasons (language, complexity, etc) we moved to a newer approach/tool: the Requests and Resources Checker (R2C) and the tools used to implement it.

3   Execution Control Level

3.1   Role and Requirements

The main role of the Execution Control Level and its main component, the R2C, is a fault protection role. Faults are inevitable, even more with complex decisional system partially based on non formal methods and tools. Yet, to be able to use such advanced decisional tools, one needs to design systems which in the worse cases prevent the system from engaging in disastrous situations. Thus, the execution control level has a "simple", yet critical, role in the architecture:

- As the interface between the decisional and the functional level, it ensures that all the requests passed to the functional level remain consistent with respect to a model of desirable or undesirable states of the system, i.e. interactions between the functional modules. For example, it is the R2C role to make sure that a request to move the robot is not issued while a picture is being taken.

- It manages the resources of the system and guarantees that any request leading to an overconsumption or inconsistent use of resources is properly handled.

- It acts synchronously with the functional level to ensure a consistent view of the state of functional modules. In one cycle, all inputs are parsed simultaneously, all outputs are produced "instantaneously" and simultaneously. This is of course an hypothesis, but it provides strong determinism to the whole checking process.

- It acts in guaranteed real-time. No request to the functional level should be delayed more than one R2C cycle before being processed.

This critical role requires the use of formal tools to validate it. Moreover for these tools to be used by the engineers developing complex autonomous systems, one needs to provide a user-friendly specification language.

3.2   State of the Art in Execution Control

Many of the concerns raised in the previous section are not new, and some robotics architectures address them in one way or another.

Indeed, some of the requirements presented above were clearly fulfilled by a previous version of the LAAS execution control layer based on KHEOPS [4]. KHEOPS is a tool for checking a set of propositional rules in real-time. A KHEOPS program is thus a set of production rules (*condition(s)* → *action(s)*), from which a decision tree is built. The main advantage of such a representation is the guaranty of a maximum evaluation time (corresponding to the decision DAG depth). However, the KHEOPS language is not adapted for resources checking and appears to be quite cumbersome to use.

Another interesting approach to prove various formal properties of robotics system is the ORCCAD system[6]. This development environment , based on the ESTEREL [2] language provides some extensions to specify robots "tasks" and "procedures". However, this approach does not address architecture with advanced decisional level such as planners.

In [12] the author presents another work related to synchronous language which has some similarities with the work presented here. The objective is also to develop an execution control system with formal checking tools and a user-friendly language. This system represents requests at some abstraction level (no direct representation of arguments nor returned values). This development environment gives the possibility to validate the resulting automata via model-checking techniques (with SIGALI, a SIGNAL extension).

In [9], the authors present the CIRCA SSP planner for hard real-time controllers. This planner synthesizes off-line controllers from a domain description (preconditions, postconditions and deadlines of tasks). It can then deduce the corresponding timed automata to control on-line the system with respect to these constraints. This automata can be formally validated with model checking techniques.

In [13] the authors present a system which allow the translation from MPL (Model-based Processing Language) and TDL (Task Description Language) to SMV a symbolic model checker language. Compare to our approach, this system seems to be more designed for the high level specification of the decisional level, while our approach focuses on the online checking of the outcomes of the decisional level.

4   R2C and the Ex^{O}Gen Tool

In this section we give a description of the R2C, the main component of the LAAS Execution Control Level. The internal model of the R2C is built using the Ex^{O}Gen tool which largely uses the G$^{en}_o$M semi formal descriptions of the underlying functional modules [7] and its namespace.

4.1   Overview

The R2C (see figure 2) is designed to support safe execution of the system. It contains a database representing the current state of the functional level (i.e. running instances of requests, resources levels, and history of requests) and – according to these information and the model checker – it calculates appropriate actions to keep the system safe.

The possible R2C actions are: to launch a request; to kill an existing request; to reject a request (and report it) and to report a request completion.

4.2   Presentation of Ex^{O}Gen

This section presents the Ex^{O}Gen system and its language used to build the main components of the R2C.

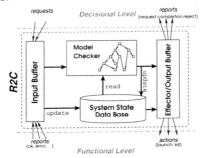

Figure 2: R2C general view.

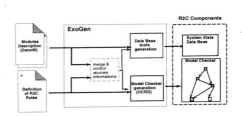

Figure 3: Ex^{O}Gen development cycle.

An Ex^{O}Gen program consists of a set of requests and resources usage descriptions. For each request of the functional level – as defined by G$^{en}_o$M – one may define the preconditions it has to verify to be executed. Those preconditions can be defined on the arguments values of the requests themselves, past requests (i.e. running requests) and states of the current system (which results from previously completed requests). Moreover , one has to specify the resources used by a particular request call when applicable.

The Ex^{O}Gen language has been specifically designed to easily represent those descriptions. For a particular application, the Ex^{O}Gen program will contain hundreds of such preconditions (which need to be verified or maintained), as well as their resources usage.

We shall now describe the Ex^{O}Gen language features:

4.2.1   Request Launching Context

Contexts are used to describe states that are either required or forbidden to launch a request. Thus, we have contexts to *prevent* request execution (**fail:**) and contexts *required* for request execution (**check:**). Moreover, these contexts can be checked *before* launching (**precond:**), and *while* the request is running (**maintain:**)

The contexts are conjunctions of predicates. We have three predicates:

Active(request(?arg)[with *cstr$^+$* **])** is true when an instance of *request* satisfying constraint(*cstr*) is currently running.

Last_Done(request(?arg):?ret[with *cstr$^+$*]) is true when the last correctly terminated instance of *request* satisfies *cstr*.

Resource tests example : *BatLevel* < 10.

The constraints can be of the following types: range of a variable, comparison of a variable with a constant value, equality of a variable with a constant. They can be defined over the arguments and the results of the requests.

Example: Consider a robot with a camera with two modes: FIXED (the robot cannot move while taking a picture) and TRAVELING (the robot must move while taking a picture). Ex⁰GEN code for TakeImg request should be[2]:

```
request Cam_TakeImg() { [...]
  fail { precond:
  Last_Done(Cam_SetMode(?m) with ?m==FIXED) && Active(Mov_Move(?_)); }
  check { maintain:
  Last_Done(Cam_SetMode(?m) with ?m==TRAVELING) && Active(Mov_Move(?_)); } }
```

4.2.2 Resources

We distinguish two types of resource:

sharable: The resource (such as battery power) is borrowed during request execution and released at the end of execution.

depletable: The resource (such as battery load) is consumed/produced by request execution[3].

To describe a resource, we have to declare its type, its name and indicate its location in the functional module description (most likely in a G^{en}$_o$M poster). Example:

```
depletable batLevel: auto(Battery_State.level);
sharable batPower: auto(Battery_State.power);
```

Resources usage is declared with: **use**(*value*) and **produce**(*value*). For instance to describe the battery usage of the Camera_takeImage request we write:

```
request Cam_TakeImg() {
  uses {
    batLevel: use(10);
    batPower: use(5);}
  [...]}
```

4.3 Internal Model Structures

To allow the real-time checking of the various constraints specified by the user, the Ex⁰GEN compiler produces a structure which is a mix of those produced by KHEOPS [8] and OBDD[4] [11]. This structures corresponds to a binary decision Directed Acyclic Graph (DAG) and guarantees a maximum execution time for checking. For example, given the request below :

```
request Cam_TakeImg() {[...]
  fail {
  preconds:
    Last_Done(Cam_SetMode(?m) with ?m==FIXED) && Active(Mov_Move(?_)); }
  [...]}
```

The Ex⁰GEN compiler will translate the $Active(x)$ predicate as: $\left(askFor(x) \wedge \neg reject(x)\right) \vee Running(x)$
The previous code is thus equivalent to the boolean formula:

$$\left(\left(\texttt{askFor(Mov\_Move())} \wedge \neg\texttt{reject(Mov\_Move())}\right) \vee \texttt{Running(Mov\_Move())}\right)$$
$$\wedge \quad \texttt{LastDone(Cam\_SetMode(}?m\texttt{)} \; with \; ?m = \texttt{FIXED)}$$
$$\Rightarrow \quad \texttt{reject(Cam\_TakeImg())}$$

The Ex⁰GEN compiler generates the OBDD shown on figure 4

In this DAG, we distinguish two types of predicates. The uncontrollable ones – LastDone, askFor, Running and resources predicates – corresponding to external events (external demands and current system state) and the controllable ones – kill, reject – which correspond to actions the R2C can perform. The principle of the R2C is to keep the overall formula true. To assess this goal the compiler

---

[2] the Cam_TakeImg and the Cam_SetMode correspond to both requests (TakeImg and SetMode) of the Cam module defined with G^{en}$_o$M.

[3] Our model remains pessimistic, as the consumption is done at the launch of the request and the production at the completion of the request.

[4] OBDD : Ordered Binary Decision Diagram.

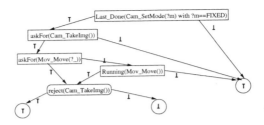

Figure 4: Example of rule DAG.

has fixed the value of controllable predicates value to true. For instance in figure 4, if the last correctly terminated instance of `Cam_SetMode` was launched with the `FIXED` argument value and the supervisor launches new instance of `Cam_TakeImg` and `Mov_Move`, then R2C will do `reject(Cam_TakeImg)` to reach the true leaf.

This DAG only deduces negative actions (i.e. `kill` and `reject`). The corresponding positive actions (respectively `launch` and `keep`) are deduced by the non-existence of the negative action. For example, if system request `rq` and R2C has not deduced `reject(rq)` then it may execute the action `launch(rq)`.

The deduction is limited to the DAG traversal so the program complexity order is proportional to its depth which is fixed at compilation.

5   Formalization of the Executive

In this section we describe how Ex°Gen constructs the structure described in section 4.3. This data structure is based upon OBDD principles with an extension to support constrained predicates keeping all well known properties of OBDD . We call it OCRD for Ordered Constrained Rule Diagram.

5.1   Definition of the OCRD

The OCRD is a binary DAG. Leaves may be true ($\top$) or false($\bot$). Each node is composed of a predicate and its attached constraint, a branch to activate when the predicate and constraint are true and a branch to activate when they are false. An OCRD node is represented as OBDD. Figure 5 presents the basic representation of a predicate. The basic constructors are :

$$\top : \qquad\qquad\qquad\qquad\qquad\qquad \to OCRD$$
$$\bot : \qquad\qquad\qquad\qquad\qquad\qquad \to OCRD$$
$$(\_, \_, \_) : \quad CPred \times OCRD \times OCRD \; \to OCRD$$

Figure 5: Representation of basics predicates.

The $CPred$ type is the association of a predicate and its constraint. Its constructor is:

$$\_ \,\|\, \_ : Predicate \times Constraint \to CPred$$

The reduction rule while building complete OCRD for a particular system is the same as OBDD : the DAG $(foo, whentrue, whenfalse)$ becomes $whentrue$ iff $whentrue = whenfalse$. The negation rule of an OCRD works like for OBDD (i.e. only the leafs are replaced by their negations).

The construction rules differ for binary operators. We now describe how we calculate the conjunction of two OCRD. The rules when one of the tree is a leaf($\top$ or $\bot$) are trivial. The conjunction of two OCRD with the same root node is: $(a, t_1, f_1) \wedge (a, t_2, f_2) = (a, t_1 \wedge t_2, f_1 \wedge f_2)$

The preceding rules are exactly the same that OBDD, the difference appears when two nodes differ. For definition of this operation we have to define some operators for the $CPred$ type :

$$\_ - \_ : \quad CPred \times CPred \; \to CPred$$
$$\_ \cap \_ : \quad CPred \times CPred \; \to CPred$$
$$\_ \prec \_ : \quad CPred \times CPred \; \to Boolean$$

$$a \,\|\, c_a \cap b \,\|\, c_b \; = \; \begin{cases} a \,\|\, (c_a \wedge c_b) & \text{if } a = b \\ \emptyset & \text{else} \end{cases}$$

$$a \,\|\, c_a \prec b \,\|\, c_b \; = \; \begin{cases} \top & \text{if } a < b \\ \bot & \text{if } a > b \\ c_a \prec_{cstr} c_b & \text{if } a = b \end{cases}$$

Where :

$$a \,\|\, c_a - b \,\|\, c_b \; = \; \begin{cases} a \,\|\, (c_a \wedge \overline{c_b}) & \text{if } a = b \\ a \,\|\, c_a & \text{else} \end{cases}$$

A complete order ($<_p$) is given on predicates. The operator $<_{cstr}$ is an order for constraints. This order is partial because $c_a <_{cstr} c_b$ is not defined for the case that $\exists x / c_a(x) \wedge c_b(x)$

Now we can define the OCRD construction rules for conjunction of $A = (a, t_a, f_a)$ and $B = (b, t_b, f_b)$

Where :

$$A \wedge B = \begin{cases} (a, t_a \wedge B, f_a \wedge B) & \text{if } a \prec b \\ (b, t_b \wedge A, f_b \wedge A) & \text{if } b \prec a \\ A \curlywedge B & \text{else} \end{cases}$$

$$\begin{aligned} (a, t_a, f_a) \curlywedge (b, t_b, f_b) &= (a \cap b, t_a \wedge t_b, f_a \wedge f_b) \\ &\wedge \quad (a - b, t_a \wedge f_b, f_a \wedge f_b) \\ &\wedge \quad (b - a, f_a \wedge t_b, f_a \wedge f_b) \end{aligned}$$

Our goal in using OCRD as an extension of OBDD, is to keep the OBDD computational and logical properties. So this data structure may be easily used for verification of temporal properties with an appropriate model checker. The complexity of operations (like "restrict", "satisfy one",... see [3]) is exactly the same as for OBDD.

It is well known that the choice of variable ordering largely influences the OBDD size (expressed by the number of its nodes). Even if OBDD minimization is an NP-hard problem, we can find some minimization techniques [5] with a lower complexity (the best is $O(n^2.3^n)$). Furthermore we have to note that this optimization will be done during compilation time. So the duration of the minimization sequence is not critical and will give some guarantees about the generated OBDD size.

6 Experimental results

We have implemented the Ex^{O}Gen tool and the resulting R2C is connected to the G^{en}$_o$M modules and the supervisor is implemented with Propice. Preliminary tests have been conducted with "toy" problems, and we are now working on an implementation on our XR4000 Nomadics: Diligent(figure 6 and 7). Our experimentation goal is twofold:

- to test the R2C it on real experiments on board our robots and see how well it behaves,
- to test it on very large problems (i.e. with numerous modules, requests and constraints to check) to see how well this approach scales (size of the OCRD, traversal time compatible with hard real time, etc).

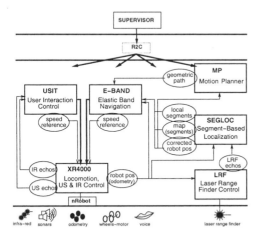
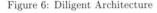

Figure 6: Diligent Architecture

Figure 7: Diligent

7 Conclusion and Future Works

We have briefly presented the LAAS architecture, its components and its integrated tools. Then the presentation focuses on the Execution Control Level of this architecture and its main component the Requests and Resources Checker (R2C). This layer of the LAAS architecture has a critical role with respect to dependability, and in particular with respect to faults protection. It must guarantee that the functional modules which "act" on the real physical system are properly controlled and do not engage in "dangerous" situations in response to "commands" coming from the decisional level.

Thus, the proposed R2C and its associated tool Ex⁰Gen offer a language in which the user can specify, using the GᵉⁿₒM namespace, the contexts in which a particular request (with constraints on its arguments) can or cannot be executed.

Our approach uses OCRD which are quite similar to OBDD. Thus the generated model used in a synchronous hypothesis context provides excellent confidence in the system states reachability. Moreover, it offers some real-time properties such as the guaranty of the maximum time taken to check new incoming requests and reports . Another interesting property inherited from OBDD is the reduction of input formula to a canonical form. The generated DAG is relatively compact (depending on predicates ordering method) and is unique. The counterpart of this reduction property is that the completeness checking of a declaration[5] is harder than with a complete representation.

We expect to add in a near future functionalities to check the modules behavior and indicates the supervisor whether services are executable or not.

Another objective is to experiment formal verification based on model-checking techniques applied to OCRD and execution control problems. Another considered extension is to add temporal information to the description of the R2C, i.e. to represent state transitions, to allow more complex validation functionalities (reachability of a state from a particular state, unreachable states, etc).

References

[1] R. Alami, R. Chatila, S. Fleury, M. Ghallab, and F. Ingrand. An architecture for autonomy. *International Journal of Robotics Research, Special Issue on Integrated Architectures for Robot Control and Programming*, 17(4):315–337, April 1998.

[2] F. Boussinot and R. de Simone. The ESTEREL Language. *Proceeding of the IEEE*, pages 1293–1304, September 1991.

[3] Randal E. Bryant. Graph-based algorithms for Boolean function manipulation. *IEEE Transactions on Computers*, C-35(8):677–691, August 1986.

[4] Adelardo A. D. de Medeiros, Raja Chatila, and Sara Fleury. Specification and Validation of a Control Architecture for Autonomous Mobile Robots. In *IROS*, pages 162–169. IEEE, 1996.

[5] Rolf Drechsler and Wolfgang Gunther. Using lower bounds during dynamic BDD minimization. In *Design Automation Conference*, pages 29–32, 1999.

[6] B. Espiau, K. Kapellos, and M. Jourdan. Formal verification in robotics: Why and how. In *The International Foundation for Robotics Research, editor, The Seventh International Symposium of Robotics Research*, pages 201 – 213, Munich, Germany, October 1995. Cambridge Press.

[7] S. Fleury, M. Herrb, and R. Chatila. Design of a modular architecture for autonomous robot. In *IEEE International Conference on Robotics and Automation*, Atlanta, USA, 1994.

[8] M. Ghallab and H. Philippe. A compiler for real-time knowledge-based systems. In *IEEE International Workshop on Artificial Intelligence for Industrial*, Hitachy City, Japan, May 1988.

[9] R. P. Goldman and D. J. Musliner. Using Model Checking to Plan Hard Real-Time Controllers. In *Proc. AIPS Workshop on Model-Theoretic Approaches to Planning*, April 2000.

[10] F.F. Ingrand, R. Chatila, R. Alami, and F. Robert. PRS: A High Level Supervision and Control Language for Autonomous Mobile Robots. In *IEEE International Conference on Robotics and Automation*, Mineapolis, USA, 1996.

[11] J.R. Burch, E.M. Clarke, K.L. McMillan, D.L. Dill, and L.J. Hwang. Symbolic Model Checking: 10^{20} States and Beyond. *Information and Computing*, 98(2):142–170, 1992.

[12] Eric Rutten. A framework for using discrete control synthesis in safe robotic programming and teleoperation. *IEEE International Conference Robotics & Automation*, pages 4104–4109, May 2001.

[13] R. Simmons, C. Pecheur, and G. Srinivasan. Towards automatic verification of autonomous systems. In *IEEE/RSJ International conference on Intelligent Robots & Systems*, 2000.

---

[5]That all reachable states are specified.

Intelligent Autonomous Systems 7
M. Gini et al. (Eds.)
IOS Press, 2002

Evaluation of Control Strategies for Multi-Robot Search and Retrieval

Paul E. Rybski, Amy Larson, Anne Schoolcraft, Sarah Osentoski and Maria Gini
Department of Computer Science and Engineering, University of Minnesota, U.S.A.
{rybski,larson,schoolcr,osentosk,gini}@cs.umn.edu

Abstract.
    We are interested in studying how environmental and control factors affect the performance of a homogeneous multi-robot team doing a search and retrieval task. In particular, we looked at the affects of target distribution (uniform or nonuniform), the number of robots, and search strategies (purposeful or random). In our experiments, purposeful search employs localization so that a robot can determine its position and remember locations of target sightings. During random search, a robot has no knowledge of either its own or a target's location. Regardless of search strategy, the robots perform the task without any explicit communication or knowledge of their teammates. Although their sensor suite is very limited, the robots are still able to complete their task. We analyzed the performance of a series of experiments and we present the results.

1   Introduction

Cooperating teams of robots have the potential to outperform a single robot attempting an identical task. Increasing task or environmental knowledge may also improve performance, but increased performance comes at a price. In addition to the monetary concerns of building multiple robots, the complexity of the control strategy and the processing overhead can outweigh the benefits. In this paper, we explore these trade-offs by comparing single robot to multi-robot team performance, as well as examining the benefits of increased intelligence in the form of environmental knowledge.

We propose a task of search and retrieval whereby robots locate, collect, and return targets to a home base. Robots are homogeneous and perform independently with a localized goal of target retrieval without the aid of communication. The task is a simplified version of minefield clearing where mines are localized using close-proximity sensors such as magnetometers, or of a search-and-rescue task where robots find and retrieve specific targets such as those dropped by air. For performance evaluation relative to strategy, the research questions we address in this paper are: How does the distribution of the targets in the environment affect performance? Is the ability to explicitly localize helpful in solving the task? How does the number of robots operating in the same area affect performance?

2   Related Work

Most research with multiple robots has focused on various forms of collaborative work as detailed, for instance, in [2, 6]. While collaboration may be essential, we are interested in studying tasks that can be done by a single robot, but where using multiple robots can potentially increase performance either by decreasing the time to complete the task or by increasing the reliability. Sample tasks include cleaning up trash, mapping a large area, and placing a distributed sensor network. For this type of task, cooperation usually requires communication among the robots [7, 9, 10]. Even simple communication has been shown to substantially increase the performance of robots when foraging, consuming, and grazing [3]. However, direct communication can be replaced by indirect communication via sensing or via the environment [1, 4].

We are interested in studying this problem from a rigorous experimental standpoint. We want to examine the kinds of unforeseen effects that are caused by the implementation of algorithms on real robots. Such details may be overlooked or be impractical to implement in a simulation study. While some of the observed effects may be unique to our hardware, we hope that by analyzing the data from real robots we can uncover cases where the performance deviates from the expected norm.

3 Robot Hardware

The robots are constructed out of LEGO Technic blocks. LEGOs were used because they are lightweight, easy to work with, and ideal for rapid prototyping. The chassis is a dual-treaded skid-steer design, allowing the robot to turn in place. Each robot is equipped with an articulated cargo bay that is capable of securely grasping a target. For obstacle avoidance, a set of bumpers are located just beyond the front of the robots' treads as well as on the back. The robots and the targets are shown in Figure 1.

Figure 1: The robots and targets

The targets that the robots attempt to locate transmit an omnidirectional stream of 40 KHz infrared light that is detectable at a range of 70 cm. Two infrared detectors are mounted on each side of the robot and two more are mounted on the front. A turret-mounted set of cadmium-sulfide (CdS) photoresistors is used to track visible-light landmarks. The on-board computer is the Handyboard, an MC68HC11-based microcontroller with 32K of RAM [8]. The software was developed in Interactive-C [11], a subset of C with multitasking capabilities.

4 Robot Software

Several parallel sensory-motor behavior processes, similar to the subsumption algorithm[5], are used to control the robot's behavior. Each process is responsible for handling one segment of the robot's control code by mapping sensors to actuators. When the sensor(s) monitored by a process are activated (e.g. when collision detection is activated by a depressed bumper), the process tries to control the actuators. Conflicts between processes running in parallel are resolved with assigned priorities.

Localization and navigation is achieved with three collinear lightbulbs that serve as both home bases and as landmarks. When a robot randomly searches for targets, it navigates by moving away then towards the light at random time intervals. For purposeful search, the robot uses localization to navigate towards a known target location. To establish a target location, the localization routine is invoked when a robot encounters a target while returning another to home base. Once the target is dropped off, the robot navigates back to where the other target was found by invoking the localization routine every 20-30 seconds and using it to correct its heading.

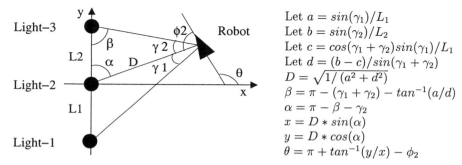

$$\text{Let } a = sin(\gamma_1)/L_1$$
$$\text{Let } b = sin(\gamma_2)/L_2$$
$$\text{Let } c = cos(\gamma_1 + \gamma_2)sin(\gamma_1)/L_1$$
$$\text{Let } d = (b - c)/sin(\gamma_1 + \gamma_2)$$
$$D = \sqrt{1/(a^2 + d^2)}$$
$$\beta = \pi - (\gamma_1 + \gamma_2) - tan^{-1}(a/d)$$
$$\alpha = \pi - \beta - \gamma_2$$
$$x = D * sin(\alpha)$$
$$y = D * cos(\alpha)$$
$$\theta = \pi + tan^{-1}(y/x) - \phi_2$$

Figure 2: Landmark localization used by the robot. The lines connecting the robot and lights represent line of sight. The origin of the coordinate system is positioned at Light-2. To avoid symmetry problems, the robot moves only in the positive X direction. The values of a, b, c & d are computed as shown on the right.

Given some assumptions about the placement of the lights in the environment, the global pose (x,y,θ) of the robot can be determined. The values of L_1 and L_2 are programmed into the robot *a priori* and are assumed never to change. The robot uses its light-tracking turret to measure the angles to the three lights with respect to its own orientation $(\phi_1, \phi_2, \text{and } \phi_3)$[1], thus $\gamma_1 = (\phi_1 - \phi_2)$ and $\gamma_2 = (\phi_2 - \phi_3)$. The angles α and β and the distance to the center landmark D are solved for and from these values, the robot's global pose (x, y, θ) can be calculated. The robot's orientation θ is measured with respect to the global x axis. Figure 2 illustrates this analytical solution. The localization method estimates the robot's position to within 25 cm and its orientation to within 5 degrees. However, if it is too close to a light, localization will fail.

5 Experimental Description

Many factors determine the effectiveness of a cooperative multi-robotic solution to a search and retrieval task. Three such factors include the physical distribution of the targets, the kinds of search strategies employed by the robots, and the number of robots used. The purpose of this work is to study how the overall performance of a robotic team is affected by altering these factors.

To solve this task, the robots started from a fixed location, searched an area for targets and returned them to one of three drop-off zones. Experiments were run with

---

[1]For the sake of clarity, only ϕ_2 is shown in the figure.

one-, two-, and four-robot configurations. The robots were not explicitly aware of each other's presence and simply treated each other as obstacles if they collided. Target locations were either distributed uniformly or nonuniformly (i.e. all placed in one far corner of the arena). Some experiments were run using localization while others were not. Without the ability to localize, a robot's search for targets was random. Figure 3 describes the experimental setup.

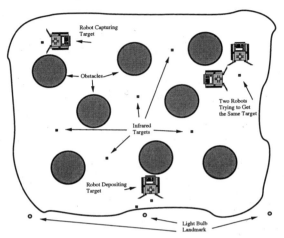

Figure 3: The experimental environment was roughly 5.4 meters on a side. All experiments contained nine targets and eight obstacles. Three lightbulbs were placed at known locations and were used to determine position and orientation. Obstacles are relatively low and do not block the robot's view of the landmarks. *Robots and targets are not shown to scale.*

6   Experimental Results

For each of the experiments, the time that a robot returned a target to a drop-off zone was recorded and averaged over five runs. Each experiment was run until all nine targets were retrieved. Table 1 shows the average time in seconds that it took to retrieve all targets. Each graph in Figure 4 shows the average time to retrieve each target when using one-, two- and four-robot teams. Results across columns differ by target distribution where left is nonuniform and right is uniform. Rows differ by use of localization. Results were obtained from experiments where no localization was used (top row), where localization was used (middle row), and where localization was used but the processing time was factored out (bottom row), making localization instantaneous.

	uniform 1 robot	uniform 2 robots	uniform 4 robots	nonuniform 1 robot	nonuniform 2 robots	nonuniform 4 robots
no localize	934	458	374	1672	1058	587
localize	1108	478	343	1911	1030	593
instant localize	986	478	323	1328	*794	*444

Table 1: Average time in seconds when the last target was retrieved. Star (*) indicates statistically significant difference at the 95% confidence level between *instant localize* and *no localize* results of the same column.

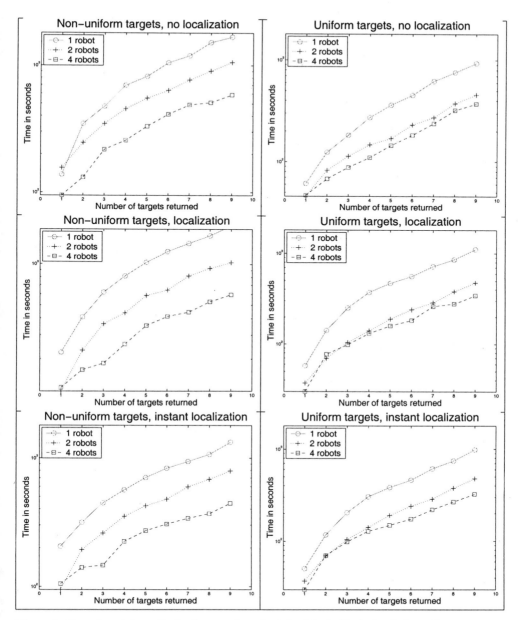

Figure 4: Experimental results for the search and retrieval task. Each chart shows the average time that it took to find each target for one, two and four robots. Experimental variables are the distribution of targets in the environment (from uniform to nonuniform), and purposeful versus random search. The time axis is shown in a base 10 logarithmic scale.

Each time a robot localizes, it must remain stationary for 18 seconds while it collects and processes the landmark data[2]. This 18-second delay had a significant effect on the overall time to complete the task, as reflected in the table. There are two reasons for factoring out localization overhead. 1) It indicates potential payoff for improvement of the localization technique, and 2) it can help determine how much overhead the system can afford while still improving task performance. In practice, instantaneous localization would be difficult to achieve but it is reasonable to assume that the 18 seconds could be significantly reduced. Varying the time it takes to localize can provide a maximum time that localization can take while still improving performance.

In looking at how localization affects performance, it can be seen from inspection of Figure 4 and Table 1 that there was no improvement when retrieving uniformly distributed targets. During the experiments, when returning a target to base, robots rarely encountered other targets, thus the ability to localize (i.e. storing the location of a found target) offered no advantage. In the experiments with nonuniformly distributed targets, there was a qualified performance improvement. Robots almost always encountered other targets when returning to base, thus a robot with localization capability could navigate directly towards the cache of targets rather than wander randomly. The computational overhead of our implementation of localization outweighed this benefit of purposeful search, but by factoring the overhead out (as described above), we can see how performance can improve.

T tests were run to determine the significance of the non-localization versus localization trials and the non-localization versus instant localization trials. Only the two- and four-robot trials with the instant localization and nonuniform target distribution were statistically significant at the 95% confidence interval (one-tailed, two-sample t test, $p = 0.0482$ and $p = 0.0291$ for the two- and four-robot cases, respectively.) All other localization results (instant or otherwise) were not statistically significant from the non-localization cases.

Additional analysis was conducted on the target search time of each robot. Table 2 illustrates the average time that it took each robot to grab a new target after returning a captured one to the base. Once again, the localization and instant localization results were compared against the no localization results for statistical significance. For this data, all three of the the instant localization with nonuniform target distributions were significant (one-tailed, two-sample t test, $p = 0.0085$, $p = 0.0032$, and $p = 0.0371$ for the one-, two- and four-robot cases.) All other localization results (instant or otherwise) were not statistically significant from the corresponding non-localization results.

	uniform 1 robot	uniform 2 robots	uniform 4 robots	nonuniform 1 robot	nonuniform 2 robots	nonuniform 4 robots
no localize	83	57	64	150	181	142
localize	96	65	79	131	143	152
instant localize	89	65	72	*88	*100	*94

Table 2: Average times in seconds for a robot to grab a new target right after a captured one has been dropped off. These values are calculated by the number of targets actually returned during the run. The differences between the starred instant localize results and the no localize results in the same column are statistically significant at the 95% confidence level.

It should be noted that the use of localization in an environment with uniformly

---

[2]This slow speed is due to the fact that Interactive-C is an interpreted language, all floating point processing on the MC68HC11 is done in emulation (there is no hardware FPU), and the Handyboard's CPU clockspeed is only 2MHz.

distributed targets actually degraded performance. This is attributed both to localization errors that would cause the robot to head off in the wrong direction, and in the multi-robot case, to navigating towards a target that was no longer present because another robot had picked it up.

To evaluate how team size affects performance, we calculated linear speed-up for each of the two- and four-robot teams. Results are shown in Table 3. We defined speed-up S_n as:

$$S_n = \frac{t_1/n}{t_n}$$

where n is the number of robots and t_n is the time it takes n robots to retrieve all targets. This is similar to a speed-up measure found in [3]. The system is said to have linear speed-up if $S_n = 1$, superlinear if $S_n > 1$, and sublinear if $S_n < 1$. As was expected, speed-up of two-robot teams was larger than the speed-up of four-robot teams.

	uniform 2 robots	uniform 4 robots	nonuniform 2 robots	nonuniform 4 robots
no localize	1.02	0.62	0.79	0.71
localize	1.16	0.81	0.92	0.81
instant localize	1.03	0.76	0.83	0.75

Table 3: Linear speed-up for two- and four-robot teams. Values of 1 indicate linear speed-up, less than 1 indicate sublinear.

7   Conclusions and Future Work

We have analyzed how the performance of a robotic team is affected by environmental factors, the number of robots, and the search strategy employed by these robots. We expected that localization would greatly assist the robots in the nonuniformly distributed environment and not so much in the uniformly-distributed environment. This turned out to not necessarily be the case due to the somewhat lengthy overhead (18 seconds) involved in localizing. We did note that if we discounted the time to localize, the robots were much faster at finding their way back to a new target once one had been dropped off. More work needs to go into the search strategies to reduce the overhead. Another hypothesis we had was that adding more robots would greatly increase the performance of the team, but continually increasing the number of robots wouldn't be as beneficial. This was shown true in our speed-up analysis. These results show that knowledge about the structure of the environment is very important when choosing a search strategy for a team of robots.

Future work will include optimizing the localization system so that it runs faster and is more accurate. This will help increase the amount of time that robots contribute to the completion of the task. Another variation to consider is a maze-like environment where the targets would be enclosed inside of small alcoves. In this case, explicit localization is expected to be extremely important. Path planning may also prove to be beneficial, if not essential, in this kind of environment. Finally, the effects of communication between the robots will be explored. Experiments will be run to analyze different kinds of communication systems and determine how much information should be shared between the robots so they can complete their task.

8 Acknowledgments

The work of Amy Larson was supported by the Doctoral Dissertation and Louise T. Dosdall Fellowships of the University of Minnesota. Sarah Osentoski, from the University of Nebraska, Lincoln, was supported by the Computing Research Association (CRA) Distributed Mentor Project. Additional partial support was provided by the Defense Advanced Research Projects Agency, Microsystems Technology Office, (Distributed Robotics) ARPA Order No. G155, Program Code No. 8H20, issued by DARPA/CMD under Contract #MDA972-98-C-0008.

We would like to thank Daniel Boley for his insights into the localization algorithm and the help that he gave us in computing a closed-form solution. We would also like to thank Joseph Djugash, Ashutosh Jaiswal, Esra Kadioglu, Elaine B. Rybski, Lorry Strother, and Harini Veeraraghavan for their help in collecting data.

References

[1] R. C. Arkin. Cooperation without communication: Multi-agent schema based robot navigation. *Journal of Robotic Systems*, 9(3):351–364, April 1992.

[2] Ronald C. Arkin and George A. Bekey, editors. *Robot Colonies*. Kluwer, 1997.

[3] T. Balch and R.C. Arkin. Communication in reactive multiagent robotic systems. *Autonomous Robots*, 1(1):27–52, 1994.

[4] R. Beckers, O. E. Holland, and J. L. Deneubourg. From local actions to global tasks: Stigmergy in collective robotics. In *Artificial Life IV*, pages 181–189. MIT Press, 1994.

[5] R. A. Brooks. A robust layered control system for a mobile robot. *IEEE Journal of Robotics and Automation*, RA-2(1):14–23, March 1986.

[6] Y. Uny Cao, Alex S. Fukunaga, and Andrew B. Kahng. Cooperative mobile robotics: antecedents and directions. *Autonomous Robots*, 4(1):7–27, 1997.

[7] E.J.P. Earon, T. D. Barfoot, and G.M.T. D'Eleuterio. Development of a multiagent robotic system with application to space exploration. In *Advanced Intelligent Mechatronics*, July 2001.

[8] Fred G. Martin. *The Handy Board Technical Reference*. MIT Media Laboratory, Cambridge, MA, 1998.

[9] Maja J Mataric. Using communication to reduce locality in distributed multi-agent learning. In *Proc. Nat'l Conf. on Artificial Intelligence*, July 1997.

[10] Lynne E. Parker. On the design of behavior-based multi-robot teams. *Journal of Advanced Robotics*, 10(6), 1996.

[11] Anne Wright, Randy Sargent, and Carl Witty. *Interactive C User's Guide*. Newton Research Labs, Cambridge, MA, 1996.

Intelligent Autonomous Systems 7
M. Gini et al. (Eds.)
IOS Press, 2002

Towards Distributed and Dynamic Task Reallocation

Wei-Min Shen and Behnam Salemi

USC Information Sciences Institute and Computer Science Department
4676 Admiralty Way, Marina del Rey, CA 90292
{shen,salemi}@isi.edu

Abstract

Distributed and dynamic task reallocation in multi-robot organization is motivated by the fact that most critical information for organizational performance must be obtained during problem solving and static techniques for task allocation cannot take the dynamic information into account. This paper presents Distributed Organizational Task Network (DOTN) as a representation for the problem, and proposes two local search techniques: trading tasks and trading responsibilities among robots, for searching the best DOTN based on global feedback. Experiments of an implemented algorithm called SOLO show a number of advantages of this approach, and illustrate some promising future research directions.

1 Introduction

In an organization of multi-robots, *task reallocation* is a process that decomposes a global mission task into subtasks and distributes the subtasks to the robots in the organization. In contrast to the problem of resource allocation, task reallocation emphasizes the task migration and organizational changes among robots, rather than allocation of resources.

Traditionally, task reallocations are often considered in a centralized and static setting. In a centralized setting, a single controller robot would gather and examine all relevant information about the current organization and mission, then decide and allocate tasks for every fellow robot. The weakness of this approach is that the accuracy of this global information is not always possible to obtain and often hard to maintain. The central controller must know the decomposition of the global task and the required capabilities and resources for each subtask. It must also know every robot's available capabilities and resources. This approach also creates a fragile bottleneck in the organization. Any failures to the controller robot will paralyze the entire organization.

Traditional approaches for task reallocation also assume that there is a known evaluation function for measuring the quality of task allocation, and this function remains unchanged during the process of problem solving. With these assumptions, a static approach [1] may solve the problem of task allocation by analyzing the evaluation function and making all necessary decisions before the problem solving starts. Similarly, researchers in coalition formation also make use of this assumption. For example, [2] assume that the values of coalitions are computable before the execution of the organization. These approaches, however, cannot take into the consideration of the dynamic aspects of the environment and unexpected changes in robot behaviors.

In real-world applications that require multi-robot collaboration, an effective solution to task reallocation must be distributed and dynamic. It must be dynamic because one cannot assume to be able to predict every outcome at the outset of a mission. A given evaluation function might be inaccurate and the capabilities and resources of robots may change. The robots must *re*-allocate the tasks during problem solving process in order to successfully carry out the given mission. A solution to task reallocation must also be distributed to eliminate the bottleneck of requiring a single robot to always obtain and maintain the correct global information for the entire organization. Without any fixed leaders, robots in an organization should negotiate among themselves to find a satisficing solution for task allocation.

Task reallocation is also closely related to the problem of self-organization, where the main objective is to decide who does what and how to collaborate with others. However, self-organization is a very diverse natural phenomenon, and a coherent and general definition is still in debate. For example, [4] defines an organization as a set of problem solvers with "information and control relationships." [3] describes an organization as a set of production systems with shared variables. [5] describes an organization as a set of "routines". [1] models an organization as a task dependent structure that includes the task units to be done, the participating (universally capable) robots, an assignment of the tasks to the robots, and a workflow structure dictates the task distribution and result assembly. Most of these definitions, although they provide valuable case studies, are not operational for task reallocation during problem solving.

This paper proposes a new approach to the problem of distributed and dynamic task reallocation based on the principles of self-organization. The approach deals with the dynamic changes in the robots and in the environment by reallocating tasks based on the performance of robots. Such reallocations are made by individual robots themselves and require no central controller robot to know the global knowledge of the organization and task progress. To focus our attention on the organizational aspects of the problem, we assume that the total communication cost in an organization is a good estimation of the quality of the task allocation and the current organizational structure. So in this paper, the organizational performance will only be measured by the costs of communication. Our approach is similar to the bottom-up approaches for self-organization [3]. However, we do not assume that robots have universal capabilities (i.e., every robot can handle every subtask), nor that the population of robots can be changed arbitrarily.

In this paper, we define a result of task reallocation as assignment of robots to a *role-graph*, and define the process of task reallocation as the optimization of these assignments, with respects to the dynamic cost function. In a role-graph, a *role* node is a set of responsibilities (or subtasks) for the given task, and a *role-relationship* edge is a commitment between roles to communicate certain types of information (such as subtasks, solutions, actions, or data). This representation separates the role requirement from the robots' capabilities, and makes the assignment of robots to roles an essential task in organization. In this paper, we assume that an initial role-graph is given to the robots, but the robots are allowed to modify the role-graph at will. As we will see later, these modifications include trading tasks and responsibilities among robots, and such modifications can affect the structure of an organization. With this representation, task reallocation is a team-learning process for adapting a role-graph and searching for an optimal robot assignment to the role-graph based on performance results during problem solving.

The rest of the paper is organized as follows. Section 2 gives a formal definition for the problem of distributed and dynamic task reallocation. Section 3 presents a solution approach based on local search with two novel heuristics for task-trading and responsibly-trading among robots. Section 4 describes the SOLO algorithm that implements the above approach. Section 5 presents the experimental results of the SOLO algorithm. Section 6 concludes the paper with future research directions.

2   Distributed and Dynamic Task Reallocation

To study the problem of distributed and dynamic task reallocation in a domain-independent fashion, it is necessary to ground the research on a rigorous computational foundation that is domain-independent and decomposable among multiple robots. Examples of such foundations include Distributed Constraint Satisfaction Problems (DCSP), Distributed Bayesian Networks, Contract Nets, and Graphical Models [6]. In this paper, we shall focus on DCSP to investigate the feasibility of the approach.

A Constraint Satisfaction Problem (CSP) is commonly defined as assigning values to a list of variables V from a respective list of domains D such that a set of constraints C over the variables is satisfied. For example, we can define an example CSP_1 as follows: $V=[x_1, x_2, x_3]$, $D=[\{1,2\}, \{2\}, \{1,2\}]$, and $C=[(x_1 \neq x_3), (x_2 \neq x_3)]$. Then a solution for CSP_1 is $(x_1,x_2,x_3)=(2, 2, 1)$. A distributed CSP is a CSP in which V, D, and C are distributed among multiple robots. A DCSP is solved if each robot solves its local portion of the CSP and the collection of all local solutions is a solution to the CSP. For instance, we can partition the above example into two parts: $V_1=[x_1,x_2]$, $D_1=[\{1,2\},\{2\}]$, $C_1=[(x_1 \neq x_3)]$ and $V_2=[x_3]$, $D_2=[\{1,2\}]$, $C_2=[(x_2 \neq x_3)]$, and assign them to two robots respectively. Then, a solution to the DCSP is $(x_1,x_2)=(2,2)$, and $(x_3)=(1)$. In the standard DCSP, however, task reallocation is not an issue because the assignment between robots and variables are given and static.

To generalize DCSP to address the problem of distributed and dynamic task reallocation, we map tasks to variables, task dependencies to constraints between variables. A task is solved if the corresponding variable is assigned a value that does not violate any involved constraints. We further introduce that (1) every task/variable has a set of required capabilities, (2) every task dependency/constraint has two responsibilities: the *supervisor* and the *subordinator*; and (3) there is a set of heterogeneous robots that collectively possess all the required capabilities. For task dependency that links two task variables, the responsibility of the supervisor is to select a value for its variable and pass the value to the subordinator. The responsibility of the subordinator is to adjust the value of its variable so that it satisfies the constraint with the supervisor's value.

Formally, the problem of distributed and dynamic task reallocation can be defined as a tuple (V, R, D, C, A), where V is a list of task/variables, R a list of required capabilities by the task/variables, D a list of value domains for the task/variables, C a set constraints, and A a set of robots with heterogeneous capabilities. The goal of this problem is to find an assignment $A \Leftrightarrow (V, C)$ that is both *complete* and *optimal*. An assignment is complete if every task and every responsibility is assigned to a *qualified* robot and no single capability is assigned to more than one task simultaneously. A robot is qualified for a task if the robot possesses the necessary capabilities required by the task. An assignment is optimal if it enables a solution to the given global problem to be found with the minimal cost. The cost of a global solution can be measured in a user-specified way. For example, it could be the total number of messages sent between robots, or the sum of computational time consumed by the participating robots. In this paper, however, we will only consider the total number of messages for communication.

To illustrate the above definitions, consider a task reallocation problem TR_1 extended from CSP_1 as follows: $V=[x_1, x_2, x_3]$, $R=[\{c_1, c_4\}, \{c_2\}, \{c_3\}]$, $D=[\{1,2\}, \{2\}, \{1,2\}]$, $C=[(x_1 \neq x_3), (x_2 \neq x_3)]$, and $A=[A_1=\{c_1, c_2, c_4\}, A_2=\{c_2, c_3, c_4\}]$. In this problem, the required capabilities for task x_1 is $\{c_1, c_4\}$, task x_2 $\{c_2\}$, task x_3 $\{c_3\}$, respectively. The robot

Table 1: All Possible Tasks Allocations for TR_1

	Task Assignment	Responsibility Assignment
O_1	$A_1:(x_1,x_2), A_2:(x_3)$	$x_1 \rightarrow x_3 \rightarrow x_2$
O_2	$A_1:(x_1,x_2), A_2:(x_3)$	$x_1 \rightarrow x_3 \leftarrow x_2$
O_3	$A_1:(x_1,x_2), A_2:(x_3)$	$x_1 \leftarrow x_3 \rightarrow x_2$
O_4	$A_1:(x_1,x_2), A_2:(x_3)$	$x_1 \leftarrow x_3 \leftarrow x_2$
O_5	$A_1:(x_1), A_2:(x_2,x_3)$	$x_1 \rightarrow x_3 \rightarrow x_2$
O_6	$A_1:(x_1), A_2:(x_2,x_3)$	$x_1 \rightarrow x_3 \leftarrow x_2$
O_7	$A_1:(x_1), A_2:(x_2,x_3)$	$x_1 \leftarrow x_3 \rightarrow x_2$
O_8	$A_1:(x_1), A_2:(x_2,x_3)$	$x_1 \leftarrow x_3 \leftarrow x_2$

A_1 has capabilities $\{c_1, c_2, c_4\}$ and is qualified for x_1 and x_2, and the robot A_2 has $\{c_2,c_3,c_4\}$ and is qualified for x_2 and x_3. Without lost of generality, we can simplify the problem by assuming that each task requires a unique capability so that the capability requirements for tasks can be embedded in the task variables. For each task that requires more than one capability, such as x_1 requires $\{c_1, c_4\}$, we create a new capability called $c_{1\&4}$ so that x_1 can be handled by $c_{1\&4}$. With this convention, TR_1 can be simplified as: $V=[x_1, x_2, x_3]$, $R=[\{c_{1\&4}\}, \{c_2\}, \{c_3\}]$, $D=[\{1,2\}, \{2\}, \{1,2\}]$, $C=[(x_1 \neq x_3), (x_2 \neq x_3)]$, and $A=[A_1=\{c_{1\&4}, c_2\}, A_2=\{c_2, c_3, c_4\}]$. Thus we can replace robots' capabilities by task variables as follows: $R=[\{x_1\},\{x_2\},\{x_3\}]$ and $A=[A_1=\{x_1, x_2\}, A_2=\{x_2, x_3\}]$.

With the above definition, we can enumerate all possible task allocations for a given problem by matching the qualification of robots with task variables and task dependency constraints. In our current example, there are eight possible task allocations listed in Table 1. For example, in the solution O_1, A_1:(x_1,x_2) indicates that the robot A_1 is assigned tasks x_1 and x_2, and A_2:(x_3) indicates that the robot A_2 is assigned x_3. The responsibility assignment $[x_1 \rightarrow x_3]$ means that for the task dependency between x_1 and x_3, the robot A_1 (who is assigned to x_1) is the supervisor while A_2 (who is assigned to x_3) is the subordinator. Similarly, for the task dependency between x_2 and x_3, $[x_3 \rightarrow x_2]$ indicates that A_2 is the supervisor and A_1 is the subordinator.

The above task allocations can be graphically represented in a Distributed Organizational Task Network (DOTN). In a DOTN, the nodes are the task variables with required capabilities, the edges are the task dependency constraints, and the direction of an edge represents the responsibilities of the involved robots in the corresponding task dependency constraint. A DOTN is complete and optimal if the robot assignment to the elements in the DOTN (nodes and edge directions) is complete and optimal. To illustrate the representation of DOTN, Figure

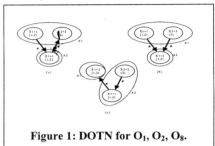

Figure 1: DOTN for O_1, O_2, O_8.

1(a), 1(b), and 1(c) shows three task allocations corresponding to O_1, O_2, and O_8, respectively. Note that O_1 and O_2 have the same task assignments $[A_1$:(x_1,x_2),A_2:$(x_3)]$, but different responsibility assignment for the dependency constrain $[x_2 \neq x_3]$. In O_1, A_1 is the subordinator and A_2 the supervisor, while in O_2, A_1 is the supervisor and A_2 the subordinator. In Figure 1(c), O_8 has a totally different task assignment: A_1 is assigned to x_1, and A_2 to (x_2, x_3).

Intuitively speaking, the solution O8 is the optimal task allocation because it is most likely for the robots to find a coherent solution for the global problem faced by the organization. This is because the tasks x_1 and x_2 are free from any dependency constraints so they are best to be assigned to different robots. Furthermore, for the only dependency constraint $[x_2 \neq x_3]$ across the two robots, x_2 has a more restricted domain than x_3, so its assignee A_2 should be given the responsibility of supervisor. In this simple illustrative example, the objective of task reallocation is to find this optimal solution O_8 during the process of solving the global DCS problem.

3   Task Reallocation by Heuristic Search

The above definition of task allocation implies that solving the problem requires a search in an enormous space for possible assignments between robots and DOTN elements (nodes and edge directions). This search can be done either exhaustively or heuristically.

The basic idea for exhaustive search is quite simple. One can find the best task allocation by going through all possible task allocations and returning the one that has the best performance. This search is complete because it guarantees to find the optimal task allocation. But the time complexity of this search makes it impractical to use. When the environment is changed, the entire search process must started over again.

To find a practical solution for distributed and dynamic task reallocation, we may use local search approaches to find an approximation of the best task allocation. Local search proceeds by incrementally improving the current task allocation based on the feedback of solving the global problem. To do so, we have to answer two questions: how to modify a task allocation, and when to apply these modifications so that they result in improvement. By definition, a task allocation can be modified by two actions: trade tasks among robots, and trade responsibilities among robots. We now discuss them in detail.

3.1   Trading Tasks among Robots

In a task allocation process, tasks with dependency constraints are partitioned into groups, and each group is then assigned to a qualified robot. In this context, we can classify a dependency constraint between two tasks as either *remote* (across robots) or *local* (within a robot). An optimal task allocation is the one that minimizes the remote dependencies between robots so that each robot can solve its own tasks in a relatively independent way.

How do we measure the dependencies between two tasks or two robots? The most straightforward way is to simply count the number of dependency relationships between tasks or robots. This is a static estimation and the best task allocation based on this measurement can be accomplished by analyzing the role-graph at the outset and partition the roles into groups to minimize the total number of dependency relationships between groups. However, this static approach does not consider the likelihood how a dependency can be satisfied. Such likelihood would depend on how easy to find solutions for the tasks involved, how wide the communication bandwidth is between robots, and many other facts. Such information is not available until the problem solving process starts.

As a first attempt for dynamic task reallocation, we estimate the dependency between two tasks x and y by the number of messages exchanged between the tasks:

$$\text{dependency}(x,y) \cong \text{the\_num\_of\_messages}(x,y).$$

When a task x is traded from robot A to robot B, some local dependencies of x may become remote, while some remote dependencies of x may become local. The purpose of such task trading is to reduce the total amount of remote dependencies among all robots.

To illustrate this point, consider the task allocation in Figure 2, where five tasks t_1, t_2, t_3, t_4, and t_5 are allocated to three robots A, B, C, and the dependencies on the links are as shown. Assume that based on the qualification of the robots, only t_2 and t_3 can migrate from A to B or C. Given these facts, the changes of remote dependencies caused by the four possible task trades can be computed as follows:

Trade t_2 from A to B: -6+3+2 = -1
Trade t_2 from A to C: 3+2 = 5
Trade t_3 from A to B: +3
Trade t_3 from A to C: -5 +3 = -2

For example, when t_2 is moved from A to B, it eliminates a remote dependency of value 6, but introduces two new remote dependencies of value 2 and 3. Among these four possibilities, the trade t3 from A to C is the most profit modification for the current task allocation.

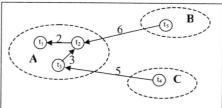

Figure 2: Trading tasks between robots.

In general, we trade a task x from a robot A to another robot B if the total remote dependencies between x and the tasks in B is higher than the total local dependencies of x in A. This trade of task will reduce the total remote dependencies in the entire task allocation. To facilitate this decision making, each robot is required to record the number of messages sent and received on each task dependency.

3.2 Trading Responsibilities among Robots

Trading responsibilities between robots is another way to modify a task allocation. The basic idea is to switch the supervisor and subordinator responsibility whenever the subordinator cannot find any possible solution to its task. The motivation of this action is that by becoming a supervisor, a robot may have more freedom to choose solutions for its task, therefore more likely to find a global solution.

In DOTN, this action switches the direction of an edge and alters the direction of information flow between roles. Such changes can affect the performance of an organization because it has been demonstrated in [8] that changes in the priority order among variables can influence the rate of problem solving in DCSP. Switching responsibilities is a special case of changing priorities.

The simple protocol of switching responsibilities between two robots, however, is too limited to be effective and can run into deadlocks. Consider a situation where both the supervisor and the subordinator cannot find solutions to their tasks, then no matter how many times they switch responsibilities, the problem can never be solved because they are constrained by other neighbors.

To overcome this problem, we extend the scope of trading responsibilities from two robots to the entire neighborhood. Inspired by the priority schema in [7], we assume that every task is assigned a global priority. If two tasks are linked by a dependency constraint, then the task with the higher priority is the supervisor. Whenever a robot fails to find a solution for its task, it will switch the priority of that task with a neighboring task that has the highest priority in the neighborhood. Different from the schema in [7], this protocol does not introduce any new priorities for the tasks yet it can prevent loop creation in DOTN.

To illustrate this protocol of trading responsibilities, assume that the tasks $[t_1,t_2,t_3,t_4,t_5]$ in Figure 2 have the priorities $[1,2,3,4,5]$. When t_2 fails to find a solution, it will switch its priority with t_5 because t_5 is the neighbor task that has the highest priority. Notice that after this switch, t_2 becomes the supervisor of all its neighbors, which includes t_6, t_3, and t_1.

4 The SOLO Algorithm

The process of task reallocation described above has been implemented as a new algorithm called SOLO, which is an extension of the Asynchronous Backtracking Algorithm [7] and the version that deals with multiple variables per agent [9]. Given a DCSP, the SOLO algorithm allows robots to reallocate the task variables assigned to them initially. The output of SOLO is a solution to the given DCSP and a near optimal task allocation among robots.

Four types of messages are used. (1) An *ok?* message is sent when a robot is proposing a new solution to its local task. When a robot Ai sends out an *ok?* message for a local task x_i, it also indicates if it is willing to give away (*giveAway?*) the task to the receiver. (2) A *nogood* message is sent when a subordinator robot finds it is impossible to find a solution for a task to satisfy all neighboring supervisors. (3) An *interested* message is sent by a qualified robot as a reply to an *ok?* message to indicate the willingness to accept the offered task. (4) A *release* message is sent by a robot to transfer a local task to an interested receiver. Among all the "interested" receivers, the offering robot will select the one that has the highest remote dependency with respect to the offered task. Since communication among robots are asynchronous, the offering robot cannot guarantee to wait for all interested parties so it is permissible to release the task to the first interested receiver.

```
when received (ok?, (Aj, xj, dj, priority, giveAway?)) do
    add (Aj, xj, dj, priority) to robotView;
    if (giveAway? & qualifedfor(xj) & profitToAccept(xj)) then send(interested, Ai, Aj, xj);
    if robotView and currentAssignments are inconsistent then checkRobotView;

when received (nogood ,xj ,nogood) do
    add nogood to the nogoodList
    if (xk, dk, priority) is contained in the nogood where xk is not in the neighbors
    then add xk to neighbors; add (Ak, xk, dk, prority) to robotView;
    checkRobotView;

when received (interested, Aj, xi) do
    if xi is in the possessedVariablesList then delete xi from possessedVariablesList; send (release, (xi, di, priority));

when received (release, (xi, di, priority)) do
    add xi to possessedVariablesList; annouce the new ownership of xi to the neighbors;

procedure checkRobotView
    if robotView and currentAssignments are consistent then
        for each xi that has a new value d do
            for each robot Ak that has a constraint with xi do
```

giveAway?=CommCost(*xi,Ak*)>localCommCost(*xi*));   send (**ok?**, (*Ai, xi, d, currentPriority(xi), giveAway?*));
else select x_k from *possessedVariablesList* that has the highest priority and violates constraint with higher priority variables;
  if no value in D_k is consistent with *robotView* and *currentAssignments*
  then record and send a nogood, the subset of *robotView* and *currentAssignments* where x_k has no consistent value;
    when the obtained nogood is new **do**
      switch the priorities between x_k and the inconsistent variable that has the highest priority in the nogood;
      $x_k = d$; where $d \in D_k$ and d minimizes the number of violations with lower priority variables;
      checkRobotView; end do;
  else $x_k = d$; where $d \in D_k$ and d is consistent with *robotView* and *currentAssignments*
    and minimizes the number of violations with lower priority variables;
    checkRobotView;

Figure 3: The SOLO Algorithm

Figure 3 illustrates the skeleton procedures for receiving messages such as *ok?*, *nogood*, *interested*, and *release*, and for checking local robot views. The algorithm starts with an initial task allocation determined randomly. Each robot assigns values to its local variables, and sends *ok?* messages to all related subordinator robots. After that, robots wait and respond to incoming messages. When an *ok?* message about a variable *x* is received, the receiver robot *A* will update its local view and send back an *interested* message if it is qualified and profit to possess *x*. When an *interested* message for a variable *x* is received, the receiver robot will relinquish the variable (by deleting *x* from its local variable list) and replies with a *release* message. The receiver of the *release* message will add the variable to it local variables list and announce the new ownership by a set of *ok?* messages.

To illustrate the SOLO algorithm in detail, let us consider our TR_1 example again. We assume that the initial task allocation is Figure 1(a). Each robot communicates these initial values via *ok?* messages. When A_1 sends an *ok?* message to A_2 for x_1's new value, it also indicates the willingness to give x_1 away because x_1 has a higher remote dependency than its local dependency. When A_2 receives this *ok?* message, it updates the *robotView* but is not interested in x_1 because the lack of qualification. After A_2 assigns a new value 2 to its local variable x_3, it sends an *ok?* message to A_1 (for A_1 is the subordinator of the constraint $x_3 \neq x_2$). This time A_1 fails to find a consistent value x_2 to satisfy the constraint of $x_3 \neq x_2$, so it performs the following actions. A_1 sends a *nogood* message $\{(x_3=2)\}$ to A_2, switches the priority value of x_2 with x_3, selects a new value 2 for x_2, sends an *ok?* message to A_2 to inform the priority switch and its willingness to give x_2 away. At this point, the task allocation becomes Figure 1(b). In this new task allocation, A_2 sends out two messages: an *interested* message to A_1 for taking x_2, and a *nogood* $\{(x_1=1),(x_2=2)\}$ message to A_1 because it fails to find a consistent value for x_3. After these messages, A_1 releases x_2 to A_2 and changes the value of x_1 to 2. At this point, all tasks are solved and the task allocation is Figure 1(c).

5 Experimental Results

We have applied the SOLO algorithm to a distributed 3-color problem. Given n variables, we first generate a random 3-color problem with $2.7n$ links (to ensure the difficulty of the problems). We then generate a set of m robots by randomly partitioning the capabilities (variables) into m even subsets. If the n/m is not an integer, then the remaining capabilities are assigned to the last robot. To make sure that robots have overlapping capabilities, we then expend each robot's capabilities by adding extra $p\%$, randomly selected different capabilities. Notice that when $p=0$, every robot has unique capabilities and there is no room for trading tasks. If $p=100$, then all robots are qualified to trade all tasks.

Table 2 lists the results of running SOLO with different number of variables (n), robots (m), and capability overlapping (p). Each data point in the table is the average for 50 randomly generated problem instances. The initial values of the variables in these trails are determined randomly. To show the effects of task trading, we have recorded the number remote and local messages, the cycles needed to solve the problem, and the number of task trading.

As we can see from the results, as the overlapping capability increases, more tasks are traded between robots, less communication is needed between robots, and the rate of converge is faster (less cycles). In general, when robots have choices for what they do, task reallocation allows them to solve the problem much more quickly than fixed task allocation. Interestingly, we notice that communication does not reduce monotonically with the capability overlapping. In the case m/n=20/8, we see an increase of communication at 30% of overlapping, before it goes down again. Further investigation is required to determine the causes of this phenomenon.

Table 2: The Effects of Task Trading			
n/m	$p=0$	$p=30$	$p=60$	$p=90$
# of remote messages			
10/4	119.0	79.3	68.1	62.0
10/8	265.1	186.7	137.4	200.8
20/8	1004.7	2394.5	1111.7	593.2
20/12	5729.7	3652.9	3038.3	1132.6
30/10	2584.0	2269.4	2490.3	2343.2
30/20	5969.4	7007.3	4236.4	348.9
50/10	2948.9	3152.1	3446.1	3310.4
100/20	6041.7	5920.6	5766.5	5718.1
n/m	# of local messages		
10/4	29.8	19.1	17.6	21.6
10/8	17.8	25.3	19.6	35.4
20/8	142.2	333.3	189.6	89.6
20/12	412.4	379.8	409.2	143.9
30/10	269.4	239.6	264.7	257.1
30/20	176.7	329.6	293.6	19.4
50/10	324.5	353.0	396.4	395.4
100/20	311.5	329.1	333.3	350.3
n/m	# of cycles for solving DCSP		
10/4	29.8	19.1	17.6	21.6
10/8	5.8	5.8	4.1	8.2
20/8	20.8	44.6	33.5	16.5
20/12	47.0	47.0	60.2	21.2
30/10	26.7	27.4	31.8	32.1
30/20	21.3	32.4	37.6	3.7
50/10	34.0	39.7	39.8	41.9
100/20	22.8	24.7	26.1	27.0
n/m	# of task trading		
10/4	0.0	0.7	1.1	0.5
10/8	0.0	0.0	2.5	2.0
20/8	0.0	0.1	2.8	3.9
20/12	0.0	3.7	5.0	4.9
30/10	0.0	1.3	3.6	5.8
30/20	0.0	7.0	8.0	6.6
50/10	0.0	7.0	8.7	7.4
100/20	0.0	7.2	9.9	10.8

6 Conclusion

This paper presents an approach to the distributed and dynamic task reallocation in multi-robot systems. Experiments with the SOLO algorithm have shown favorable performance in graph coloring problems. This work also suggests a number of future research directions in task reallocation. More factors other than the number of messages must be considered to better estimate the dependencies between tasks and robots. Applications of the approach to real-world problems that involve robots are also necessary. An even more challenging problem is to deal with the changes in the environment where solutions to tasks are non-stationary. We are grateful that this work is in part supported by the US AFOSR contract F49620-01-1-0020.

References

1. So, Y.-P., E.H. Durfee. An organizational self-design model for organizational change. in National Conference on Artificial Intelligence. 1993.

2. Shehory, O., S. Kraus. Formation of overlapping coalitions for precedence-ordered task-execution among autonomous robots. in International Conference on Multiple Agent Systems. 1996.

3. Ishida, T., L. Gasser, M. Yokoo, Organization Self-Design of Distributed Production Systems. IEEE Transactions on Knowledge and Data Engineering, 1992. **4**(2):123-134.

4. Durfee, E.H., V.R. Lesser, D.D. Corkill, Trends in Cooperative Distributed Problem Solving. IEEE Transactions on Knowledge and Data Engineering, 1989. **1**(1):63-83.

5. Levitt, B., J.G. March, Organizational Learning. Annual Review of Sociology, 1988. **14**:319-340.

6. Jordan, M., ed. Learning in Graphical Models. 1998, MIT Press.

7. Yokoo, M., E.H. Durfee, T. Ishida, K. Kuwabara, The Distributed Constraint Satisfaction Problem: Formalization and Algorithms. IEEE Transactions on Knowledge and Data Engineering, 1998. **10**(5):673-685.

8. Armstrong, A., E. Durfee. Dynamic prioritization of complex agents in distributed constraint satisfaction problems. The International Joint Conference on Artificial Intelligence. 1997.

9. Yokoo, M., K. Hirayama. Distributed Constraint Satisfaction Algorithm for Complex Local Problems. The International Conference for Multiple Agent Systems. 1998.

Intelligent Autonomous Systems 7
M. Gini et al. (Eds.)
IOS Press, 2002

Cognition and Periodic Control of Autonomous Robot Using Net-list Evolution

Tomokazu Shindo, Koji Nishikawa, Hiroshi Yokoi and Yukinori Kakazu
Autonomous System Engineering, Hokkaido University
North 13, West 8, Kita-Ku, Sapporo, 060-8628, Japan
shindo@complex.eng.hokudai.ac.jp

Abstract. In this paper, the Tripod Robot is controlled by a logic circuit generated by Net-list evolution. The Net-list chromosome is a circuit model that abstracted a logic circuit structurally. The Tripod Robot is one of the soft mechanics robots, and it has SMA actuators and it is characterized by its symmetry and reconfigurability. The generation of the adaptive behavior of the Tripod Robot is discussed through the experiment of using a simulator.

1. Introduction

In this paper, Net-list evolution [1] is used as one of the methodologies to develop adaptive logic circuits for autonomous robots. Evolvable Hardware (EHW)[2] is a methodology to develop logic circuit as hardware adaptively using reconfigurable devices called Field Programmable Gate Array (FPGA). In general approach, the internal structure of FPGA restrains functional evolution of the logic circuits as a phenotype. Therefore another method is proposed that abstracts a logic circuit functionally using Hardware Description Language (HDL) [3]. On the other hand, Net-list is a circuit model that abstracts a logic circuit structurally, so it doesn't have those limitations.

The Tripod Robot to which Net-list evolution is applied is a soft mechanics robot that employs Shape Memory Alloy (SMA) in an actuator [4]. In order for this robot to gain aimed movements, it is necessary to generate a suitable motion pattern, particularly for a rhythm. The period of the rhythm should be changed according to environment. Moreover, this robot has a symmetrical structure with a point, and the function is not distributing to it geometrically. It is difficult to control the above objective by the conventional approaches. Therefore Net-list evolution is applied for the controlling the Tripod Robot.

2. Net-list Evolution

First of all, logic circuits are abstracted structurally into Net-list chromosomes. A logic circuit can be abstracted into basic logical elements such as AND, OR and NOT. Net-list consists of n pieces of logical elements and theirs connections. The logical element called a logical cell consists of a pair of inputs, a function and an output. A k-th logical cell C_k is defined by

$$C_k = (i_{kl}, o_k, f_k) \qquad (0 \le k < n, l = 1,2) \qquad (1)$$

where i_k is a set of two inputs, o_k is an output and f_k is a function of the logical cell. And Net-list is a free-link network that doesn't have any limitations of connections between the logical cells. Therefore the inputs i_{kl} of each logical cell can be the output of any logical cell including itself as following equation.

$$i_{kl} \in \{o_i \mid 0 \le i < n\} \tag{2}$$

The kind of the logical cell's functions f is seventeen: D flip-flop and all sixteen kind of Boolean functions having two inputs and an output. D flip-flop also has two inputs, one is a clock signal and the other is a D value. The output of the logical cell is defined by the inputs and the function as following equations.

$$for\ 0 \le f \le 15, \quad o_k = \frac{f_k \cdot 2^{i_{k1} + 2i_{k2}}}{2^{i_{k1} + 2i_{k2}}} \tag{3}$$

$$for\ f = 16, \quad o_k = DFF(i_{k1}, i_{k2}) \tag{4}$$

Therefore, the chromosome coded for Net-list is represented by three integers; function, input-1 and input-2. Fig.1 shows a cording of the circuit into Net-list chromosomes as an example. As shown in this figure, the input cells (k=0,1) are used only for inputs of the circuit. Therefore, they don't have genes. And the output cells (k=18,19) are used only for outputs, so they don't have gene of function and input-2. There are some logical cells that don't work for output signal such as the cell of C_6 in this figure. The genes of these unused cells are left in the chromosomes as they are, so that the whole circuit has redundancy. It is expected that the redundancy produces the effect to the genetic operation.

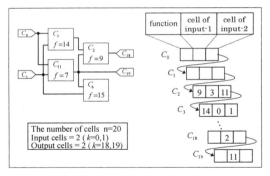

Fig.1: Coding of Net-list3. Experiment

To generate any Net-list chromosomes automatically, Evolutional Programming is applied to it. When generating the next generation, the operations of elite reproduction, selection using roulette strategy, mutation and new individual generation are carried out.

3. Experiment

Fig.2: Picture of Tripod robot and its structure

3.1. Objective

Tripod robot, shown in Fig.2, is characterized in terms of two points. One is that Shape Memory Alloy (SMA) is used for actuators. A SMA actuator consists of a SMA and a heater. If the SMA is heated, the shape of the SMA changes to straight. Because the heat is necessary as medium for SMA actuator to work, there is time delay between movements and control commands. The other is that Tripod robot has a unique structure that three legs are arranged in a point symmetric position. The controller must operate the robot that doesn't have the geometric relation of front and rear, right and left. Therefore, the rhythm of the motion pattern for controlling the Tripod Robot is complicated more than the usual animal type robot of the line symmetrical structure. In addition, the SMA actuators are arranged between one leg and the other. That is, it is complex to move only one leg. Therefore, the control of the robot using conventional methods is difficult.

3.2. Settings

Tripod simulator is set up as follows. The heater of each actuator heats SMA according to control command. The amount of heating per unit time is different from the one of cooling. If the temperature of SMA becomes higher than a threshold, the spring constant of the SMA changes from k_{low} to k_{high} and the legs attached in the both ends move. A rotation of each leg causes the movement of the center of the gravity. Movements and rotation of Tripod robot in this simulator come from a relation between a movement of robot's center of the gravity and a friction with the floor. The sensors of the Tripod robot to detect the light source are three solar panels attached in the same angle as the actuators.

The interface between the Tripod simulator and an evolved logic circuit is set up as follows. All three sensory reading are encoded into 3-bit values (1-bit value for each sensor) as input from Tripod robot to the circuit. And also the logic circuit inputs 1-bit clock pulse. The outputs from the circuit are 3-bit values for the switch of each SMA actuator of the robot.

The robot starts the central point of the 500×500 grids environment. The light source as a target is put on the point on the circumference of which the center point is initial position of the robot and the radius is 200-grids. The fitness of the generated circuit is estimated by following two steps. First, the distance d of approaching toward light source is estimated by following equation.

$$d = 200 - \sqrt{(L_x - G_x)^2 + (L_y - G_y)^2} \tag{5}$$

In this equation, (L_x, L_y) is a position of the light source and (G_x, G_y) is a final position of the robot. Then, the distances of four trials are added as fitness. The position of light source is decided randomly at first and then shifted every $\pi/2$ for every trial. The period of the clock pulse is set up shorter than period of leg's movement.
The other condition of the experiment is shown in Table.1.

Table.1: Parameters of the experiment

Parameters of Tripod Simurator		Parameters of Net-list Evolution	
Maxmum Temperature of SMA	200	The number of logical cells	100
Minimum Temperature of SMA	0	Population Size	100
Up temperature per unit time	12	Elite Population	10% of population size
Down temperature per unit time	9	New generated population	10% of population size
Threshold of temperature	100	Mutation rate	3% per logical cell
The ratio of the spring constant	$k_{high} : k_{low} = 2:1$		

3.3. Result

Fig.3 shows the changes of the maximum and average fitness of the evolved circuit. After about the 170th generation, the Tripod Robot acquired the movement that can move toward all directions.

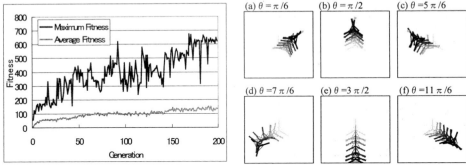

Fig.3: Changes of fitness value   Fig.4: Trajectory of Tripod Robot

Fig.4 shows the trajectory of the Tripod Robot controlled by the circuit of the 200th generation. In this figure, the trial number is increased to 6 in order to confirm the adaptability of the circuit. Tripod Robot could move toward all directions by generating the rhythm that reflect its structure and given environment. In the Fig.4, the movements of the robot are divided into two patterns. When the light source as a target located upper field of the figure such as (a)(b)(c), the robot moved toward the target after rotating so that upper leg turned to the target. On the other hand, when the target located lower field such as (d)(e)(f), the robot moved toward the target directly. In this way, it is considered that the generated circuit absorbed the characteristic of a robot's point symmetrical structure, and carried out functional specialization appropriately.

4. Conclusion

In this paper, Net-list Evolution was used for generating the logic circuit. Through the experiment using Tripod simulator, it was confirmed that the generated circuit reflected the structural characteristics of the autonomous robot in order to control periodically.

The next step is applying the methodology to real Tripod robot and multi-Tripod robots connected each other.

References

[1] Tomokazu Shindo, Hiroshi Yokoi and Yukinori Kakazu, "Adaptive Logic Circuit Based on Net-list Evolution", Journal of Robotics and Mechatronics, pp.144-149, Vol.12, No.2, Apr., 2001

[2] T. Higuchi, H. Iba, B. Manderick, "Evolvable Hardware", in Massively Parallel Artificial Intelligence, edited by H. Kitano, J. Hendler, pp398-421, MIT Press, 1994

[3] Hitoshi Henmi, J.Mizoguchi, K.Shimohara, "Development and Evolution of Hardware Behaviors", Lecture Notes in Computer Science Towards Evolvable Hardware, Vol.1062, pp.250-265

[4] Koji Nishikawa, Hiroshi Yokoi and Yukinori Kakazu, "Multi-agent Control System for the Symmetric Body Robot", 3rd European Workshop on Advanced Mobile Robots (EUROBOT'99), Proceedings, pp.25-32, 1999

Intelligent Autonomous Systems 7
M. Gini et al. (Eds.)
IOS Press, 2002

Real-Time Obstacle Avoidance by an Autonomous Mobile Robot using an Active Vision Sensor and a Vertically Emitted Laser Slit

Seydou SOUMARE*, Akihisa OHYA** and Shin'ichi YUTA*

Intelligent Robot Laboratory, University of Tsukuba
**PRESTO, JST / University of Tsukuba*
1-1-1 Tennodai, Tsukuba, 305-8573 JAPAN
{soumare,ohya,yuta}@roboken.esys.tsukuba.ac.jp

Abstract: *This article presents a real-time vision-based obstacle detection and avoidance method in an indoor environment for an autonomous mobile robot. In this study, we propose a scenario in which 3-Dimensional obstacles mainly building our real world environment, are detected by scanning a vertically emitted laser slit paired with an active stereo vision system. We present a simple and effective obstacle avoidance algorithm and experimental results to illustrate the advantages of our approach.*

Keywords: *Mobile robot, Obstacle avoidance, Laser range sensor, Active stereo vision.*

1 Introduction

For an autonomous mobile robot performing a navigation-based task in a vague environment, to detect and to avoid encountered obstacles is an important issue and a key function for the robot body safety as well as for the task continuity. Obstacle detection and avoidance in a real world environment - that appears so easy to humans - is a rather difficult task for autonomous mobile robots and is still a well-researched topic in robotics.

In many previous works, a wide range of sensors and various methods for detecting and avoiding obstacles for mobile robot purpose have been proposed. Good references related to the developed sensor systems and proposed detection and avoidance algorithms can be found in [1-11].

Based on these developed sensor systems, various approaches related to this work can be grouped into two categories. The first one tends to use ultrasonic sensors for their simple implementation and fast obstacle detection, but they show great accuracy and reliability limits when it comes to detect obstacles having a 3-Dimensionally complicated shape[1][2][3][4]. On the other hand, we have the vision-based sensor systems, which can be divided into two subgroups of sensor systems: stereo vision and laser range sensors. The former one applies with good reliability to the detection of 3-Dimensional objects but reveals to be deficient in term of speed and towards weakly textured obstacles[5][6][7][8]. The latter one, when applied as an horizontally emitted laser range sensor is efficient only towards 2-Dimensional obstacles[9][10]. We have also, 2-Dimensional laser range finder sensor which can efficiently detect 3-Dimensional obstacles but is poorly characterized in real-time detection[11]. From this background, the

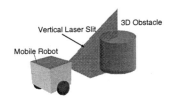

Figure 1: The vertical emission of the laser slit.

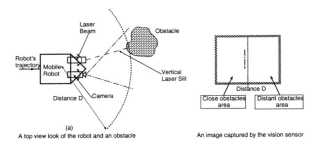

Figure 2: Obstacle detection principles.

main contribution of this work is the realization of an accurate, reliable and real-time obstacle detection and avoidance conducted in an indoor environment by an autonomous mobile robot using an active vision sensor coupled with a vertically emitted laser slit.

In this paper, we introduce first our obstacle detection strategy, using an active stereo vision system paired with a vertical laser slit and its implementation. Next, we outline the control of the viewing field of the robot during its navigation. Last, we present briefly the obstacle avoidance algorithm applied in this work and yet show the experimental results of an obstacle avoidance action that took place in a corridor.

2   Obstacle detection

2.1   Principles

In this study, we use a laser range sensor to detect obstacles. A laser range sensor consists of coupling a vision sensor with an -usually- horizontally emitted laser slit. It enables an intelligent robot to get the distance to any obstacle present in its 2-Dimensional vicinity, thanks to a simple and speedy image processing[12]. However, emitting horizontally the laser slit does not procure the height information of obstacles such as tables, chairs and many others present in an indoor environment. In this work, in order to obtain the height information of any object obstructing the robot path, regardless of its shape, we emit our laser slit vertically as illustrated in figure 1.

As shown in figure 2(a), the laser range sensor components are set with the mobile robot. From the laser beam projector is emitted a vertical laser slit which is reflected when it encounters an obstacle, and is captured by the vision sensor(figure 2(b)). On an

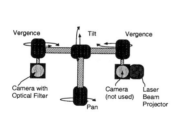

(Mechanism of the Active Vision Sensor)

Figure 3: Autonomous mobile robot equipped with active stereo vision and vertically emitted laser projector.

image captured by the camera, an obstacle appears as a vertical segment. Its horizontal coordinates taken from the image left border correspond to the obstacle position viewed from the robot center. For each captured image, computing an obstacle position from the robot center will not fulfill our seek for a real-time obstacle detection, since it is a time-consuming image processing. Instead, we fix a threshold distance under which obstacles on the mobile robot's path would be detected quickly and accurately. For instance, if we suppose that the image's vertical center line represents a given threshold distance D from the robot center, then a vertical segment representing any obstacle that appears on the left-side of our vertical line threshold, we can say that such an obstacle should exist in the immediate vicinity of the robot, within a distance D from its center. On the other hand, if the segment happens to be on the right-side of the threshold line, we deduce that the obstacle is located beyond a distance D from the robot position. Thus, by checking constantly the image's left-side region, we enable our sensor to detect efficiently and quickly any obstacle standing in its front neighborhood.

However, since the vision sensor and the vertical laser slit, making our laser range sensor, are sensing in only one direction, which is insufficient to know the presence of obstacles standing on the robot's path. The above explained detection method is basically not relevant enough and needs to be ameliorated. The main improvement brought to this detection method is to change the meeting point of the laser beam projector and the vision sensor, in other words to scan our sensor throughout all the mobile robot front environment thanks to actuators (active vision).

Therefore, scanning horizontally a vertical laser slit throughout the robot front environment and capturing and checking constantly the image's left region would enable a robot equipped with such sensor to detect quickly and efficiently any obstacle present in its front neighborhood.

2.2 Sensor system hardware

In this research, the authors used an active stereo vision system[13], set on the top of our YAMABICO robot[14], the intelligent mobile robot jointly used in this study (figure 3). The mechanism of this sensor system is mainly made up of four degrees

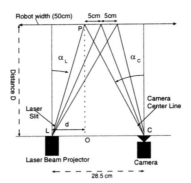

Figure 4: Obstacle detection range related to the vergences.

of freedom: pan, tilt and two independent vergences. Under each vergence, a camera is fixed as illustrated in figure 3. We only used the robot's right hand side camera through an optical filter which cuts visible lights and paired it with an infrared laser beam projector set with the robot's left vergence. Below, we outline that activating both vergences enables our laser range sensor to scan effectively and quickly the mobile robot vicinity.

2.3 Strategy and implementation

The above introduced method has been implemented on our YAMABICO robot. Then, once our laser range sensor was defined and set with our mobile robot, we had to define quantitatively its range of detection consisting of a 3-Dimensional area in which obstacles must be detected reliably and speedily. We defined such an area with features, such as the detection distance of an obstacle, the robot's height and width.

Firstly, the detection distance of an obstacle, viewed from the robot center, relies deeply on the intersection of the camera center line and the vertical laser slit. Recursively, this will depend on the rotational angles of both vergences, as our laser range sensor's components are activated by these actuators. In figure 4, α_L and α_C represent respectively the rotational angle of each vergence. The distance D of an obstacle location from the robot is related to the rotational angles of each vergence, as follows.

$$D = LC/\left(\tan\alpha_L + \tan\alpha_C\right) (1) \quad \text{for any } \alpha_L, \alpha_C \text{ (with LC=28.5 cm)}$$

Moreover, another critical issue here is to precise how wide the robot should search for possible obstacles on its path. We recall that the need of scanning our sensor throughout the robot's path, is due to the vertical emission of the laser slit in one direction which reveals to be ineffective and not enough to secure the robot's body safety against any sided obstacle near the path. Therefore, the mobile robot must scan constantly its own width on the path in order to continue safely its navigation-based task. In this implementation, our YAMABICO robot has a 50 cm width that we slice into ten segments of 5 cm each. We then, orient periodically our sensor on each segment thanks to the vergences. In figure 4, we first set the laser beam projector(left vergence) on one segment and doing so will direct the vision sensor(right vergence) on the same vergence according to these following expressions.

$$\alpha_L = \tan^{-1}(d/D) \ (2) \quad \text{and} \quad \alpha_C = \tan^{-1}((LC - d)/D) \ (3)$$

d : distance to any segment from the laser beam projector.

In the above expressions, both α_L and α_C angles are related to D, the distance of an obstacle from the robot. The capital issue here is to precise how distant -D- our sensor should detect obstacles standing on the robot's path which generally consists of a mix of lines and curves. In our implementation, we set the distance D dynamically in relation with the path complexity and the robot's velocity relative to the path. For instance, if we suppose that the mobile robot follows a linear path, our sensor is then able to search far ahead any obstacle present on the route. However, when the path is circular, the detection distance D should be small enough to enable our sensor to detect any obstacle standing on the forthcoming circular path. Path complexity in this study is basically acknowledged by setting high velocity for linear paths and relatively slow velocity for circular paths. This yields us to have the obstacle detection distance D proportional to the robot velocity on a given path. In addition, by default, when our robot is navigating a linear path at 20 cm/s of speed, the detection distance is set to 2 m. The distance is set to 1 m for a velocity of 10 cm/s, characterizing a circular path.

In this section, we detailed the main lines of our obstacle detection strategy and its implementation. However, as we will discuss in the next section, the effective detection of obstacles is performed by image processing, which is an important issue here to fulfill our seek for an accurate, reliable and real-time obstacle detection in a real-world environment.

2.4 Image processing

Previously we showed that any possible obstacle standing in the robot's front environment would show up as a vertical segment in the left region of the image captured by our sensor(refer to figure 2(b)). Here, we outline that the use of an optical filter to filtrate only the laser slit and to cut all visible lights enables our sensor to capture nearly binarized images, which consist of an uniformly black textured background with a white textured vertical segment, in case of obstacle presence. Doing so simplifies greatly the image processing to be conducted in this study in matter of difficulty, and mostly helps our seek for a real-time obstacle detection.

Based on such knowledge, the image processing performed here, consists simply of scanning from right to left, top to bottom the left region of any captured image, with the help of an horizontal line and confronts all met pixels with a constant threshold value. For a single image, it takes about one tenth of a second (0.1s)[1] to detect an obstacle presence in the robot vicinity. Therefore, as we scan our sensor in ten different directions ahead of the robot, it costs nearly a second to our sensor to scan fully and detect accurately any obstacle's presence in the robot's front environment.

3 Orienting dynamically the active vision sensor on the robot's path.

Once the mobile robot equipped with such sensor system, is navigating a defined path, particularly a circular one, the equipped sensor's orientation on the path becomes a matter of great importance here, since the robot must anticipatorily detect and avoid obstacles on the path. In this study, such navigation based-problem is solved by controlling the field scanned by our sensor. Such control is performed by directing

---

[1]Vision Module CPU : Transputer T-805 (Clock:20MHz)

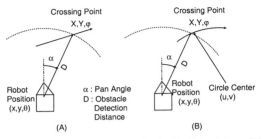

Figure 5: Pan orientation on the robot's path.

dynamically the meeting point of our sensor's components on the robot's path, thanks to the pan degree of freedom of the active vision sensor(refer to figure 3).

We define the robot's path as a succession of of lines and curves. And we basically control the field viewed by our sensor by computing dynamically the intersection of the path(line or curve) with a circle defined by its center, the robot's position on the path and its radius, the obstacle detection distance. Figure 5 shows such computational process and α represents the needed pan angle to redirect dynamically the active vision sensor on the forthcoming path.

4   Obstacle avoidance

4.1   Obstacle avoidance algorithm

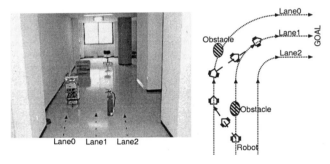

Figure 6: Obstacle avoidance action.

Once our mobile robot equipped with such sensor is able to detect stationary obstacles located on its path, the issue of avoiding those obstacles and continuing safely its navigation-based task naturally comes up. Historically, there are numerous obstacle avoidance methods adapted to a variety of sensors. Theoretically, we have methods based on avoiding obstacles accordingly to their shape, which demand a great environment knowledge of the sensors. Another major avoidance strategy is to avoid obstacles regardless of their shape and to consider for avoidance space only the navigable free space. Since the sensor developed for this work does not procure shape information

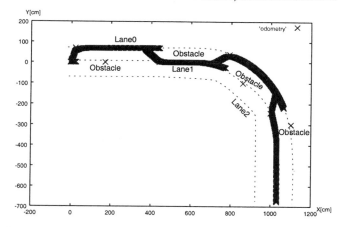

Figure 7: Experimental results of an obstacle detection and avoidance action in a corridor.

for the detected obstacles, we adapt the latter avoidance strategy to our sensor in an avoidance action taking place in an indoor corridor.

A view of such obstacle avoidance action is illustrated in figure 6. The basic and simple idea behind our obstacle avoidance algorithm is to enable our robot, after detecting objects obstructing partially its path to change dynamically its path in the free navigable space, in order to avoid obstacles, repeatedly until it reaches the goal of its navigation. For that purpose, we define offline a set of lanes parallel to the robot's path(center lane) and change dynamically the path on an adjacent lane when an obstacle is encountered on the path. This yields us to a simple to implement and effective obstacle avoidance action.

4.2 Experiments

For experiment purpose, we used a corridor(figure 6) having a width of 260 cm, in which our robot can navigate safely on three distinct lanes, parallel to each other. In this indoor environment, the robot's path is about 15 m of length and consists in a linear, circular and linear trajectory. We deliberately set 3-Dimensional obstacles on different lanes, and conducted our real-time obstacle detection and avoidance action in such an environment. During the experiment, the odometry of the robot was recorded and is plotted in figure 7.

Based on these results, we can deduce that an intelligent mobile robot equipped with the sensor developed in this study, is able to perform a reliable and real-time 3-Dimensional obstacle detection and avoidance in a real-world indoor environment.

5 Conclusion and future work

In this paper, the authors dealt with a real-time obstacle detection and avoidance method performed by an autonomous mobile robot using an active vision sensor and a vertical laser slit. After detailing our detection strategy and implementation of the developed sensor with our robot, we also introduced the basic algorithm of our avoidance method. We confirmed the originality of our approach with results of an obstacle avoidance experiment in an indoor corridor.

Reducing obstacle detection failures in well-lighted environment, due to the sunlight for instance, and improving our avoidance algorithm for moving obstacles are the topics of our future work.

References

[1] J.Borenstein, "Real-time obstacle avoidance for fast mobile robots", *IEEE Trans. Syst. Man Cybern.*, vol.19, pp.1179-1187, 1989.

[2] J.Borenstein and Y.Koren, "Histogramic in motion mapping for mobile robot obstacle avoidance", *IEEE Trans. Robot. Automat.*, vol.7, pp.535-539, 1991.

[3] K.Higashijima, H.Onda and T.Ogasawara, "A study on Estimation Wire Obstacle using Ultrasonic Sensors and Planning for Avoidance", *Proc. IEEE/RSJ IROS 98*, vol.3, pp.1622-1627, 1998.

[4] A.Ohya, A.Kosaka and A.Kak, "Vision-based navigation by a mobile robot with obstacle avoidance using single-camera vision and ultrasonic sensing", *IEEE Trans. Robot. Automat.*, vol.14, pp.969-978, 1998.

[5] K.Yoshida, S.Yokota and I.Watabe, "A Real-time Motion Stereo Vision Based on Brightness Difference", *Proc. IEEE/RSJ IROS 99*, vol1, pp.77-82, 1999.

[6] K.Onoguchi, N.Takeda and M.Watanabe, "Obstacle Location Estimation Using Planar Projection Stereopsis Method", *IEEE ICRA 97*, vol.3, pp.2553-2558, 1997.

[7] T.Nakamura and M.Asada, "Stereo Sketch:Stereo Vision-based Target Reaching Behavior Acquisition with Occlusion Detection and Avoidance", *Proc. IEEE ICRA 96*, vol.2, pp.1314-1319, 1996.

[8] M.Kumano, A.Ohya and S.Yuta, "Obstacle Avoidance of Autonomous Mobile Robot using Stereo Vision Sensor", *Intl. Symp. Robot. Automat.*, pp.497-502, 2000.

[9] N.Okada and E.Kondo, "3-Dimensional Object Model Construction from Range Images taken by a Range Finder on a Mobile Robot", *Proc. IEEE/RSJ IROS 98*, vol.3, pp.1853-1858, 1998.

[10] R.M.Taylor and P.J.Probert, "Range Finding and Feature Extraction by Segmentation of Images for Mobile Robot Navigation", *Proc. IEEE ICRA 96*, vol.1, pp.95-100, 1996.

[11] J.Vandorpe, H.Van Brussel and H.Xu, "Extraction Dynamic Map Building for a Mobile Robot using Geometrical Primitives Produced by a 2D Range Finder", *Proc. IEEE ICRA 96*, vol.1, pp.901-908, 1996.

[12] J. Iijima, S. Asaka and S. Yuta, "Searching Unknown Environment by a Mobile Robot Using Range Sensor–An Algorithm and Experiment–", *Proc. IEEE/RSJ IROS 89*, vol.1, pp.46-53, 1989.

[13] A.Ohya, D.Kawai and S.Yuta, "Autonomous mobile robot navigation using Active Stereo Vision", *Intelligent Autonomous Systems 6, E.Pagello et al(Eds) Press* pp.1041-1048, 2000.

[14] S.Yuta, S.Suzuki and S.Iida, "Implementation of a small size experimental self-contained autonomous robot–sensors, vehicle control and description of sensor based on behavior–", *Experiment. Robot. II*, R.Chatilla et al. ,Eds. Springer-Verlag, 1993, pp.344-358.

Intelligent Autonomous Systems 7
M. Gini et al. (Eds.)
IOS Press, 2002

Global Locomotion from Local Interaction in Self-Reconfigurable Robots

K. Støy[1], W.-M. Shen[2] and P. Will[2]

kaspers@mip.sdu.dk {shen,will}@isi.edu

[1]The Maersk Mc-Kinney Moller Institute for Production Technology
University of Southern Denmark, Campusvej 55, DK-5230 Odense M, Denmark
[2]Information Sciences Institute, University of Southern California
4676 Admiralty Way, Marina del Rey, CA 90292, USA

Abstract. We present a general distributed control algorithm for achieving locomotion of a self-reconfigurable robot. In this algorithm each module continuously performs a cyclic sequence of actions with a period T. When a specified fraction of this period d has elapsed a signal is sent to all child modules. Upon receiving this signal the child module resets its action sequence making it delayed d compared to its parent. The algorithm is minimal and robust to loss of synchronization signals and change in the number of modules. We show in three different experiments that the algorithm can be used to implement a caterpillar, a sidewinder, and a rolling wheel gait in a real self-reconfigurable robot consisting of eight modules.

1 Introduction

Reconfigurable robots are robots made from a possibly large number of independent modules connected to form a robot. If the modules from which the reconfigurable robot is built are able to connect and disconnect without human intervention the robot is a self-reconfigurable robot. Examples of physically realized self-reconfigurable robots can be found in [8, 6, 9, 15, 13, 11, 7].

Several potential advantages of self-reconfigurable robots over traditional robots have been pointed out in literature:

- **Versatility**. The modules can be combined in different ways making the same robotic system able to perform a wide range of tasks.

- **Adaptability**. While the self-reconfigurable robot performs its task it can change its physical shape to adapt to changes in the environment.

- **Robustness**. Self-reconfigurable robots consist of many identical modules and therefore if a module breaks down it can be replaced by another.

- **Cheap production**. When the final design for the basic module has been obtained it can be mass produced and thereby keep the cost of the individual module low.

Self-reconfigurable robots can solve the same tasks as traditional robots, but as Yim et al [15] point out; in applications where the task and environment are given a priori it is often cheaper to build a special purpose robot. Therefore, applications best suited for self-reconfigurable robots are applications where some leverage can be gained from the special abilities of self-reconfigurable robots. The versatility of these robots make them suitable in scenarios where the robots have to handle a range of tasks. The robots can also handle tasks in unknown or dynamic environments, because they are able to adapt to these environments. In tasks where robustness is of importance it might be desirable to use self-reconfigurable robots. Even though real applications for self-reconfigurable robots still are to be seen, a number of applications have been envisioned [11, 15]: fire fighting, search and rescue after an earthquake, battlefield reconnaissance, planetary exploration, undersea mining, and space structure building. Other possible applications include entertainment, service robotics, and payload management.

The potential of self-reconfigurable robots can be realized if several challenges in terms of hardware and software can be met. In this work we focus on one of the challenges in software: how do we make a large number of connected modules perform a coordinated global behavior? Specifically we address how to design algorithms that will make it possible for self-reconfigurable robots to locomote efficiently. In order for a locomotion algorithm to be useful it has to preserve the special properties of these robots. From the advantages and applications mentioned above we can extract a number of guidelines for the design of such a control algorithm. The algorithm should be distributed to avoid having a single point of failure. Also the performance of the algorithm should scale with an increased number of modules. It has to be robust to reconfiguration, because reconfiguration is a fundamental capability of self-reconfigurable robots. Finally, it is desirable to have homogeneous software running on all the modules, because it makes it possible for any module to take over if another one fails.

It is an open question if a top-down or a bottom-up approach gives the best result. We find that it is difficult to design the system at a global level and then later try to distribute it, because often properties of the hardware are ignored and a slow robotic system might be the result. Therefore, we use a bottom-up approach where the single module is the basic unit of design. That is, we move from a global design perspective to a bottom-up one where the important design element is the individual module and its interactions with its neighbors. The global behavior of the system then emerges from the local interaction between individual modules. A similar approach is also used by Bojinov et al [1].

2 Related Work

In the related work presented here we focus on control algorithms for locomotion of self-reconfigurable robots.

Yim et al [14, 15] demonstrate caterpillar like locomotion and a rolling track. Their system is controlled based on a gait control table . Each column in this table represents the actions performed by one module. Motion is then obtain by having a master synchronizing the transition from one row to the next. The problem with this approach is that the amount of communication needed between the master and the modules will limit its scalability. Another problem is the need for a central controller, since it gives the system a single point of failure. If there is no master it is suggested that the modules can be assumed to be synchronized in time and each module can execute its column of actions open-loop. However, since all the modules are autonomous it is a questionable assumption to assume that all the modules are and can stay synchronized. In order to use the gait control table each module needs to know what column it has to execute. This means that the modules need IDs. Furthermore, if the configuration changes or the number of modules changes the table has to be rewritten.

Shen, Salemi, and others propose to use artificial hormones to synchronize the modules to achieve consistent global locomotion. In earlier versions of the system a hormone is propagated through the self-reconfigurable system to achieve synchronization [11]. In later work the hormone is also propagated backwards making all modules synchronized before a new action is initiated [12, 10]. This synchronization takes time $O(n)$ where n is the number of modules. This slows down the system considerably, because it has to be done before each action. Also, the entire system stops working if one hormone is lost. This a significant problem, because a hormone can easily be lost due to unreliable communication, a module disconnecting itself before a response can be given, or a module failure. In fact, the system has n-points of failure which is not desirable. The earlier version is better in this sense, but still performance remains low because a synchronization hormone is sent before each action.

In our system all modules repeatedly go through a cyclic sequence of joint angles describing a motion. This sequence could come from a column in a gait control table, but in our implementation the joint angles are calculated using a cyclic function with period T. Every time a module has completed a given fraction d of the period a message is sent through the child connectors. If the signal is received the child module resets its action sequence making it delayed d compared to the parent. This way the actions of the individual module are decoupled from the synchronization mechanism resulting in a faster and more reliable system. Furthermore, there is no need to make changes to the algorithm if the number of modules changes.

3 General Control Algorithm

We assume that the modules are connected to form a tree structure, that a parent connector is specified, and that this connector is the only one that can connect to child connectors of other modules. Furthermore, we assume that the modules can communicate with the modules to which they are connected.

The algorithm is then used by specifying three components. The first component is a cyclic action sequence $A(t)$. This sequence describes the actions that each module is to repeat cycle after cycle. The second, is the period T of this cycle. The third, is a delay d. This delay specify the fraction of a period the children's action sequences are delayed compared to their parents. The skeleton algorithm looks like this:

```
t = 0
while(1) {
    if (t=d) then <send signal to child connectors>
    if <signal received from parent> then t=0
    <perform action A(t)>
    t = (t+1) modulus T
}
```

Ignoring the first two lines of the loop, the module repeatedly goes through a sequence of actions parameterized by the cyclic counter t. This part of the algorithm alone can make a single module repeatedly perform the specified sequence of actions. In order to coordinate the actions of the individual modules to produce the desired global behavior the modules need to be synchronized. Therefore, at step $t = d$ a signal is send through all child connectors. Note that it does not matter if a child module is actually connected or not. If a child receives a signal it knows that the parent is at $t = d$ and therefore sets its own step counter to $t = 0$. This enforces that the child is delayed d compared to its parent.

From the time the modules are connected it takes time proportional to d times the height of the tree for all the modules to synchronize. To avoid problems with uncoordinated modules initially we make sure the modules do not start moving until they receive the first synchronization signal. After the start-up phase the modules stay synchronized using only constant time.

4   Experimental Setup

To evaluate our algorithm we conducted several experiments using the CONRO modules shown in Figure 1. The CONRO modules have been developed at USC/ISI [3, 5]. The modules are roughly shaped as rectangular boxes measuring 10cm x 4.5cm x 4.5cm and weigh 100grams. The modules have a female connector at one end and three male connectors located at the other. Each connector has a infra-red transmitter and receiver used for local communication and sensing. The modules have two controllable degrees of freedom: pitch (up and down) and yaw (side to side). Processing is taken care of by an onboard Basic Stamp 2 processor. The modules have onboard batteries, but these do not supply enough power for the experiments reported here and therefore the modules are powered through cables. Refer to http://www.isi.edu/conro for more details and videos of the experiments reported later.

5   Experiments

In general, it is a problem how to report performance of a specific part of a self-reconfigurable system because there is such a tight coupling between hardware and software. In this work we choose to report the length of our programs as a measure of the complexity of the control algorithm. This metric is used to support our claim that this control system is minimal. We also report the speed of the locomotion patterns, but this should only be considered an example, the reason being that in our system the limiting factors are how robust the modules physically are, how powerful the motors are, and how much power we can pull from the power source. To report a top speed is not meaningful before we run the robot autonomously on batteries.

5.1   Caterpillar Locomotion

We connect eight of our modules in a chain and designate the male opposite the female connector to be the parent connector. We then implement the algorithm described above with the following parameters.

$$
\begin{aligned}
T &= 180 \\
pitch(t) &= 50^\circ sin(\tfrac{2\pi}{T}t) \\
yaw(t) &= 0 \\
d &= \tfrac{T}{5}
\end{aligned}
\tag{1}
$$

The motor control of our modules makes the motor go to the desired position as fast as possible. This means that way-points have to be specified to avoid jerky motion. The period T can be used to control the number of way-points and therefore the smoothness and speed of the motion. The action sequence is an oscillation around 0° with an amplitude of 50° and the yaw joint is kept straight. Each module is delayed one fifth of a period compared to its parent.

The modules are connected and after they synchronize a sine wave is traveling along the length of the robot. Refer to Figure 1. This produces caterpillar like locomotion at a speed

Figure 1: A CONRO module (left). A snapshot of caterpillar like locomotion (right).

of 0.13km/h. Note, that it is easy to adjust the parameters of this motion. For instance, the length of the wave can be controlled using the delay. The program is simple. The main loop contains 16 lines of code excluding comments and labels. The initialization including variable and constant declaration amounts to 18 lines of code.

5.2 Sidewinder Locomotion

We now turn our attention to a locomotion pattern similar to that of a sidewinding snake. A detailed mathematical analysis of this motion pattern has been reported in [2]. Here we just use the intuition that by having modules moving to one side lifted and those moving to the other touching the ground a sidewinder like motion is achieved. The result can be seen in Figure 2. The sidewinder moves at 0.24km/h. The main loop and the initialization contain respectively 19 and 17 lines of code. The parameters used are:

$$
\begin{aligned}
T &= 180 \\
pitch(t) &= 20°cos(\tfrac{2\pi}{T}t) \\
yaw(t) &= 50°sin(\tfrac{2\pi}{T}t) \\
d &= \tfrac{T}{4}
\end{aligned}
\tag{2}
$$

5.3 Rolling Track Locomotion

If we maintain that each module can only have one parent, but remove the assumption that the structure forms a tree we include loops as structures that can be handled. The rolling track is an example of such a configuration. However, this poses a problem to our algorithm. In the previous experiments we have exploited the assumption that the modules form a tree to implicitly find a conductor. The conductor being the root of the configuration tree. This is a simple mechanism that guarantees that there is one and only one conductor. In a loop configuration this is not the case.

One solution to this problem is to introduce IDs. In our implementation we just make the modules pick a random number and use that as ID. It is not guaranteed to find a unique conductor, but it is a simple solution that works in most cases. The short comings of this approach can easily be avoided if each module has a unique serial number.

The synchronization part of the algorithm now works as before, but it is combined with a simple well-known distributed leader election algorithm [4]. The signals from parent to child now contains a number which is the ID of the module originally sending the signal. Upon receiving a signal a module compares the signal's number to its ID. If it is higher the

Figure 2: A snapshot of sidewinder like locomotion (left) and the rolling track (right).

module is synchronized and the signal and its ID is propagated along with the synchronization signal. Otherwise, the module consider itself the conductor and ignores the signal. After the system has settled the module with the highest ID dictates the rhythm of the locomotion pattern. The leader election algorithm runs continuously which means that the system quickly synchronizes if modules are replaced. The advantage of combining the algorithms is that there is no need to detect if the conductor fails.

We used this algorithm to implemented the rolling track which can be seen in Figure 2. The rolling track is the fastest gait and achieves a speed of 0.50km/h. The program is now a little more complex and the main loop and initialization contain respectively 35 and 28 lines of code. The parameters for the eight module rolling track is:

$$
\begin{aligned}
T &= 180 \\
pitch(t) &= \begin{cases} 60°(1 - sin(\frac{2\pi}{T}t)) & if\ t =< \frac{T}{2} \\ 60° & if\ t > \frac{T}{2} \end{cases} \\
yaw(t) &= 0 \\
d &= \frac{T}{4}
\end{aligned}
\tag{3}
$$

Unlike the sidewinder and the caterpillar this control algorithm only works with 8 modules, because of the physical constraint. It might be possible to make a more general solution by making $pitch(t)$ and d a function of the number of modules. The number of modules in the loop could be obtained by the conductor by including a hop count in the signal.

6   Handling a General Configuration

We saw in the previous section that we had to introduce IDs to find a unique conductor in a configuration that contains loops. Introducing the ID mechanism unfortunately ruins the opportunity to use the synchronization algorithm to automatically find a conductor in a tree structure. In fact, the loop algorithm will fail in this situation unless the module with the highest ID also happens to be the root. In order to make a general algorithm the synchronization signal has to be propagated both upwards and downwards in the tree.

7   Discussion

We have presented a general control algorithm and presented examples of how it can be used to achieve three different locomotion patterns. We will now discus some of the properties of this control algorithm.

An important issue in the design of control algorithms for self-reconfigurable robots is if the algorithm scales with the number of modules. The presented algorithm is only initially dependent on the number of modules, because it decides how long time it takes for the synchronization signal to be propagated through the system. After this start-up phase the time it takes to keep the modules synchronized is independent of the number of modules implying that the algorithm scales. Furthermore, all modules run identical programs making it easier to manage the development process when programming systems consisting of many modules.

The modules of the robot are only loosely coupled through the synchronization signal and therefore the system is highly robust to changes in the number of modules. In fact, the caterpillar can be divided in two and both parts still work. If they are reconnected in a different order they will quickly synchronize to behave as one long caterpillar again. This also implies that the system is robust to module failure. If a module is defect and it can be detected this module can be ejected from the system and the remaining modules when reconnected can continue to perform. Finally, if a synchronization signal is lost it is not crucial for the survival of the system. If a signal is lost it just means that the receiving module and its children will be synchronized a period later.

In the algorithm the synchronization signal is only sent once per period. This means that in order for the modules to stay synchronized the time to complete a period has to be the same for all modules.In the experiments presented here the cycles take the same amount of time, but in more complex control systems where other parts of the control system use random amounts of computation time this can not be assumed to be true. This problem can easily be handle by using timers. Even though timers are not precise enough to keep modules synchronized over a long period of time they can be used for this purpose.

8 Future Work

Our future work will go along two lines. Can the algorithm handle more complex locomotion patterns? We suspect it can be achieved by using different delays through different connectors. For instance, in a multi-legged robot where the head is the conductor, the synchronization signal could travel along the spine modules. When a spine module receives a signal it can first propagate it to the left leg and then the right before propagating the signal to the next spine module.

Another issue is that if the self-reconfigurable robot is to locomote automatically in a real complex environment the control algorithm has to be able to take feedback from the environment into account. A first step in this direction could be to mount sensors on the side of the caterpillar robot and use these to control the yaw joint of the modules.

9 Conclusion

We have presented a general control algorithm for self-reconfigurable robots. The algorithm has the following properties: distributed, scalable, homogeneous, and minimal. We have shown how the algorithm easily can be used to implement a caterpillar and a sidewinder like locomotion pattern. Furthermore, we have seen that with the introduction of IDs in the modules it is possible to handle loop configurations. We have demonstrated this using the rolling track as an example. Finally, we have pointed out interesting lines for future research.

10 Acknowledgements

This work is supported under the DARPA contract DAAN02-98-C-4032, the EU contract IST-20001-33060, and the Danish Technical Research Council contract 26-01-0088.

References

[1] H. Bojinov, A. Casal, and T. Hogg. Emergent structures in modular self-reconfigurable robots. In *Proceedings of the IEEE int. conf. on Robotics & Automation*, volume 2, pages 1734–1741, San Francisco, USA, 2000.

[2] J.W. Burdick, J. Radford, and G.S. Chirikjian. A 'sidewinding' locomotion gait for hyper-redundant robots. In *Proceedings of the IEEE int. conf. on Robotics & Automation*, pages 101–106, Atlanta, USA, 1993.

[3] A. Castano, R. Chokkalingam, and P. Will. Autonomous and self-sufficient conro modules for reconfigurable robots. In *Proceedings of the 5th int. Symposion on Distributed Autonomous Robotic Systems*, pages 155–164, Knoxville, Texas, USA, 2000.

[4] E. Chang and R. Roberts. An improved algorithm for decentralized extrema-finding in circular configurations of processes. *Communications of the ACM*, 22(5):281–283, 1979.

[5] B. Khoshnevis, B. Kovac, W.-M. Shen, and P. Will. Reconnectable joints for self-reconfigurable robots. In *Proceedings of the IEEE/RSJ int. conf. on Intelligent Robots and Systems*, Maui, Hawaii, USA, 2001.

[6] K. Kotay, D. Rus, M. Vona, and C. McGray. The self-reconfiguring robotic molecule. In *Proceedings of the IEEE int. conf. on Robotics & Automation*, pages 424–431, Leuven, Belgium, 1998.

[7] S. Murata, E. Yoshida, H. Kurokawa, K. Tomita, and S. Kokaji. Self-repairing mechanical systems. *Autonomous Robots*, 10(1):7–21, 2001.

[8] A. Pamecha, C. Chiang, D. Stein, and G.S. Chirikjian. Design and implementation of metamorphic robots. In *Proceedings of the ASME Design Engineering Technical conf. and Computers in Engineering conf.*, pages 1–10, Irvine, USA, 1996.

[9] D. Rus and M. Vona. A physical implementation of the crystalline robot. In *Proceedings of the IEEE int. conf. on Robotics & Automation*, pages 1726–1733, San Francisco, USA, 2000.

[10] B. Salemi, W. Shen, and P. Will. Hormone controlled metamorphic robots. In *Proceedings of the IEEE int. conf. on Robotics & Automation*, pages 4194–4199, Seoul, Korea, 2001.

[11] W.-M. Shen, B. Salemi, and P. Will. Hormone-based control for self-reconfigurable robots. In *Proceedings of the int. conf. on Autonomous Agents*, pages 1–8, Barcelona, Spain, 2000.

[12] W.-M. Shen, B. Salemi, and P. Will. Hormones for self-reconfigurable robots. In *Proceedings of the int. conf. on Intelligent Autonomous Systems*, pages 918–925, Venice, Italy, 2000.

[13] C. Ünsal and P.K. Khosla. Mechatronic design of a modular self-reconfiguring robotic system. In *Proceedings of the IEEE int. conf. on Robotics & Automation*, pages 1742–1747, San Francisco, USA, 2000.

[14] M. Yim. *Locomotion with a unit-modular reconfigurable robot*. PhD thesis, Department of Mechanical Engineering, Stanford University, 1994.

[15] M. Yim, D.G. Duff, and K.D. Roufas. Polybot: A modular reconfigurable robot. In *Proceedings of the IEEE int. conf. on Robotics & Automation*, pages 514–520, San Francisco, USA, 2000.

Intelligent Autonomous Systems 7
M. Gini et al. (Eds.)
IOS Press, 2002

Autonomous Distributed Control of Traffic Signal Network

Masao Sugi, Hideo Yuasa, and Tamio Arai
Department of Precision Engineering, The University of Tokyo, Japan
E-mail: {sugi, yuasa, arai}@prince.pe.u-tokyo.ac.jp

Abstract. A new method is proposed for controlling a large number of traffic signals, which are represented by nonlinear oscillators, each governed by a reaction-diffusion equation at a node of a graph. The behavior of each signal is determined only by the states of its neighbors, with the system as a whole being also organized globally. In this report, both offsets between neighboring signals and phase splits of them are controlled according to the traffic condition. The simulation results are exemplified to show the validity of the method.

1 Introduction

This paper deals with control of a large number of traffic signals by the decentralized method, which provides a contrast to the conventional method based on off-line planning and centralized control scheme [1]. In the conventional method, traffic volume is estimated by on-the-spot survey; using this stationary traffic volume, a kind of quasi optimization problem is obtained. A solution of this optimization determines the behavior of signals, which are controlled by a central computer. In general, the traffic condition varies day by day, or even within a day (e.g., the main traffic flow often turns its direction). To cope with this, the central computer prepares several modes of signals' behavior. However, this method cannot follow dynamic changes of traffic volume, nor unexpected traffic conditions caused by road constructions or accidents. Its centralized architecture makes it impossible to cover wide area; the traffic condition becomes more unstationary and uncertain as the area to be controlled becomes larger. It is therefore necessary to develop a new method to overcome the limitations of the conventional one.

In this respect, we propose a method for controlling a large number of traffic signals in a distributed manner. This method will enable to handle a wide area with dynamic environments effectively. Signals are modeled as nonlinear coupled oscillators. The behavior of each oscillator is governed by a reaction-diffusion equation on a graph [5]. Each signal determines its split (ratio of the green light duration for each direction) and offset (difference between the onset times of green lights of neighboring signals) from its local traffic conditions.

In section 2, the traffic signal control is outlined. The related research is also mentioned. The oscillator model and the control method are described in section 3. Some simulation results are discussed in section 4. We conclude this paper in section 5.

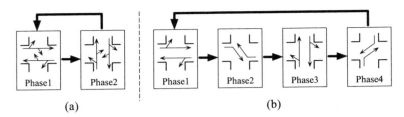

Figure 1: Examples of phase cycle systems. (a) 2-phase cycle, (b) 4-phase cycle

2 Related research

2.1 · Overview of traffic signal control

In the traffic signal control, the phase in the signal cycle and three parameters, the cycle length, the split and the offset, are determined to maximize efficiency of traffic flows.

A signal in an intersection changes its state cyclically among several patterns (Note that a *one-to-one* correspondence is held between "a signal" and "an intersection". In traffic engineering, a set of signal apparatuses in an intersection is called as "a signal"). In each pattern, the green light, allowing to pass through the intersection, is assigned to one or more traffic flows. These patterns are referred to as *phases*. The number of phases in one cycle and the choice of flow sequences in each phase have many alternatives. Figure 1 shows two examples of them, a 2-phase cycle and a 4-phase cycle, respectively. *Cycle length* means the period of a signal. *Split* is the ratio of duration of each phase to the cycle length. *Offset* (or *phase offset*) denotes the difference between the onset times of the phases of neighboring signals.

There are two schemes in the traffic control; the isolated control and the coordinated control. The former deals with only one signal, and controls cycle length and split of the signal. Its neighbors are not taken into account. The latter aims at coordinating multiple signals neighboring one another on the same road, and controls the offsets among them, along with the cycle lengths and the splits. This can be expected to achieve more effective performance than the isolated control. Extension of the coordinated control from one single road to network is often called the wide area control. As is mentioned in the introduction, the wide area control in dynamic environments is difficult for the conventional off-line and centralized control method.

2.2 Research on wide area signal control

There are several studies that deal with control of a large number of traffic signals by the decentralized method. In Refs. [3, 4], the multi-agent reinforcement learning is applied to control the split of each signal. This approach is suitable only for a stationary environment. The offsets and cycle lengths are not controlled. Reference [6] introduces a distributed fuzzy controller to regulate three parameters. This method, however, based on the empirically determined fuzzy rules, is difficult to analyze, and even stability is not guaranteed. In Ref. [2], a traffic network is modeled using nonlinear coupled oscillators. The offsets are controlled through the phase lock of oscillators, but Ref. [2] also reported that the system sometimes falls out of control due to drift problems. Therefore, wide area control of traffic signals by the distributed method is

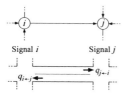

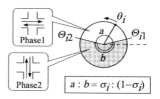

Figure 2: Traffic volume and oriented graph model

Figure 3: Traffic signal model

still a problem to be developed.

3 Autonomous distributed signal control

3.1 Preassumptions

In the present method, automobiles are assumed to go straight in an intersection and turn neither right nor left. The roads are limited to be bidirectional two-lane road, and all the intersections are crossroads. The cycle system of the signals is a 2-phase cycle: "Phase1" for the traffic flows from east and west and "Phase2" from north and south, as shown in Fig. 3.

Each signal can measure the traffic volume (i.e., number of automobiles) from each of four directions, which is shown in Fig. 2. The traffic volume is defined by,

$$q = \frac{1}{q_{max}\tau} \int_{t}^{t+\tau} Q(t')\,dt', \tag{1}$$

where q_{max} is the maximum volume, $Q(t)$ is the instantaneous traffic volume at time t, and τ is the time duration of the green light. Equation (1) normalizes the traffic volume into $q \in [0, 1]$.

Along with the traffic-volume values from all directions, each signal is to recognize the values *to* all directions. In Fig. 2, two neighboring signals i and j share the information on $q_{i \leftarrow j}$ (flow volume from j to i) and $q_{j \leftarrow i}$ (volume from i to j).

3.2 Traffic signal model

In this paper, the state of a signal i is described by two variables: the phase angle $\theta_i \in [0, 2\pi)$ and the split $\sigma_i \in [0, 1]$, which is shown in Fig. 3. Here the split σ_i is defined as,

$$\sigma_i = \frac{\tau_{\text{phase1}}}{\tau_{\text{phase1}} + \tau_{\text{phase2}}}, \tag{2}$$

where τ_{phase1} and τ_{phase2} are the time duration of "Phase1" and "Phase2 ", respectively.

The actual phase of a signal i is determined using the phase angle θ_i and the split σ_i. Let us first define the phase switch points Θ_{i1} and Θ_{i2}, shown in Fig. 3, i.e.,

$$\Theta_{i1} = (\frac{1}{2} - \sigma_i)\pi \pmod{2\pi}, \qquad \Theta_{i2} = (\frac{1}{2} + \sigma_i)\pi \pmod{2\pi}. \tag{3}$$

Then the phases are determined by Θ_{i1} and Θ_{i2};

$$\begin{array}{ll} \text{Phase1:} & \sin\theta_i \geq \sin\Theta_{i1}, \quad \theta_i \neq \Theta_{i2}, \\ \text{Phase2:} & \sin\theta_i \leq \sin\Theta_{i2}, \quad \theta_i \neq \Theta_{i1}. \end{array} \tag{4}$$

The network of traffic signals is modeled as an oriented graph, which is shown in Fig. 2. A signal (identical with an intersection here) is represented as a node, and a link between two nodes indicates that the corresponding signals are connected by a road with two lanes (Note that all of the roads are bidirectional. Consequently, the direction of a link does not represent that of the traffic flow). Here, we define a function on a link (i, j) between two nodes i and j,

$$\text{sign}(i, j) = \begin{cases} 1, & \text{if } i \text{ is end-point of the link } (i, j), \\ -1, & \text{if } i \text{ is origin of the link } (i, j). \end{cases} \tag{5}$$

Using this function, the differences, $\rho_{(i,j)}$ and $\phi_{(i,j)}$, of splits and phase angles, respectively, between neighboring signals i and j, are defined as,

$$\rho_{(i,j)} = \text{sign}(i, j)(\sigma_i - \sigma_j), \tag{6}$$
$$\phi_{(i,j)} = \text{sign}(i, j)\{(\theta_i - \xi(i, (i, j))) - (\theta_j - \xi(j, (j, i)))\}, \tag{7}$$

where $\xi(i, (i, j))$ is a function defined on a node i and a link (i, j),

$$\xi(i, (i, j)) = \begin{cases} \theta_{i1}, & \text{if Phase1 of signal } i \text{ gives green light to the link } (i, j), \\ \theta_{i2}, & \text{if Phase2 of signal } i \text{ gives green light to the link } (i, j). \end{cases} \tag{8}$$

It is noted that the reciprocal relations, $\rho_{(i,j)} = \rho_{(j,i)}$ and $\phi_{(i,j)} = \phi_{(j,i)}$, hold in the above formulation.

3.3 Split control

We introduce a gradient system, which is shown below, to the dynamics of the split σ_i of signal i,

$$\frac{d\sigma_i}{dt} = -\frac{\partial}{\partial \sigma_i}(W_0 + W_1), \tag{9}$$

where W_0 and W_1, the potentials of the gradient system of σ_i, are given as,

$$W_0 = \sum_i \alpha \left\{ \sigma_i - \frac{q_{i \leftarrow w(i)} + q_{i \leftarrow e(i)}}{\sum_{j \sim i} q_{i \leftarrow j}} \right\}^2, \tag{10}$$

$$W_1 = \sum_i \sum_{j \sim i} \beta (q_{i \leftarrow j} + q_{j \leftarrow i}) \cdot (\rho_{(i,j)})^2, \tag{11}$$

where $w(i)$ and $e(i)$ represent the east and west neighbors of i, respectively, $j \sim i$ denotes that j is a neighbor of i, α and β are parameters to control the speed of convergence. The potential W_0 is determined by the local traffic condition alone. It attains the minimum value at $\sigma_i = (q_{i \leftarrow w(i)} + q_{i \leftarrow e(i)})/(\sum_{j \sim i} q_{i \leftarrow j})$ for all signal i. This minimum value is the ratio of the east and west traffic volume to the total one, and is based on the definition of the split, with the time durations in Eq. (6) being replaced by the traffic volumes. The potential W_1 is affected by the neighboring signals, having a role of smoothing the splits of the neighbors. It attains the minimum value at $\sigma_j = \sigma_i$ for all signal pair i, j.

Since Eqs. (10) and (11) contain the traffic volumes, the system can adapt to the changes of the traffic conditions by moving its minimum value. Since the traffic volumes can also be time-varying, the convergence of the split σ_i is not guaranteed in this dynamics. However, if the traffic volume is almost stationary, the split will converge into the minimum value of the potential.

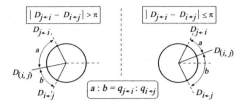

Figure 4: Relations among $D_{(i,j)}$, $D_{i \leftarrow j}$ and $D_{j \leftarrow i}$

3.4 Offset control

As is the case with the split control, we use a gradient system of the phase angle θ_i to control offsets as,

$$\frac{d\theta_i}{dt} = \omega - \frac{\partial V}{\partial \theta_i}. \tag{12}$$

The potential V is given as,

$$V = \sum_i \sum_{j \sim i} \left\{ -\gamma(q_{i \leftarrow j} + q_{j \leftarrow i}) \cos(\phi_{(i,j)} - D_{(i,j)}) \right\}, \tag{13}$$

where γ is a parameter to control the speed of convergence, and ω is the angular acceleration common to all signals. Here $D_{(i,j)}$, which is a target difference of $\phi_{(i,j)}$ to be attained, is represented by,

$$D_{(i,j)} = \begin{cases} 0, & (q_{i \leftarrow j} = q_{j \leftarrow i} = 0) \\ \frac{q_{i \leftarrow j} D_{i \leftarrow j} + q_{j \leftarrow i} D_{j \leftarrow i}}{q_{i \leftarrow j} + q_{j \leftarrow i}}, & (|D_{i \leftarrow j} - D_{j \leftarrow i}| \leq \pi) \\ \frac{q_{i \leftarrow j} \{D_{i \leftarrow j} + \text{sign}(i,j)\pi\} + q_{j \leftarrow i} \{D_{j \leftarrow i} + \text{sign}(j,i)\pi\}}{q_{i \leftarrow j} + q_{j \leftarrow i}} + \pi, & (|D_{i \leftarrow j} - D_{j \leftarrow i}| > \pi) \end{cases} \tag{14}$$

where $D_{j \leftarrow i}$ and $D_{i \leftarrow j}$ denote the target differences of $\phi_{(i,j)}$, being related to flow from i to j and from j to i, respectively. These are simply defined by the maximum velocity of the automobiles v_{max} and the distance between two signals L_{ij}, as,

$$D_{i \leftarrow j} = -\text{sign}(i, j)\frac{\omega L_{ij}}{v_{max}}, \quad D_{j \leftarrow i} = -\text{sign}(j, i)\frac{\omega L_{ij}}{v_{max}}. \tag{15}$$

Equation (14) means that $D_{(i,j)}$ is the weighted average of $D_{j \leftarrow i}$ and $D_{i \leftarrow j}$ with regard to the traffic volume from j to i and that of the opposite direction, which is shown in Fig. 4. Equation (15) means that $D_{j \leftarrow i}$ is the product of the angular acceleration ω and the necessary time for the automobiles travelling from signal i to j at the maximum speed. It is noted that the differences between signals' phase angles are equal to the actual offsets multiplied by the angular acceleration ω. In fact, if $q_{j \leftarrow i} \neq 0$ and $q_{i \leftarrow j} = 0$, the target difference of $\phi_{(i,j)}$ is $D_{(i,j)} = D_{j \leftarrow i}$. The actual offset in this case is given as $D_{j \leftarrow i}/\omega = -\text{sign}(j,i)L_{ij}/v_{max}$, which is widely used for offset optimization in conventional signal control methods.

Equations (2), (7), and (8) indicate that $\phi_{(i,j)}$ is a function of θ_i, θ_j, σ_i, and σ_j. Because σ_i is also nonstationary, Eq. (12) is not a gradient system of θ_i. However, in the case of $\frac{d\Theta_{i1}}{dt} (= -\pi \frac{d\sigma_i}{dt}) \ll \frac{d\theta_i}{dt}$, the terms of σ_i and σ_j in Eq. (12) can be ignored, thus we can consider Eq. (12) as a gradient system of θ_i. If the traffic volume is almost stationary, the difference of the phase angles will converge, as is the case with the split control.

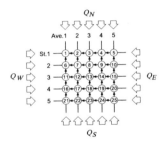

Figure 5: Simulation environment

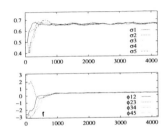

Figure 6: Time evolutions of splits and phase angle differences

Table 1: Simulation parameters

Signal		Road and car	
ω	$\pi/60$ (rad/s)	L_{ij}	120 (m)
α	0.002	v_{max}	14 (m/s)
β	0.002	a^+	0.15 (m^2/s)
γ	$\omega/8.0$	a^-	0.5 (m^2/s)

4 Simulation

4.1 Traffic flow model

We use a discrete traffic model. The velocity v_i of an automobile i is controlled into a target velocity $\tilde{v}_i$, as,

$$\frac{dv_i}{dt} = \begin{cases} a^+, & (v_i < \tilde{v}_i) \\ 0, & (v_i = \tilde{v}_i) \\ -a^-, & (v_i > \tilde{v}_i) \end{cases} \qquad (16)$$

where constants a^+ and a^- represent the acceleration and deceleration of automobiles, respectively. Target velocity $\tilde{v}_i$ depends on $d_i = x_{i-1} - x_i$, which is the distance between the automobile i and the one in front of i;

$$\tilde{v}_i = \tilde{v}(d_i) = \begin{cases} v_{max}, & (d_i > \frac{(v_{max})^2}{2a^-}) \\ \sqrt{2d_i a^-}. & (d_i \leq \frac{(v_{max})^2}{2a^-}) \end{cases} \qquad (17)$$

where v_{max} is the maximum velocity of automobiles.

4.2 Simulation environment

The environment used in this simulation is shown in Fig. 5. This is a square road network with 5 parallel roads for each direction, and the total number of signals is 25. The traffic volume of each of four directions is assumed to be uniform in the five roads. The values of parameters are shown in Table 1.

4.3 Case 1: Static condition

In this subsection, the traffic volume of each direction is fixed stationary. The arrival frequency of automobiles (number of automobiles per unit time) from each direction

Table 2: Simulation result (case 1): necessary time to pass through

	Present method (s)	Non-control (s)
Q_N	258.9	164.8
Q_S	184.6	166.1
Q_W	123.3	189.0
Q_E	136.1	166.5
total	142.6	179.6

Table 3: Simulation result (case 2): necessary time to pass through

	Present method (s)	Non-control (s)
Q_N	150.1	186.1
Q_S	148.2	166.6
Q_W	134.2	186.7
Q_E	163.6	166.8
total	146.8	179.5

is set to be $Q_N = 0.074$ (1/s), $Q_S = 0.029$ (1/s), $Q_W = 0.294$ (1/s), and $Q_E = 0.098$ (1/s). The initial value of the phase angle θ_i is random for each signal i, and that of split is $\sigma_i = 0.5$. The simulation time is 4200 (s).

Figure 6 shows the time evolutions of splits and phase angle differences of the five signals no. 1–5 on the street no. 1 in Fig. 5. At $t = 1000$ (s), we find that the values of both split and phase difference are approximately homogenized over all signals, $\sigma \simeq 0.67$ and $\phi \simeq 0.37$.

Table 2 shows the necessary time for the automobiles from four directions to pass through the simulation environment. Table 2 also shows the comparison of the result of the present method with the non-control case, where the splits and the offsets are fixed at $\sigma = 0.5$ and $\phi = 0$ for all signals. The present method allows the major traffic flow to pass through faster. The minor flow requires much more time than the case of non-control, but in total, the mean time over all directions is reduced down to approximately 80%.

4.4   Case 2: Dynamic condition

In this subsection, the frequency of car arrival will be changed. For $0 < t \leq 2000$ (s), the frequencies of arrival are set to be $Q_N = 0.074$ (1/s), $Q_S = 0.029$ (1/s), $Q_W = 0.294$ (1/s), and $Q_E = 0.098$ (1/s). Then for $2000 < t \leq 4000$ (s), the frequencies are set to be $Q_N = 0.294$ (1/s), $Q_S = 0.098$ (1/s), $Q_W = 0.029$ (1/s), and $Q_E = 0.074$ (1/s). As the initial values, the splits are set to be 0.5 and the phase angles are randomized. Other parameters are shown in Table 1, which is the same as the case of the previous subsection.

Table 3 shows the necessary time for automobiles in the present method, in comparison with the case of non-control ($\sigma = 0.5$, $\phi = 0$). This method can achieve a high performance, even in the dynamic traffic conditions.

Figure 7 shows the number of automobiles existing in the environment at time t, comparing with the case of non-control. At $t = 2000$ (s), the frequencies of arrival are changed. The number of automobiles is increased, since the offsets at that moment are suitable for the previous traffic conditions. At approximately $t = 2700$ (s), the phase angle differences are converged, forming new offset patterns appropriate for the current traffic condition. The number of automobiles for the present controlled case fluctuates

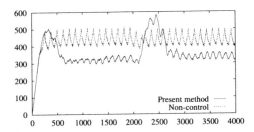

Figure 7: Number of automobiles in the environment at time t

within the range of approximately 300–350, while the corresponding range for the case of non-control is approximately 400–500. Thus, more efficient traffic flow is realized in the present controlled case. It is strongly suggested that the present method can quickly adapt to the changes in the traffic conditions.

5 Conclusion

This paper proposed a new method for controlling a large number of signals. Signals, being modeled as nonlinear coupled ocsillators with the dynamics by a gradient system, control their own splits and phase angles only by the local traffic conditions and the states of the neighbors. The global coordination, i.e., the coordinated offset/split control, is also realized by such local interactions.

The results of the simulation have shown a high stability in a stationary environment and a high adaptability in a dynamic environment. This method also achieved a high performance in both stationary and dynamic environments.

As a future work, we will handle the control of cycle lengths, along with splits and offsets. At the same time, the environment will be advanced to more practical one; traffic flows with left/right-turning, non-square signal layouts, and so on. Online estimation of the optimal offset is also a subject to be developed.

References

[1] Shogo Kawakami and Hiroshi Matsui: *Traffic Engineering*, Morikita Shuppan Inc., 1987 (*in Japanese*).

[2] Kousuke Sekiyama, Jun Nakanishi, Isao Takagawa, Toshimitsu Higashi and Toshio Fukuda: Self-organization of Offset Pattern in Urban Traffic Signal Network, *Proc. 13th SICE Symposium on Decentralized Autonomous Systems*, pp. 45–50, 2000 (*in Japanese*).

[3] Sadayoshi Mikami and Yukinori Kakazu: Self-Organized Control of Traffic Signals Through Genetic Reinforcement Learning, *Proc. IEEE Intelligent Vehicles*, pp. 113–118, 1993.

[4] Tadanobu Misawa, Haruhiko Kimura, Sadaki Hirose and Nobuyasu Osato: Multiagent-based Traffic Control with Reinforcement Learning, *Trans. IEICE*, D-I, Vol.J83-D-I, No.5, pp. 478–486, 2000 (in Japanese).

[5] Hideo Yuasa and Masami Ito: Internal Observation System and a Theory of Reaction-diffusion Equation on a Graph, *Proc. 1998 IEEE Int'l Conf. Systems, Man, and Cybernetics*, pp. 3669–3673, 1998.

[6] Jee-Hyong Lee and Hyoung Lee-Kwang: Distributed and Cooperative Fuzzy Controllers for Traffic Intersections Group, *IEEE Trans. Systems, Man, and Cybernetics* C, Vol.29, No.2, pp. 263–271, 1999.

Intelligent Autonomous Systems 7
M. Gini et al. (Eds.)
IOS Press, 2002

Fault-Tolerance for Autonomous Decentralized Multi-Legged Robot System

Takanori Suzuki
The University of Tokyo
Tokyo, Japan
suzuki@prince.pe.u-tokyo.ac.jp

Shinkichi Inagaki
The University of Tokyo
Tokyo, Japan
inagaki@prince.pe.u-tokyo.ac.jp

Hideo Yuasa
The University of Tokyo & RIKEN–BMC
Tokyo, Japan & Nagoya, Japan
yuasa@prince.pe.u-tokyo.ac.jp

Tamio Arai
The University of Tokyo
Tokyo, Japan
arai@prince.pe.u-tokyo.ac.jp

Abstract. Autonomous decentralized multi-legged robot is an applicative example of autonomous decentralized system. As one of the advantages of autonomous decentralized system, we propose a fault-tolerance algorithm in the case of an autonomous decentralized multi-legged robot. Moreover, we create the potential function in each leg which is required in our algorithm and verified by simulation. Finally, the effectiveness of our algorithm is verified, especially by applying it to a motor failure.

1 Introduction

Animals have a variety of walking patterns (Gait). Walking patterns which we deal with are a set of timing to move each leg. Various walking patterns which appear in animals are considered to be self-organically made from the neural oscillator group called CPG (Central Pattern Generator) [1](Figure 1). Walking patterns are formed only of the interaction between the connected oscillators (Figure 2).

Recently, many researchers have been working on the autonomous decentralized multi-legged robot system (we call ADMRS) [2, 3]. In general, it is well known that the autonomous decentralized system has high fault-tolerance [4]. However, the system design is very difficult because it is necessary to decide actions of the subsystem by using information from the neighbor parts, and quite few researchers have been studying on the fault-tolerance of ADMRS.

An adaptive algorithm of the autonomous decentralized system [4] was proposed, as one of the researches on the fault-tolerance of ADMRS. It is the algorithm to adjust the feature of the oscillator on the CPG network by which the interaction between each oscillator is reduced (Figure 3(a)). Besides, The Principle of Energy Consumption

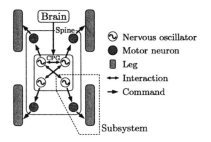

Figure 1: Concept figure of legs' control by CPG

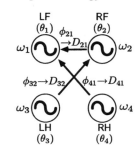

Figure 2: Example of a CPG Network

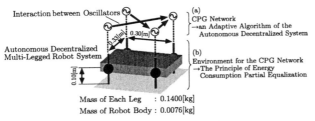

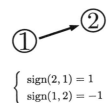

$$\begin{cases} \text{sign}(2,1) = 1 \\ \text{sign}(1,2) = -1 \end{cases}$$

Figure 3: CPG network and environment   Figure 4: Sign Function

Partial Equalization [3] was also proposed to partly equalize the consumption energy of each leg. In this case, the load of failure is distributed, and an environmental adjustment is achieved (Figure 3(b)). Here, "environment" is the outer space for the CPG network. The former adapts only to the changes inside the CPG network, whereas the latter shows the adaptive algorithm for the changing environment.

The legs of the robot are controlled by CPG. Therefore, it is necessary to reflect the influence of the failure in CPG, and to convert it into actions of CPG for achievement of the robotic fault-tolerance. However, a fault-tolerance algorithm considering the failure of the robot's physical structure has not been studied so far.

In this research, we propose a fault-tolerance algorithm of ADMRS. To approach it, we pay attention to the consumption energy in each leg. Here, we define "failure" as the condition of concentrating the load to certain legs, and "fault-tolerance" as the that of distributing the load to all the legs.

This paper is organised as follows. In Sec. 2, we will describe how the algorithm achieves fault-tolerance in ADMRS. In Sec. 3, we will describe the physical simulation to configure the potential function of the fault-tolerance algorithm. In Sec. 4, the effectiveness of fault-tolerance algorithm will be verified by using the potential function mentioned in Sec. 3. In Sec. 5, we will describe a real robot to verify the fault-tolerance algorithm. In Sec. 6, we will draw a conclusion and mention about the future projects of the fault-tolerance algorithm.

2   Fault-Tolerance Algorithm

Firstly, in our fault-tolerance algorithm, we measure the influence of failure as consumption energy difference among legs. Secondly, the oscillators' action changes in order to reduce the influence. Then, the oscillation pattern of CPG changes. Finally, the walking pattern changes to a better one according to the failure.

2.1   Oscillator's Action

In this paper, we use the Ito's [4] CPG model of the gradient system. In the model, a periodic movement of the leg is expressed as a oscillator phase.

Each leg moves depending on each oscillator's phase. θ_i is the phase of oscillator i. At this time, phase difference ϕ_{ij} between oscillator i and j is defined as

$$\phi_{ij} = \text{sign}(i,j)(\theta_j - \theta_i) \ . \tag{1}$$

Here, $\text{sign}(i,j)$ is the function (Figure 4) which shows the direction of oscillators' connection. With this definition, the phase difference between the oscillators is uniquely given. In short, the equation $\phi_{ij} = \phi_{ji}$ is formed.

The walking pattern is subscribed by the phase difference between oscillators. Here, we express the desired phase difference between oscillator i and j as D_{ij}. Therefore, it has to settle phase difference ϕ_{ij} between oscillator i and j to phase difference D_{ij} in the target walking pattern (Figure 2). Thus, we designed the oscillator's action with the gradient system, so that the phase difference between the oscillators converges to the desired phase difference.

Consequently, the potential function V_θ of the oscillators can be defined as

$$V_\theta(\phi) = \sum_i \sum_{j \sim i} \tau_\theta (D_{ij} - \phi_{ij})^2 . \tag{2}$$

Here, τ_θ is a common parameter which decides the adjustment speed of the oscillators. $j \sim i$ shows the set of oscillators in which oscillator j connects to oscillator i, and ϕ shows a set of desired phase difference ϕ_{ij}, that is, $\phi = \{\phi_{ij}\}$. For example, in Figure 2, $j \sim 2$ means the set of oscillator 1 and 3. The oscillator i's interaction f_i which is received from the connected ones is defined as

$$f_i = -\frac{\partial V_\theta(\phi)}{\partial \theta_i} \tag{3}$$

$$= \tau_\theta \sum_{j \sim i} 2 \, \text{sign}(i, j)(D_{ij} - \phi_{ij}) \tag{4}$$

so that the phase difference ϕ_{ij} can converge to the desired phase difference D_{ij}.

Here, the behavior of phase θ_i of each oscillator can be shown as Equation (5) where ω is the angular velocity of the oscillator, and it decides the walking speed of the robot.

$$\dot{\theta}_i = \omega + f_i . \tag{5}$$

2.2 The Principle of Energy Consumption Partial Equalization

The Principle of Energy Consumption Partial Equalization is used to measure the working stress (Interaction F_i is differing from the interaction f_i) on each leg of the robot. The algorithm is assumed that the consumption energy of the entire robot is minimized, when the consumption energy of each leg is equal to the mean consumption energy of the connected legs. Then the algorithm conducts the load-distribution. Interaction F_i is defined as

$$F_i[\boldsymbol{D}] = \frac{\sum_{j \sim i} e_j[\boldsymbol{D}]}{\deg(i)} - e_i[\boldsymbol{D}] \tag{6}$$

so that the consumption energy on the each leg becomes the mean value of the connected legs. Here, $\deg(i)$ shows the number of oscillators connected with oscillator i, and $\boldsymbol{D}$ shows a set of desired phase difference D_{ij}, that is, $\boldsymbol{D} = \{D_{ij}\}$. $e_i[\boldsymbol{D}]$ shows consumption energy on leg i at the walking pattern $\boldsymbol{D}$. Using the different sets of D_{ij}, energy evaluation function $e_i[\boldsymbol{D}]$ in each leg can be measured by repeating hundreds of experiments. [ ] shows the discrete function, and it means the consumption energy is measured in the discrete walking pattern $\boldsymbol{D}$.

2.3 Adaptive Algorithm of the Autonomous Decentralized System

In addition, we introduce the adaptive algorithm. This algorithm updates $\boldsymbol{D}$ to minimize the potential function $V_D[\boldsymbol{D}]$ (Figure 5). Considering the walking is the periodic movement, we design the potential function as follows to show the balance of consumption energy of one walking cycle.

$$V_D[\boldsymbol{D}] = \sum_i \int_T \sum_{j \sim i} \tau_v \{F_j[\boldsymbol{D}]\}^2 \, dt \tag{7}$$

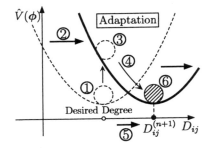
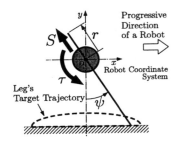

Figure 5: Update of a desired phase difference   Figure 6: Control method of every leg

Here, τ_D is a parameter to decide the adjustment speed of target phase difference $\boldsymbol{D}$, $V_D[\boldsymbol{D}]$ is the integration value of the square sum of interaction $F_i[\boldsymbol{D}]$ in the walking cycle T. The algorithm follows these steps:

Step 1: First of all, the measured potential function $V_D[\boldsymbol{D}]$ is linearized around the minimum value to obtain the potential function $\hat{V}(\phi)$ of an oval.

Step 2: Before the failure, ϕ is converged to $\boldsymbol{D}$ which makes $\hat{V}(\phi)$ minimized (Figure 5. ①).

Step 3: The shape of potential function $\hat{V}(\phi)$ changes when the failure occurs ($\hat{V}(\phi) \to \hat{V}'(\phi)$, Figure 5.②). Here, it is assumed that potential function $\hat{V}'(\phi)$ can be calculated numerically by the potential function $\hat{V}(\phi)$ before the failure. Thus, we deal with the "failure" which likes the failure leg's is certain multiple of the normal one.

Step 4: However, it keeps settling ϕ to $\boldsymbol{D}$ by Equation (4), (5) (Figure 5.③).

Step 5: Then, the inclination $-\dfrac{\partial \hat{V}'(\phi)}{\partial \phi}$ is obtained by $\hat{V}'(\phi)$ (Figure 5.④).

Step 6: Then, $\boldsymbol{D}$ is updated by Equation (8) (Figure 5.⑤).

$$D_{ij}^{(n+1)} = D_{ij}^{(n)} - \int_T \frac{\partial \hat{V}'(\phi)}{\partial \phi_{ij}}\, dt \tag{8}$$

Step 7: After some updates, ϕ converges to $\boldsymbol{D}$ which minimize $\hat{V}'(\phi)$ (Figure 5.⑥).

3 Physical Model Simulation

3.1 Physical Model and Calculation of Consumption Energy

The robot model of the physical model simulator is shown in Figure 3. Each leg has two degrees of freedom, virtical sliding and rotation, and its movement is limited in the sagittal plane. In this robot model, the body is very light compared with legs. There are three reasons for this design. The 1st reason is that the actual robot's body is an aluminum frame (Figure 7). The 2nd reason is to amplify the influence of leg failure by making legs heavier than the body. The final reason is to reduce the calculation time.

According to the oscillation pattern of CPG, each leg position is controlled to follow the target trajectory (Figure 6) in each leg coordinate system.

Concretely, phase θ decides a target toe position $\boldsymbol{x}$ uniquely. In order to follow the target toe position, leg length r and shake angle ψ is fed back to output force S and torque τ of the motor. Here, S is the force which works in the direction of shortening the leg, and τ is the torque which rotates the leg.

Using these values, the consumption energy $e_i[\boldsymbol{D}]$ on leg i par one walking cycle is calculated by Equation (9).

$$e_i[\boldsymbol{D}] = \int_T \left(S\dot{r} + \tau\dot{\psi} \right) dt \qquad (9)$$

3.2 Simulation Environment

We used the physical model simulator, "DADS" (Dynamic Analysis and Design System: Product by LMS International http://www.lms.be/), as the simulation environment.

The differential equation which describes the multi-legged robot walking is stiff with the geometrical restraint by the legs on the ground, and dynamic with sending off the leg to forward. Therefore, we selected the "ODE45 (Dormand-Prince method)" as a integration method to solve the differential equation. We set the "relative tolerance" of integration as 5.0×10^{-3}, and "absolute tolerance" of integration as 1.0×10^{-6} [J]. About 6.5 hours are required to calculate the walking simulation of 36 seconds of this robot model on this condition by using the computer, CPU: Pentiumn III 1[GHz], Memory: 512[MByte], and OS: Microsoft Windows NT 4.0 Service Pack 6a.

The initial phase of the oscillator is provided to achieve the target phase difference in each simulation in order that we can reduce the time to settle the oscillation pattern. Moreover, the robot is dropped from the height 0.10[m] at the same time as the simulation starts to lose an unnatural friction of the leg and the ground. However, it is thought that the influence of robot's dropping is lost in 8[s], because the consumption energy of each leg is steady. Therefore, as the consumption energy $e_i[\boldsymbol{D}]$ of each leg, we use the average value of the data from 8[s] to 36[s] at the simulation time. The angular velocity of the oscillator is 1.7[rad/s]. Thus, there are approximate 8 walking cycles used in one simulation.

3.3 Measurement of Potential Function

The potential function $V_D[\boldsymbol{D}]$ is shown in Figure 9. The axes in the figure show the difference of the desired phase difference ($\boldsymbol{D}$) between "walk" and other walking patterns. Here, "walk" is the walking pattern which the legs are move in order of Leg4 (Right Hindlimb) → Leg2 (Right Forelimb) → Leg3 (Left Hindlimb) → Leg1 (Left Forelimb) → Leg4 ⋯ (Figure 8), and the phase difference between each oscillator is $D_{21} = -\pi$, $D_{32} = \pi/2$, and $D_{41} = -3\pi/2$ relatively. Each axis shows the gap of the phase difference from the walk gait to the desired phase difference of the oscillator. The density of each coordinate indicates the value of potential function $V_D[\boldsymbol{D}]$. Total 215 points whose value of each axis is the following value:

$$\left\{ -\frac{\pi}{6}, -\frac{\pi}{9}, -\frac{\pi}{18}, \pm 0, \frac{\pi}{18}, \frac{\pi}{9}, \frac{\pi}{6} \right\}$$

Figure 7: Real robot of simulation model

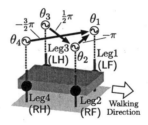

Figure 8: Robot Legs' No.

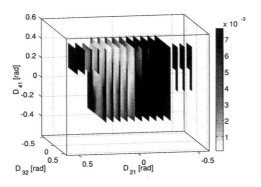

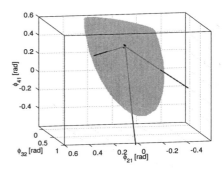

Figure 9: Potential function V_D for expressing consumption energy's balance before the failure

Figure 10: The potential function $\hat{V}(\phi)$ of equal estimation value

are measured. The minimum value of the potential function is

$$(D_{21}, D_{32}, D_{41}) = \left(-\pi, \frac{\pi}{2} + \frac{\pi}{9}, -\frac{3}{2}\pi + \frac{\pi}{9}\right) \quad . \tag{10}$$

The discrete potential function $V_D[\boldsymbol{D}]$ is 3-dimensional interpolated in $[-0.6, 0.6]$ of each axis. The measured potential function is linearized around the minimum value to obtain the potential function $\hat{V}(\phi)$ of an oval figure (Figure 10). Here, each axis of Figure 10 shows the principal direction of the oval potential function. The unit vector of principal direction, variance and contribution rate in the (D_{21}, D_{32}, D_{41}) coordinate system to each axis are shown in Table 1.

The priority should be given to the adjustment of the phase difference for the direction with the steepest inclination. Thus, in this robot, D_{21} is the most important direction to decentralize the consumption energy in legs. In a word, the phase difference between the right leg and the left leg has a large influence on the consumption energy of the robot.

4 Fault-Tolerance Simulation

We assume that it is a failure when the use of the motor electric supplied power is more than ordinary.

We suppose a situation in which the consumption energy in the right forelimb (Leg2) is 1.05 times of motor's work. It is necessary to assume not the control part failure (Calculation part: CPG), but the motor failure (Driving part). A 4-legged robot becomes impossible to walk only one leg's motor fails. Moreover, the acceptable consumption energy range is not so large. Because, over-enlarging the difference between the failed leg and the normal one is supposed to cause the significant difference between ϕ which

Table 1: Unit vector, variance and contribution rate in each principal direction

	1st Principal	2nd Principal	3rd Principal
Unit Vector of Principal Direction	−0.1446	−0.3756	0.9154
	−0.4010	0.8680	0.2928
	−0.9046	−0.3247	−0.2761
Variance	0.0390	0.0249	0.0072
Contribution Rate	0.5485	0.3502	0.1013

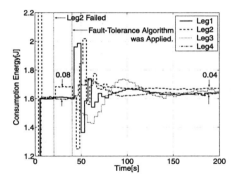

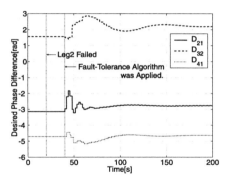

(a) Consumption energy on legs

(b) Desired phase difference between Oscillators

Figure 11: Simulation Results of applying the Fault-Tolerance algorithm

makes the potential function $\hat{V}'(\phi)$ minimize and D of Equation (10). So, we suppose the failure in which the consumption energy in the leg is 1.05 times of original one.

The result of applying the fault-tolerance algorithm to this failure is shown. The consumption energy of each leg is shown in Figure 11(a), and the target phase difference between each set of oscillators is shown in Figure 11(b). The minimum value of the potential function after Leg2 failure is

$$(D_{21}, D_{32}, D_{41}) = \left(-\pi + \frac{\pi}{9}, \frac{\pi}{2} + \frac{\pi}{9}, -\frac{3}{2}\pi - \frac{\pi}{9}\right) \quad . \tag{11}$$

We change the target phase difference between oscillators to reduce the influence of the failure of Leg2 (see Sec. 2). As a result, the difference of the consumption energy among the legs is decreased from 0.08[J] to 0.04[J] after applying the fault-tolerance algorithm. Here, [J] (Joule) is assumed to be a unit which shows the consumption energy in the simulation environment. Therefore, it is confirmed that the influence of the leg failure is reflected in the CPG network, and the propposed effectiveness of fault-tolerance algorithm is verified.

However, the influence of the consumption energy in the failure leg is only 1.05 times of the original one, which is too little in the case of a real robot. Moreover, it is not possible to adapt the walking pattern into failures such as no movement. Therefore, it is important to verify the falut-tolerance algorithm by using a real robot.

5 Real Robot Experiment

In Sec. 4, we used a 4-legged robot model shown in Figure 3 to verify the effectiveness of the fault-tolerance algorithm. However, it is very difficult to make an autonomous decentralized 4-legged robot walk. A 4-legged robot becomes impossible to walk only with one leg failure. However, it is possible for a hexapodal robot to keep static walking, if only one leg fails and does not move, because the gravity center and ZMP[5] (Zero Moment Point) are included in the support polygon that made of the legs on the ground. Therefore, a more-than-6 legged robot is suitable for the verification of the fault-tolerance algorithm.

To verify the effectiveness of the fault-tolerance algorithm in a real robot, ADMRS (named "NEXUS") [6] shown in Figure 12 is developed. Each leg is about 0.9[kg]. MPU (Micro-Processing Unit: H8S/2655, 20[MHz]) is used for the control of legs. The

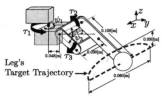

Figure 12: Real Robot NEXUS: W0.32, H0.23, L0.37[m]

Figure 13: Real Robot NEXUS on the treadmil

Figure 14: Control method of every leg

serial communication between MPUs is conducted. The phase difference, the target phase difference, and the consumption energy in legs are communicated as information. The structure of each leg is shown in Figure 14. The same as in the simulation, the joint angles of the legs are controlled to follow the target trajectory. Moreover, the consumption energy in each leg is measured as the electrical current of each motor, by using the analog to digital conversion.

6 Conclusion

In this paper, a fault-tolerance algorithm considering of a robot structure was presented. The potential function of the robot were proposed by the physical model simulation. Furthermore, the effectiveness of the algorithm was confirmed by the simulation. We are now going to make an algorithm with a real robot. Now, the fault-tolerance algorithm is installed in NEXUS, and we are experimenting to verify the algorithm by making NEXUS walk on the treadmil (Figure 13).

As one of our future projects, we are going to propose the fault-tolerance algorithm by using the symmetry of the walking pattern and CPG network. Moreover, the proposal of the fault-tolerance algorithm in other CPG models, for example the Hamilton system [6] and etc. are considered to be interesting themes.

References

[1] Keir Pearson. The Control of Walking. *Scientific American*, Vol.235, pp.72–74,79–86, 1976.

[2] K. Tsujita, K. Tsuchiya, A. Onat, S. Aoi and M. Kawakami. Locomotion Control of a Multipod Locomotion Robot with CPG Principles. Proc. of The Sixth International Symposium on Artifical Life and Robotics, Vol.2, pp.421–426, 2001.

[3] Tadashi Odashima, Hideo Yuasa, Zhi-wei Luo, Masami Ito. Emergent Generation of Gait Pattern for a Myriapod Robot System Based on Energy Consumption, *Journal of the Robotics Society of Japan (In Japanese)*, Vol.17, No.8, pp.1149–1157, 1999.

[4] Satoshi Ito, Hideo Yuasa, Zhi-wei Luo, Masami Ito, Dai Yanagihara. A mathematical model of adaptive behavior in quadruped locomotion, *Biological Cybernetics*, Vol.78, pp.337–347, 1998.

[5] Miomir Vukobratović. Legged Locomotion Robots, *The Nikkan Kogyo Shimbun(In Japanese)*, 1975.

[6] Shinkichi Inagaki, Hideo Yuasa, Takanori Suzuki, Hideki Kumagai, Tamio Arai. Gait Pattern Generation of Autonomous Decentralized Multi-Legged Robot System — CPG Model and Development of Hardware —, *Proceeding of 14th SICE Symposium on Decentralized Autonomous Systems (In Japanese)*, 2C2–3, 2002.

Intelligent Autonomous Systems 7
M. Gini et al. (Eds.)
IOS Press, 2002

Active Artifacts: for New Embodiment Relation between Human and Artifacts

Kazunori TERADA

Synsophy Project, Communications Research Laboratory

2-2-2, Hikaridai, Seika-cho, Soraku-gun, Kyoto, 619-0298, Japan

E-mail: kazuno-t@crl.go.jp[†]

Toyoaki NISHIDA

Department of Information and Communication Engineering,

School of Engineering, The University of Tokyo

7-3-1 Hongo, Bunkyo-ku, Tokyo 113-8656, Japan

Abstract. The *active artifact* provides new embodiment relation between human and artifacts. In the active artifact paradigm, an artifact perform its task by itself. In this paper we argue the implicit embodiment communication between active artifact and human and propose an architecture to realize such communication. In order to perform such communication, we introduce a concept of *affordance distance*. We employ utility function in order to express the affordance distance. We have built an *autonomous mobile chair* as an example of active artifacts. In order to show the validity of our proposed approach, experiments have been carried in the computer simulation.

1 Introduction

In recent years, there have been many efforts towards designing autonomous robot which improves the quality of our life [5][2][4][7]. Humanoid robots may be most suitable for laboring instead of human in our everyday work, since our living areas are designed with human usage. Humanoid robots are aimed at performing tasks requested from human. If we want to take a rest, we can request a humanoid robot to bring a chair. However, there is a serious problem in the humanoid paradigm. The problem is that the humanoid has to equip sophisticated interface, for example, natural language processing system and speech interface, so that it can communicate to a human to understand what the human desires. It is difficult for artificial system to understand the natural language. The reason causing this problem is that a humanoid robot has various functions and has to perform various work like a human. In order to avoid this problem, we propose *active artifacts*. The active artifact is an everyday artifact such as furniture which has simple function, but it actively performs its task by itself. The same concept of active artifact is seen in Things That Think project proposed by MIT Media Laboratory. The project attempts to embed computational capabilities into everyday object such as toys, clothing, furniture, and even balloons. Both concepts of active artifact and Things That Think are large one including various research issues. In this paper, we focus on the implicit embodiment communication issue between human and active artifact.

The active artifact does not need equip so sophisticated interface for communication, because communication required to perform simple task is limited. Therefore, the active artifact should understand only a few specific signal. For example, an automatic door, which is an existing active artifact, should detect only whether human is approaching or not. Furthermore, exchange of the explicit signal like language does not need for communication between human and active artifact, because the task is determined by embodiment relation between human and artifact. Instead of the explicit communication, implicit embodiment communication is important for task execution. In this paper we argue the implicit embodiment communication and propose an architecture to realize such communication.

2   Embodiment communication between human and artifact

Some psychological researches have concluded that more than 65 percent of the information exchanged during a face-to-face interaction is expressed through nonverbal information in human-human communication[1]. There are some classifications of nonverbal information[8][6]. We introduce typical classification; 1)gesture, 2)head movements, 3)facial expression, 4)eye contact and gaze, 5)kinesics, body language, 6)proxemics, 7) haptics, use of touch, and 8)silence and breathing. On the other hand, these classes can be mainly divided into two modes; (1)explicit mode and (2) implicit mode. In the explicit mode, connection is established between sender and receiver of information, and signals with certain meaning or intention are exchanged. In the implicit mode, connection is not established and a signal with the clear meaning is not exchanged.

Each artifact has its own function. For example, a chair has the function to sit down. If we want to take a rest, we can request an active chair to come. In order to tell our intention of sitting, we firstly use explicit communication channel, for example gesture or language. Once the intention is understood by chair, the chair moves to us. While the chair moves, the embodiment relation between human and chair changes as the chair moves. Both the chair and human should have to optimize the embodiment relation in order to complete the sitting task. It can be said that implicit communication is to acquire the embodiment information mutually and execute task collaboratively between human and active artifact.

3   Affordance distance

We focus on the embodiment relation between human and artifacts in order to control the active artifacts. In the following section, we describe an architecture with which the artifact can realize its function autonomously. Firstly, we consider a description method of functional relation between human and artifact. In the field of ecological psychology, there is a term of *affordance*[3]. Affordance means a property of environment which is appeared by interaction between human and environment. For example a chair has affordance to sit down, and it is not appeared unless human generate an action to sit down. Gibson said that it is important that the agent perceive the relative relation between of surfaces between agent's body and environment, and such relation is described relatively by agent's body as a standard. Based on the concept of affordance, we consider a description method of functional relation between human and artifact. As mentioned by Gibson, physical relation, especially, surface relation, is important for agent to behave in physical world. In order to describe relation between a surface of agent's body and a surface of environment, we introduce a concept of *affordance distance*. Affordance distance is a value and defined as following description.

- Affordance distance will be minimum value at the end of action sequence, i.e., a tactile state.

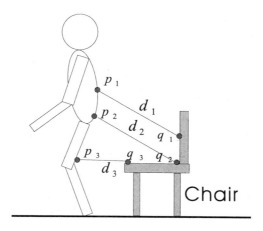

Figure 1: Distance of affordance

– Affordance distance reduces as the agent goes away from the tactile state.

– Affordance distance is defined between a point of artifact's body and a point of human.

For example, in the case of chair drawn in figure 1, the affordance distance will be minimum value when the human sit down, and the affordance distance will reduce as the human goes away form chair. Affordance distance is not a Euclid distance but corresponds to the action cost which is required when the artifact moves to a state when it touches to human, because an optimal action path to minimize the distance between two points is not necessary shortest distance. The optimal action path depends on locomotive ability of artifact and relative angle of two surfaces.

3.1 Calculation and minimizing of the affordance distance

The purpose of artifact is to search and move a point where the affordance distance is minimum. In this subsection we describe a calculation method of affordance distance and control method of artifact. We employ utility function in order to express the affordance distance. Utility function is widely used in a research area of autonomous agents. Utility function can represent a distance to the goal state considering the locomotive ability of agent.

The utility value of each state is calculated by following equation.

$$U(s) = R(s) + \max_a \sum_{s'} M_{ss'}^a U(s') \tag{1}$$

where M_{ij}^a is the transition probability of reaching state j if action a is taken in state i. $M_{ss'}^a$ is obtained by repetition of same actions from a same state;

$$M_{ss'}^a = \frac{n_{s'}}{n_s^a} \quad . \tag{2}$$

where n_s^a is the number of times of action a at state s, and $n_{s'}$ is the number of reaching state s'. where $R(i)$ is a reward function which returns a value of reward in state i. We give a reward when a certain point of artifact's body touches a certain point of humans body appropriately, i.e., the affordance distance is minimum. The tactile condition is;

$$(x, y, z)_p = (x, y, z)_q \wedge (\theta, \phi, \varphi)_p = -(\theta, \phi, \varphi)_q \tag{3}$$

where $(x, y, z)_p$ and $(x, y, z)_q$ is the coordinates of point of humans body and artifact's body respectively, and $(\theta, \phi, \varphi)_p$ and $(\theta, \phi, \varphi)_q$ is an angle of normal vector of point of humans body and artifact's body respectively.

In our method we use a simple iterative algorithm called *value iteration* in order to calculate the utility value of each state. Value iteration procedure is performed by following equation.

$$U_{t+1}(s) \leftarrow R(s) + \max_a \sum_{s'} M_{ss'}^a U_t(s') \tag{4}$$

Where $U_t(s)$ is utility value of s after t times iteration. It is know that the utility value converges to the correct value when $t \to \infty$.

Given an utility function U and if a state transition holds Markov property, optimal policy for Markov decision problem is calculated as follows;

$$f(i) = \arg \max_a \sum_j M_{ij}^a U(j) \tag{5}$$

where $\arg \max_a f(a)$ returns the value of a with the highest value for $f(a)$.

4 Experiment

In this section we describe our experimental system and simulation result.

4.1 Autonomous mobile chair

We have built an autonomous mobile chair as an example of the active artifact. We remodeled some parts of an alminium chair so that it can move around (See Figure 2). The chair has five legs radiating in all directions, and each leg has a caster wheel which can rotate in horizontal plane freely. In our system, we replace two caster wheels with powered wheels each of which is fixed to the leg so that it could not rotate in horizontal plane. The autonomous mobile chair equips motion capture system made by Ascension Technology which enable to measure the position and orientation of its body. The motion capture system employs pulsed DC magnetic-field transmitting and sensing technology to measure the position and orientation of miniaturized sensors attached to measuring equipment. The autonomous mobile chair is controlled by Linux PC connected via RS232C cable. A subject in this experiment also has to carry motion sensor so that the autonomous mobile chair may know a reaching point.

4.2 Simulation results

In order to show the validity of our proposed method, experiments have been carried in the computer simulation. The purpose of the autonomous mobile chair is to contact its back reclining on humans back.

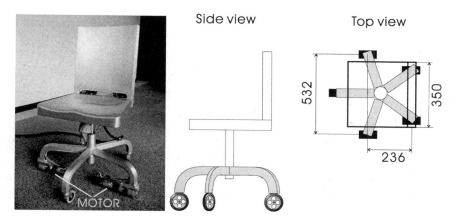

(a) Two caster wheels are replaced with powered wheel.

(b) The modeling of autonomous mobile chair.

Figure 2: Autonomous mobile chair.

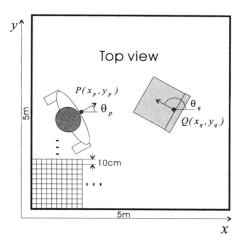

Figure 3: Modeling of the environment and state space. The task of the autonomous mobile
chair is making Q reach P.

utility value

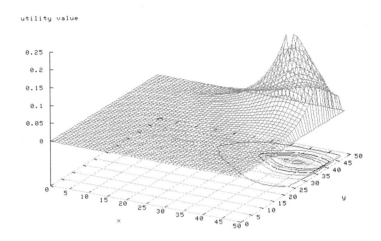

Figure 4: Utility function. The goal state is $(40, 40, 0)$. The utility value is mean value of 16 states of θ.

4.2.1 Modeling of the autonomous mobile chair and state space

In order to perform computer simulation, the environment and the autonomous mobile chair was modeled as follows. The environment is floor of $5m$ square. The size of chair and arrangement of wheels is as Figure 2(b). The state space is constructed as follows (See Figure 3). We assume that the dimension of the autonomous mobile chair, and the environment is 2 for the simplification of the simulation, that is, height is disregarded and the agent can move only on the 2D plane. As a result, the dimension of the state space turned into 3 dimensions; (x, y, θ). The floor is divided into 50×50 grid. The angle of a normal vector of the surface is discritized into 16 steps. As a result, the number of the state is $50 \times 50 \times 16$. We also assume that we can send 6 action commands to the autonomous mobile chair. The command is executed by specification of speed for each motor. Commands are $A1(-V, V)$, $A2(0, V)$, $A3(V, V)$, $A4(V, 0)$, $A5(V, -V)$, and $A6(-V, -V)$. In this experiment V is $0.3m/sec$. We define an action unit as a segment of the movement until a change of state is observed.

4.2.2 Calculation of utility value

Firstly, transition probability model is calculated. Each of six action command is executed from 100 uniform points in a grid, i.e., n_s^a in equation (2) is 100. This operation is performed for each 16 angles. A transition probability mode of a certain one grid is applied to other grids since transition probability is equal in every grid.

Next, the utility function is calculated. The goal point, i.e., the contact point, is set to $(40, 40, 0)$. The reward given in goal point is 1. Figure 4 shows the utility function

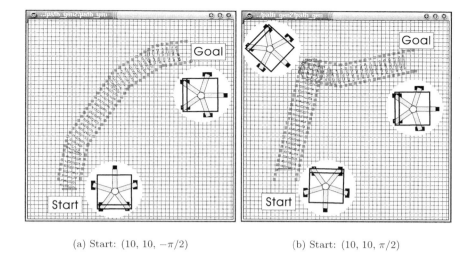

(a) Start: $(10, 10, -\pi/2)$ (b) Start: $(10, 10, \pi/2)$

Figure 5: Generated path.

calculated using the dynamic programming algorithm.

4.2.3 Reaching experiments

In order to show the validity of calculated utility function, we have reaching experiments of two cases. One case, i.e., case (a) is that the state state of autonomous mobile chair is $(10, 10, -\pi/2)$, and the other case, i.e., case (b) is $(10, 10, \pi/2)$. Figure 5 shows generated path of two cases. In both cases, the autonomous mobile chair generates reaching path appropriately and reaches the goal. In case (a), the autonomous mobile chair change the direction gradually as it reaches the goal. In case (b), the autonomous mobile chair backs once and changes the direction, and then it moves to the goal.

5 Discussion and Conclusion

In this paper we proposed the active artifact as anti-humanoid paradigm and discussed the embodiment aspects of communication. Implicit communication is to acquire the embodiment information mutually and execute task collaboratively between human and active artifact. In order to realize such implicit communication, we introduce a concept of *affordance distance*. We employ utility function in order to express the affordance distance. We have built an *autonomous mobile chair* as an example of active artifacts. In order to show the validity of our proposed approach, experiments have been carried in the computer simulation. Experimental result shows that the autonomous mobile chair can generate appropriate path to the goal by means of acquired utility function. Therefore, it can be said that affordance distance is suitable for realize implicit communication.

In the future work, an experiment with real autonomous mobile chair will be performed. Furthermore, we intend to argue the explicit communication between human and active artifact, and build complete useful system of active artifact.

References

[1] Michael Argyle. *Bodily Communication*. Methuen & Co., 1988.

[2] Rainer Bischoff and Tamhant Jain. Natural communication and interaction with humanoid robots. In *Second International Symposium on Humanoid Robots*, pages 121–128, 1999.

[3] James J. Gibson. *The Ecological Approach to Visual Perception*. Houghton Mifflin Company, 1979.

[4] Kazuo Hirai, Masato Hirose, Yuji Haikawa, and Toru Takenaka. The development of honda humanoid robot. In *IEEE International Conference on Robotics and Automation*, pages 1321–1326, 1998.

[5] Toshihiro Matsui, Hideki Asho, and Futoshi Asano. Map learning of an office conversant mobile robot, jijo-2, by dialogue-guided navigation. In *International Conference on Field and Service Robotics*, pages 230–235, 1997.

[6] Miles L. Patterson. *Nonverbal behavior: A functional perspective*. Springer-Verlag New York. Inc., 1983.

[7] Nicholas Roy, Gregory Baltus, Dieter Fox, Francine Gemperle, Jennifer Goetz, Tad Hirsch, Dimitris Margaritis, Michael Montemerlo, Joelle Pineau, Jamieson Schulte, and Sebastian Thrun. Towards personal service robots for the elderly. In *Workshop on Interactive Robots and Entertainment*, 2000.

[8] Marjorie Fink Vargas. *Louder than Words: An introduction to nonverbal communication*. Iowa State University Press, 1986.

Intelligent Autonomous Systems 7
M. Gini et al. (Eds.)
IOS Press, 2002

Analysis of the Adaptability of an Evolved Neurocontroller with Neuromodulations

Seiji Tokura

Dept. of Electrical Engineering, Nagoya University,
Nagoya 464-8603, Japan
tokura@cmplx.cse.nagoya-u.ac.jp

Akio Ishiguro, Hiroki Kawai

Dept. of Computational Science and Engineering, Nagoya University,
Nagoya 464-8603, Japan
{ishiguro, hiroki}@cmplx.cse.nagoya-u.ac.jp

Peter Eggenberger

Artificial Intelligence Laboratory, University of Zurich,
CH-8057 Zurich, Switzerland
eggen@ifi.unizh.ch

Abstract. One of the serious drawbacks in the Evolutionary Robotics approach is that evolved agents in simulated environments often show significantly different behavior in the real environment due to unforeseen perturbations. This is sometimes referred to as the gap problem. In order to alleviate this problem, we have so far proposed Dynamically–Rearranging Neural Networks(DRNN) by introducing the concept of neuromodulations. In this study, in order to quantitatively analyze the adaptability of evolved neuromodulatory effect a state–transition map technique was employed. Through this analysis, it is recongnized that the DRNN outperforms a standard neural network in terms of adaptability.

1 Introduction

Recently, the Evolutionary Robotics(ER) approach has been attracting a lot of attention in the field of robotics and artificial life [1]. In contrast to the conventional approaches where designers have to construct controllers in a top–down manner, the methods in the ER approach have significant advantages since they can autonomously and efficiently construct controllers by taking *embodiment* (e.g. physical size and shape of robots, sensor/motor properties and disposition, etc.) and the *interaction dynamics* between the robot and its environment into account.

In the ER approach, artificial neural networks are widely used to construct controllers for autonomous mobile agents. Because they can generalize, are non–linear and noise–tolerant [1, 2, 3, 4]. Another advantage of neural network–driven robots is that a neural network is a *low–level description* of a controller. More precisely, it directly maps sensor readings onto motor outputs. Although the ER approach has the above advantages, there still exist several issues that can not be neglected. One of the most serious issues is known as the *gap problem*: evolved agents in simulated environments often show significantly different behavior in the real environment due to unforeseen perturbations. Therefore, it is highly indispensable to develop a method which enables the evolved controllers to adapt not only to specific environments, but also to environmental perturbations.

Conventionally, to realize this requirement, many authors have been using techniques such as fitness–averaging [5] and/or adding noise [6, 7] in the evolutionary process.

However, we speculate that these *passive* approaches will not be essential solutions for the above requirements. What is important for robust controllers is dynamically to change the function of controller and to adapt the action of robot for each situation[8, 9].

In order to satisfy this requirement, we have so far proposed *Dynamically–Rearranging Neural Networks* (hereafter, DRNN) by introducing the concept of *neuromodulations* (hereafter, NMs) with a diffusion–reaction mechanism of signaling molecules called *neuromodulators* [10, 11]. However, a detailed analysis of the evolved DRNN remains uninvestigated.

In this study, we analyze the evolved DRNN and carry out a quantitative comparison of DRNN and standard *monolithic* neural networks(SNN) in which synaptic weights and neuron's bias are simply the targets to be evolved. In order to visualize how the evolved neuromodulation process works to increase the adaptability, a *state–transition map* technique is implemented. By using this visualization technique, the way of evolved neuromodulation mechanism contributes to increase the adaptability against environmental perturbations is observed.

2 Basic Concept of DRNN

The basic concept of our proposed DRNN is schematically depicted in Figure 1. As in the figure, unlike conventional neural networks, we assume that each neuron can potentially diffuse its specific (i.e. genetically–determined) type of NMs according to its activity, and each synapse has receptors for the diffused NMs. We also assume that each synapse independently interprets the received NMs, and changes its properties (e.g. synaptic weights). In the figure, the thick and thin lines denote the connections being strengthened and weakened by NMs, respectively. The way these changes are exerted on the synapses is also genetically–determined.

It should be stressed that the functions such as *situation evaluation, behavior generation* and *behavior regulation* can all be embedded within a *monolithic* neural network. This is the core of our concept. Due to this remarkable feature, we expect that whole genetic information can be reduced rather than straightforward approaches.

In summary, in contrast to the conventional ER approach that evolves synaptic

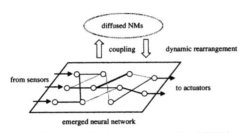

Figure 1: Basic concept of the DRNN.

weights and neuron's bias of neurocontroller, in this approach we evolve the following mechanisms:

- Diffusion of NMs (when and which types of NMs are diffused from each neuron?)

- Reaction to the diffused NMs (how do the receptors on each synapse interpret the received NMs, and modify the synaptic property?)

3 Analysis of the Evolved Agents

3.1 Task: a peg–pushing problem

In this study, we use a *peg–pushing problem* as a practical example. Here the task of the robot is to push the peg toward a light source.

The robot is equipped with six infra-red sensors which tell the distance to the detected peg in the frontal side, three light detectors each of which has different heading angle (left, center and right), and two independently–driven DC motors.

3.2 The controller

As mentioned previously, the aim of this study is not only to analyze the evolved DRNN controller but also to quantitatively compare the adaptability of DRNN with that of SNN. In this study for ease of analysis we use a feedforward neural network with three layers for DRNN and SNN controllers. Both of these controllers consist of nine sensory neurons (i.e. six infra-red, and three light direction sensors) and four motor neurons (i.e. CW/CCW motor neurons for the right and left motors) in the network. The significant difference between these neurocontrollers is the number of hidden neurons: the DRNN controller has five hidden neurons, whilst the SNN controller has 12 hidden neurons. This is because it is preferrable to set the length of genome for these neurocontrollers to be nearly identical in order to exclude the difference of the search space. For detailed information about how these neurocontrollers were evolved, see [11, 12].

3.3 Results

The DRNN and SNN controllers were evolved under the condition in which initial relative angles to the peg and the light source with respect to the heading direction of the robot are selected at random from -21 [deg.] to +21 [deg.] Figure 2 depicts typical trajectories of the best evolved agents under the condition experienced during their evolved process. From the figures, it is recognized that both evolved agents can successfully carry the peg toward the goal.

However, as mentioned earlier, the evolved agent should not only be adapted to the given environments during the evolutionary process, but also be able to act robust against environmental perturbations. Without this we can not seamlessly transfer the evolved agents from the simulated to the real world. In the following, we show the comparison of adaptability of both methods.

Figure 3 shows resultant trajectories of the same controllers under a newer-experienced condition during the evolved process. To put it more concretely, the relative angle to the peg with respect to the robot as initial condition was set to be more than +21 deg.

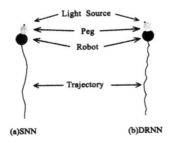

Figure 2: Resultant trajectories of the best evolved agents under the condition experienced during their evolutionary processes.

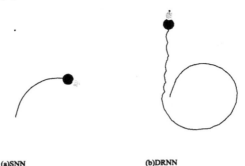

Figure 3: The resultant trajectories under the unexperienced situations.

As the figure indicates in contrast to the DRNN controller the SNN controller can not successfully push the peg toward a light source.

Now a question arises. Why can the DRNN have higher adaptability than the SNN? How does the DRNN increase the adaptability? In order to visualize the adaptability of the DRNN and the SNN, here we introduce a technique using a type of *state–transition map*. Figure 4 schematically illustrates the state–transition map used in this study. The vertical axis of this map(Figure 4[b]) denotes the relative angle to the light source with respect to the heading direction of the robot(see θ_L in the figure [a]). The horizontal axis represents the relative angle to the peg(see θ_P in the figure [a]). Each arrow in the map was drawn in the following way: we first put the evolved robot at a certain state and then recorded the resultant state after one step. For example, the arrow from A to B in figure (b) means that the robot will move to state B when it is put on state A. Note that in this task the robot should reach line C (i.e. the diagonal line) for successful peg–pushing toward the light source.

Figure 5 represents the resultant state–transition map of the evolved SNN. This map is obtained under the same initial condition experienced during the evolutionary process. The solid line denotes the trajectory of the state transition. As the figure indicates the trajectory is moving around line C. This implies that the controller can successfully push the peg toward the light source. However, it should be noted that the directions of the arrows in some regions are totally inappropriate (i.e. heading away from line C). Another important thing to be noticed is that there exist some regions that contain no arrows (i.e. *empty region*). We frequently observed that such regions are very easily created through the evolutionary processes. Consequently it is often observed that the robot can not reach the line C under the conditions never experienced in the evolutionary process. We will give an example to show how these regions influence the behavior of the controller. The solid line in Figure 6 is the resultant trajectory under the different initial condition never experienced in the evolutionary process. As in the figure the trajectory can not reach and stay on line C. In other words the SNN controller can not cope with such environmental perturbation. In sum, monolithic neurocontrollers tend to be very specific to the given environments, and it is extremely difficult to evolve SNN controllers which possess appropriate sets of state–transition vectors.

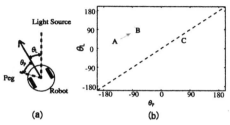

Figure 4: Explanation of the state–transition map used for the sake of visualization.

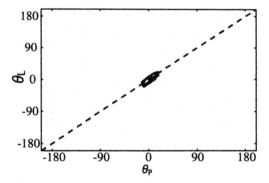

Figure 5: State–transition map of the SNN and the trajectory under the initial experienced situation in the evolutionary process.

As already mentioned, in contrast to SNN controllers the significant feature of evolved DRNN controllers is that they can modulate their own circuits' properties (e.g. synaptic weights) in real time during performance. Figure 7 depicts how the evolved DRNN controller dynamically modulates its synaptic weights during the peg–pushing task. Due to this dynamic modification of the synaptic weights, the DRNN controller can change its function, and therefore the state–transition map of the evolvement DRNN will dynamically change according to the current situation. Figure 8 shows the transition of the state–transition maps of the evolved DRNN controller. These maps are obtained under different initial conditions never experienced in the evolved process. As in the figure, the robot can successfully visit the desired states(i.e. the diagonal line). Interestingly, the figure indicates the direction and the length of the arrows are dynamically changed according to the situation in order to eventually reach and stay on the diagonal line. This is the strength of the neuromodulation mechanism.

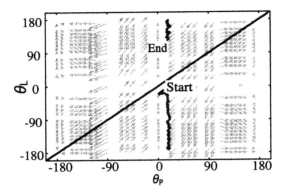

Figure 6: State-transition map of the DRNN and the trajectory under the initial unexperienced situation in the evolutionary process.

Due to lack of space, we need mention here only that evolving *the way of regulation*(i.e. neuromodulation) instead of simply evolving synaptic weights is much more powerful if one wants to create adaptive neurocontrollers.

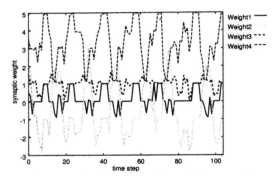

Figure 7: An example of the transition of the synaptic weights.

4   Conclusions and Further works

In this study, we analyzed the evolved DRNN and investigated how the neuromodulation process works to increase adaptability against environmental perturbations. In order to visualize the neuromodulatory effects on the adaptability of the evolved controllers, we introduced a technique based on a state–transition map. We observed that the network's property changed dynamically in order to generate appropriate behavior according to the current situation. We expect such a *polymorphic* property is essential for any adaptive neurocontrollers.

Currently, we use a simple direct encoding scheme to evolve the DRNN. However, both the network architecture and the way of diffusion and reaction to NMs in the DRNN are closely related not only to the given task but also to the embodiment and the interaction dynamics within the environment. Thus it is preferable to automatically

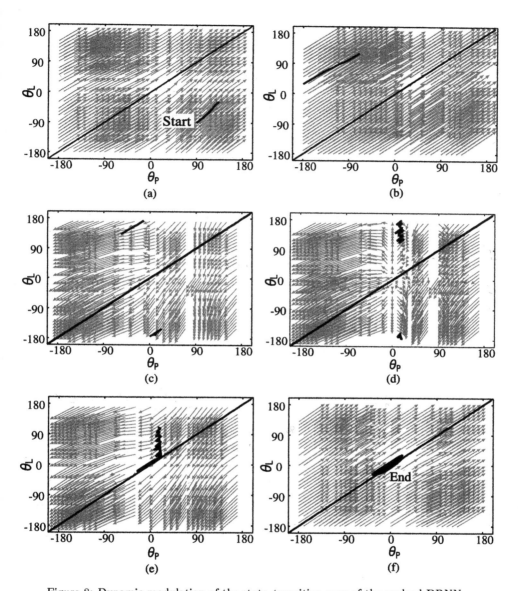

Figure 8: Dynamic modulation of the state–transition map of the evolved DRNN.

determine these parameters through the evolutionary process. In order to exclude possible presuppositions on these parameters we are currently investigating the concept of the *developmental process.*

Acknowledgements

This research was supported in part by Research Fellowships of the Japan Society for the Promotion of Science for Young Scientists, and a grant from the Japanese Ministry of Education, Culture, Sports, Science and Technology(No.C:12650442).

References

[1] Floreano, D. and Mondada, F. "Automatic creation of an autonomous agent: Genetic evolution of a neural-network driven robot", *Proc. of the 3rd International Conference on Simulation of Adaptive Behavior*, MIT Press, pp.421–430, 1994.

[2] Nolfi, S. Parisi, D. "Learning to adapt to changing environments in evolving neural networks", *Adaptive Behavior*, **5**-1, pp.75–98, 1997.

[3] Beer, R. Chiel, J. and Sterling, L. "An artificial insect", *American Scientist*, **79**, pp.444–452, 1989.

[4] Ackley, D. Littman, M. "Interactions Between Learning and Evolution", *Artificial Life II*, Addison-Wesley, pp.487–509, 1992.

[5] Reynolds, C. W. "An Evolved, Vision-Based Model of Obstacle Avoidance Behavior", *ARTIFICIAL LIFE III*, pp.327–346, 1994.

[6] Jacobi, N. Husbands, P. and Harvey, I. "Noise and the Reality Gap: The Use of Simulation in Evolutionary Robotics", *Third European Conf. on Artificial Life (ECAL95), Advances in Artificial Life*, Springer, pp.704–720, 1995.

[7] Miglino, O. Lund, H. H. and Nolfi, S. "Evolving Mobile Robots in Simulated and Real Environments", *Artificial Life 2*, pp.417–434, 1995.

[8] Floreano, D. and Urzelai, J. "Evolutionary robots with on-line self-organization and behavioral fitness", *Neural Networks 13*, pp.431–443, 2000.

[9] Husbands, P. Philippides, A. Smith, T. and O'Shea, M. "The Shifting Network: Volume Signalling is Real and Robot Nervous Systems", *Advances in Artificial Life(ECAL2001)*(Eds. J.Kelemen and P.Sosik), (Lecture Notes in Artificial Intelligence 2159), Springer, pp.23-36, 2001.

[10] P. Meyrand, J. Simmers and M. Moulins, "Construction of a pattern-generating circuit with neurons of different networks", *NATURE*, Vol.351, 2 MAY pp.60-63, 1991.

[11] P.Eggenberger, A.Ishiguro, S.Tokura, T.Kondo, T.Kawashima and T.Aoki, "Toward Seamless Transfer from Simulated to Real Worlds: A Dynamically-Rearranging Neural Network Approach", *Advances in Robot Learning*(Eds. J.Wyatt and J.Demiris), Lecture Notes in Artificial Intelligence 1812, Springer, pp.44-60, 2000.

[12] S.Tokura, A.Ishiguro, H.Kawai, P.Eggenberger, "The Effect of Neuromodulations on the Adaptability of Evolved Neurocontrollers", *Advances in Artificial Life(ECAL2001)*(Eds. J.Kelemen and P.Sosik), (Lecture Notes in Artificial Intelligence 2159), Springer, pp.292-295, 2001.

Intelligent Autonomous Systems 7
M. Gini et al. (Eds.)
IOS Press, 2002

FIDO Rover System Enhancements for High-Fidelity Mission Simulations

Edward Tunstel, Terry Huntsberger, Ashitey Trebi-Ollennu, Hrand Aghazarian,
Michael Garrett, Chris Leger, Brett Kennedy, Eric Baumgartner and Paul Schenker

Jet Propulsion Laboratory, California Institute of Technology
4800 Oak Grove Drive, Pasadena, CA 91109 USA
{firstname.lastname}@jpl.nasa.gov / http://fido.jpl.nasa.gov

ABSTRACT

The FIDO (Field Integrated Design & Operations) rover is an advanced mobility platform and research prototype for future Mars surface missions planned by NASA. It was used in a recent field trial aimed at realistic simulation of the 2003 NASA Mars Exploration Rovers (MER) mission. This paper reports the key system enhancements made to improve the utility of the FIDO rover as a platform for high-fidelity, *physical* simulation of Mars surface missions. We describe the approach taken to improve mission-relevant fidelity of the rover system in support of resource modeling, contingency sequencing, and added rover functionality as proposed for the actual mission. This is followed by a brief overview of autonomy technologies demonstrated in the field trial, infused into the flight mission, and advanced relative to the current mission baseline.

1. Introduction

The first autonomous planetary rover, named Sojourner, was deployed on Mars in the summer of 1997 as part of the payload on the NASA Mars Pathfinder lander. Sojourner demonstrated the viability of exploring planetary surfaces using mobile robot technology; its mission was limited to minimal scientific surface exploration confined to an area in close proximity to the lander. In 2003, NASA plans to launch a follow up Mars mission that will use two autonomous rovers to explore distinct regions of the planet's surface. These rovers will have greater mobility and autonomy than Sojourner since they are expected to traverse up to 100 meters each Martian day (*sol*) and to conduct exploration independent of a surface lander.

During the spring of 2001, JPL conducted an extended field trial in the southern Soda Mountains of California's Mojave Desert to physically simulate the mission operations approach planned for one rover of the MER mission. The autonomous vehicle used for the field trial was JPL's FIDO rover, a MER prototype. The FIDO rover and its associated end-to-end field testing infrastructure [12] was used to simulate 20 sols of MER-like operational sequences in a complex geological setting analogous to the Martian surface. A team of mission scientists and engineers collaborated with FIDO robotics technologists to conduct the field trial via satellite at JPL, 180 miles from the desert test site, without prior knowledge of the site location (except for aerial imagery typical of Mars orbital observations). As such, the field trial was conducted as if it were a real Mars mission.

Specific objectives, approach, and results of the field trial are reported in [14]. This paper describes enhancements made to increase the fidelity of the FIDO rover as a platform for realistic physical simulation of MER operations scenarios. In Section 2, we provide a brief overview of the system and discuss salient differences between the FIDO rover and the rovers designed for the MER mission. Enhancements to the rover system are then outlined in Section 3. Section 4 highlights the breadth of technologies and related advancements successfully demonstrated in the field, followed by concluding remarks.

2. FIDO Rover Overview and MER Comparison

The JPL FIDO *concept* represents a functional architecture and infrastructure for the development, rapid prototyping, and testing of mobility platforms and advanced robotic technologies. The FIDO rover is but one of several autonomous mobility platforms developed at JPL using the FIDO system architecture [12, 13], and it is the predecessor of similar rovers under development at NASA including the JPL Athena Software Development Model [9] and Rocky 8 rovers, and the NASA Ames K9 [5] rover. In form and function, FIDO is a terrestrial model for the rovers designed to accomplish the objectives of the MER mission (see Fig. 1), and is used for end-to-end mission concept testing and validation via annual terrestrial field trials.

FIDO field trials are conducted like remote science exploration missions using a *semi-autonomous* surface mobility approach. The prefix "semi" connotes remote planning, command-sequencing and visualization of rover activity sequences and related data products by an Earth based science-engineering team, all under extreme time delay and intermittent communication afforded by daily uplink/downlink cycles of deep space networks. In the context of science exploration, the FIDO rover does not actually decide *what* to do or *how* to conduct exploration (yet). Rather, the FIDO rover is equipped with intelligent algorithms and software, supported by advanced robotic hardware, necessary to perform autonomous execution of commanded directives and achievement of specified high-level goals that originate with mission operators. More detailed technical descriptions and specifications of the critical FIDO subsystems, mission operations tools, and capabilities can be found in [1, 2, 4, 11, 12].

2.1 FIDO and MER Rovers: Functional Contrasts and Similarities

The FIDO rover is similar in function and capabilities to the MER rovers, although the MER rovers are about 1.5 times larger in size and 2.5 times as massive. Solar panels and onboard batteries provide power for each vehicle. From a systems-level viewpoint, there are subtle functional differences between the rovers' design and configuration in the areas of mobility and sensing for navigation and control. Both designs employ the JPL 6-wheel rocker-bogie suspension and are compatible with respect to motor control and mobility performance, as well as sensing implementations for visual, inertial, and celestial navigation.

Less subtle differences exist with regard to robotic mechanism designs, the most apparent being the mast, which carries stereo imaging systems for terrain surveys and navigation planning, and an infrared point spectrometer for measuring mineral composition of surface materials from a distance. The FIDO mast arm is deployable to variable heights above the solar panel up to 2 meters above ground at full extent; the MER mast has a fixed height of 1.3 meters above ground after a one-time deployment.

Figure 1. FIDO rover prototype (left) and artist's rendering of MER design (right).

In addition, it is apparent from Fig. 1 that the FIDO mast is located at the rear of the vehicle in contrast to the frontal placement of the MER mast, and that the solar array configurations are very different for the two rover designs. This results in different near-field obscuration patterns (due to the solar panels) for the various mast-mounted instruments. To compensate for some of these differences when simulating MER, the FIDO mast is deployed at the MER mast height during the field tests. Both rover configurations include a robotic arm beneath the frontal area of the solar panel that carries a suite of instruments used for in situ science investigation of surface materials. The FIDO instrument arm carries a color microscopic imager and a spectrometer model for this purpose. The MER instrument arm will carry a microscopic imager, two types of spectrometers, and a tool for abrading rock surfaces.

Relative differences in onboard computing implementations are not significant from the viewpoints of system functionality and remote operations. However, for completeness it is worth noting that significant differences often exist between Earth-based prototypes and rovers bound for space flight. The FIDO onboard computer is a PC/104 266Mhz, Pentium-class CPU running the VxWorks 5.4 real-time operating system. The MER onboard computer will be quite different due to the meager availability of processors that are qualified for space flight (radiation-hardened/tolerant). MER vehicles will carry a 20MHz Rad6000, a radiation-hardened processor, also running VxWorks [9]. The FIDO software is written in ANSI-C and organized as a three-layer architecture. The bottom layer handles low-level hardware device drivers, while the top layer handles application software including motion and instrument command sequences, obstacle avoidance, and path-planning. The middle layer provides the abstraction between higher-level software and the hardware dependencies; it is also responsible for all motion-control functions, stereo vision processing, instrument interfaces, forward and inverse kinematics, etc.

3. FIDO Rover System Enhancements

Although important differences in size and instrumentation exist between the FIDO and MER rover designs, the similarities of the two are significant enough to maintain that the same types of challenges exist in commanding FIDO operations in complex terrain on Earth, as are expected for MER on Mars. Be that as it may, additional enhancements to the FIDO rover system were required to improve its fidelity relative to functionality, resource modeling, and operational command sequencing. In this section, we highlight efforts made towards increasing mission relevance and physical simulation fidelity relative to the baseline plan for MER mission operations.

3.1 Wheel-Based Soil Excavation

The FIDO rover is capable of autonomously executing a large variety of robotic activities ranging from instrument pointing and fine-positioning to long-range navigation in complex terrain. In early field test activities [4, 12], its wheels have been used exclusively for mobility. The MER mission scientists desire to use rover wheels for augmented functions related to soil mechanics experiments, as was done previously with Sojourner [10]. As such, it was necessary to implement an approach to use the cleats on the FIDO rover wheels as tools for excavating soil.

Prior to the field trials of spring 2001 the FIDO rover was used to evaluate the feasibility of this approach for exposing near subsurface soil features for observation using its science instruments. Through a series of preliminary tests in hard- and soft-packed soils, techniques for soil *trenching* were evaluated. Trenching was accomplished by rotating the front-left wheel backward while all other wheels maintained fixed positions. After trenching, FIDO backs up to deploy its arm-mounted instruments on the trench and acquire science data. Reverse wheel rotation facilitates backing the rover wheel out of the trench with minimal re-coverage and subsequent disturbance of the excavated soil. Rotating a FIDO rover wheel (20 cm diameter) for six revolutions was sufficient to break through the duricrust of hard-packed soil to depths of 2.5-4.0 cm. In very soft soil, trench depths of one wheel radius could be achieved after three wheel revolutions. A representative soil trench with wheel tracks is shown in the left of Fig. 2 as indicated by the white rectangle. This view is as seen from the frontal body-mounted cameras after driving backward away from the trench. Subsequently, the rover drives forward to deploy the microscopic imager, on the end of its instrument arm, inside the trench as shown in the middle image. The right image of Fig. 2 shows the resulting extreme close-up image of the soil inside the trench.

This preliminary testing revealed a need to improve the strategy for wheel-based trenching as successive trenching attempts would produce wheel motor stall conditions. It was determined that this approach to excavating near sub-surface soil requires constant monitoring and rest periods between short trenching cycles to allow the wheel motor to cool. To achieve this, an improved approach was formulated based on a *progressive trenching* motion sequence. This involves progressive rotation of the wheel through a desired range of motion, interrupted by timed periods of halted motion during which the motor winding was allowed to cool. The sequence of intermediate rotations and cool times was designed based on the motor thermal models to reduce the risk of motor failure. Execution of this sequence results in safer loads during the trenching operation. As a backup, the pre-existing FIDO onboard software provides fault detection and protection against motor stalls, thus mitigating the risk of ultimate motor failure.

This evaluation defined a technical approach and operational baseline for shallow sub-surface trenching followed by remote and in-situ visual analysis. On this basis, soil trenching using a single wheel was elected as a new FIDO rover functionality to be field tested during the MER-FIDO field trial. Additional motor thermal modeling and analysis, supported by motor current monitoring, are required to characterize and further improve the approach.

Figure 2. Left: Soil trench dug using wheel (rover's-eye view); Middle: Placement of the arm-mounted microscopic imager; Right: microscopic image of soil.

3.2 Resource Model Development

The various differences in actual hardware, software, and operations approaches used by the FIDO system versus that planned for the MER mission represent important incompatibilities with regard to resource tracking for rover operations. It was necessary to develop resource models to best account for appropriate mappings between the MER baseline (as known) and FIDO end-to-end systems.   To achieve a viable mission simulation, analogue resource models were required for predicting command execution times and associated telemetry data volumes corresponding to a sol's worth of activity sequences. The FIDO onboard power system was upgraded to enable software-controlled power switching of instruments and devices, thus facilitating power monitoring and resource management. However, only modest attention was given to modeling energy utilization for the field trial due to its minimal impact given the test operations timeline — a compressed version of the actual mission operations timeline. The field test timeline allowed for only 2 hours of rover sequence execution per sol simulated. For this duration, it was considered unrealistic for FIDO to consume as much modeled energy as a MER rover would during an actual mission timeline, for which the execution time is about three times longer. From the MER mission operations point of view, the more important issue was to familiarize test participants with the concept of dealing with resource constraints.

The MER mission plan defines several types of Martian sols according to the primary rover activity to take place on a given sol. The categories include sols dedicated to the following activities: panoramic imaging; short (2-10m) and long (>10m) traverses to approach science targets; long drives of at least 80m; remote science using instruments on the mast; and in situ investigations using instruments on the arm.   Sequences of rover commands required to autonomously execute each of these sol types were predefined for the MER mission. Based on these MER sol definitions and information about the expected resource utilization by the MER rovers, functionally equivalent FIDO rover sols were defined. Numerous isolated tests were performed with the FIDO rover to record execution times and estimated energy utilization for all relevant FIDO commands/sequences.   In addition, models were developed to compute the telemetry data volume associated with FIDO commands (i.e., the expected number of bits of rover state data, images, and/or spectra). This data formed the basis for modeling how long the FIDO system would take to execute equivalent MER sol activities, how much telemetry would be generated and transmitted, and how much energy might be consumed. Overall, 52 resource models were created for FIDO/MER rover command execution. These models were integrated into the mission operations tools with appropriate software modifications for automated downlink processing, uplink planning, and report generation [2]. For each sol, command sequence builders utilized the models to efficiently plan, generate, and verify sets of rover activities that complied with constraints on allotted execution time, available communications bandwidth, and predicted energy budgets over the entire mission.

3.3 Contingency Sequencing

Resource modeling enhancements ensure that feasible command sequences are uplinked to the rover. However, in the inevitable event that autonomous execution of a full sol of activity falters, the rover must be smart enough to fail cognizantly [6]. That is, it must be able to detect failures when they occur. A cognizant failure capability was provided by augmenting the FIDO rover onboard software to support execution of contingency sequences in the event of an unexpected occurrence. Specific engineering and science contingency sequences were defined to be executed in response to failed completion of uplinked commands involving science data acquisition, arm operations, traversal, target approach, and trenching. Engineering contingencies typically return critical rover state information for immediate Direct-to-Earth (DTE) transmission, followed by later transmission of imagery expected to reveal some aspect of the problem via UHF, or orbital relay, communications. Science contingencies are designed to return critical rover state information and remote science data for immediate DTE transmission, while corresponding panoramic imagery is transmitted during the next UHF opportunity.

In each case, detection of a failed command sequence is followed by a contingency sequence that provides the necessary telemetry for mission operators to diagnose the situation, while keeping the rover in a safe mode. Following successful diagnosis, recovery is usually possible via uplink of a corrective sequence of commands. This enhancement not only facilitates ground-based recovery from detected failures, it also allows sequence builders to anticipate potential problems and build-in contingencies that would reduce the risk of losing valuable science data or endangering the rover. A richer instantiation of cognizant failure will be implemented on the MER rovers using language constructs and conditional statements for ground-based and onboard sequencing.

4. Field-Demonstrated Technology

In this section we briefly highlight some of the robotics technologies recently developed at JPL and demonstrated in the 2001 MER-FIDO field trial. They were fully integrated with the FIDO architecture, verified in prior field tests and in the JPL MarsYard (an outdoor rover test facility), and improved for utilization in the field trial. The following technologies enhanced FIDO rover autonomy and capability, thereby contributing to a successful MER mission simulation. Fig. 3 shows related scenes of field trial activity.

- *Autonomous on-board arm collision avoidance* software was demonstrated for safe instrument arm placement to diagnose potential arm interaction with the rover and the terrain. The algorithm automatically builds a terrain model from stereo range images and requires no human interaction [8]. The algorithm was also used to supplement mission operations tools as a means to efficiently build safe instrument arm sequences.
- *Onboard Extended Kalman Filter (EKF) state estimation* was demonstrated which fuses wheel odometry, CCD-based sun sensor estimates of absolute heading, and full inertial measurement data (rate sensing and attitude) for accurate rover localization. It provided valuable information used by mission operators for localizing the rover within aerial views of the desert test site. Errors of less than 1% of distance traveled have been reported for this technique [3].
- *Long-range autonomous navigation* was demonstrated with onboard hazard detection and avoidance control using a local path planning algorithm called *DriveMaps* [7].

Figure 3. Remote science, autonomous traverse, and in situ science in the field.

Retrospective evaluations of field trial operations and results often reveal functional limitations or the apparent need for investigating alternative technology options. As an example, the current structure and flow of the planned MER mission operations activities imposes a limitation on the minimum number of sols required to approach a specific target and place instrument arm devices onto it. Successful accomplishment of this sequence of activities can take 2-4 sols under the present strategy. After the 2001 field trial, this limitation was addressed by the FIDO team through implementation and field validation of an automatic target approach sequence that can be accomplished using a single high-level command, and in a single sol. The autonomous sequence uses a visual servoing technique to facilitate navigation to the designated science target while tracking the target location using homography transforms; this is followed by placement of the microscopic imager using automatic focusing based on wavelet texture features. Details of the related algorithms are reported in [7] along with additional algorithmic options for key autonomous mission functions such as high-level navigation and path planning for longer traverses, and onboard execution monitoring for fault detection and diagnosis. Information on these developments may also be found at the JPL FIDO web site (http://fido.jpl.nasa.gov).

One of the objectives of the FIDO concept is to infuse advanced technologies into flight system implementations to increase autonomy and capabilities for planned and future missions. The most recent successes in this regard are the infusion of the onboard arm-collision avoidance technique into the MER flight software and the infusion of the operations planning tool [1] into the MER ground data system tool-suite. The FIDO rover localization and motion control technology are also being considered for implementation on the MER mission. In particular, FIDO algorithms and software for EKF-based state estimation and velocity synchronized all-wheel drive (for improved 6-wheel odometry) are undergoing evaluation by the MER flight software team.

Finally, a number of insights and lessons learned resulted from the 2001 field trial experience. Many of them pertain to desirable features and improvements related to the mission operations tools and user interfaces. There were two specific insights pertaining to the FIDO rover system that provide guidance to further increase fidelity relative to MER mission simulation. The first highlighted the utility of an autonomous capability to perform (and react to) onboard checking of science data quality in order to avoid occasional loss of a sol (due to return of poor quality data) or the risk of data loss. The second called for onboard telemetry and memory management. This insight expresses the notion that there should be a method by which the rover can reveal what requested telemetry it *thinks* it has acquired and transmitted, as well as what science data it has stored in onboard memory. Attention to these issues will be paid in the course of future work.

5. Summary and Conclusion

This paper described enhancements made to the JPL FIDO rover system to increase its fidelity for realistic physical simulation of the NASA 2003 Mars Exploration Rovers mission. An additional functionality for soil trenching using a rover wheel was described, as well as measures taken to develop resource models and a facility for contingency sequencing. Specific autonomous rover technologies verified in the field were highlighted in addition to technology advances relative to the MER mission baseline.

Realistic physical simulations such as the 2001 MER-FIDO field trials are valuable rehearsals as well as proving grounds for proposed rover mission operations. They provide opportunities to test sequences in realistic settings, train mission personnel on how to use autonomous rovers to conduct remote field-based science, and identify technologies that require additional development and/or evaluation. FIDO field experience to date has shown that these terrestrial system analogues reduce mission risk, providing cost-efficient integrated technology development, testing and evaluation within a flight-relevant environment, with direct flight participation.

Acknowledgments
The research described in this paper was performed at the Jet Propulsion Laboratory, California Institute of Technology, under contract with the National Aeronautics and Space Administration. The authors would like to thank all of the people affiliated with the FIDO team for their contributions to this effort. The collaboration and support provided by the Mars Exploration Rovers Project and personnel is gratefully acknowledged, with special regard to John Callas of the MER Science Office at JPL, and the MER Principal Investigators, Steve Squyres of Cornell University and Ray Arvidson of Washington University, St. Louis.

References

1. Backes, P.G. et al, "Sequence Planning for the FIDO Mars Rover Prototype," submitted to *Journal of Geophysical Research*, January 2001.
2. Backes, P.G. and J.S. Norris "Automated Rover Sequence Report Generation," *IEEE Aerospace Conference*, Big Sky, MT, March 2001.
3. Baumgartner, E.T., H. Aghazarian, and A. Trebi-Ollennu, "Rover Localization Results for the FIDO Rover," *SPIE Photonics East Conference*, October 2001.
4. Baumgartner, E.T., "In-Situ Exploration of Mars Using Rover Systems, *Proceedings of the AIAA Space 2000 Conference*, AIAA Paper No. 2000-5062, Long Beach, CA, September 2000.
5. Bresina, J.L. et al, "K9 Operations in May '00 Dual-Rover Field Experiment," 6th *Intl. Symp. on AI, Robotics and Automation in Space*, Montreal Canada, Paper No. AM010, June 2001.
6. Gat, E. and G. Dorais, "Robot Navigation by Conditional Sequencing," *IEEE Intl. Conference on Robotics and Automation*, 1994.
7. Huntsberger, T. et al, "Rover Autonomy for Long Range Navigation and Science Data Acquisition on Planetary Surfaces", submitted to *IEEE Intl. Conference on Robotics and Automation*, Washington, DC, May 2002.
8. Leger, C., "Efficient Sensor/Model Based On-Line Collision Detection for Planetary Manipulators", submitted to *IEEE Intl. Conference on Robotics and Automation*, Washington, DC, May 2002.
9. Maimone, M.W., J. Biesiadecki and J. Morrison, "The Athena SDM Rover: A Testbed for Mars Rover Mobility," 6th *Intl. Symp. on AI, Robotics and Automation in Space*, Montreal Canada, Paper No. AM026, June 2001.
10. Moore, H.J. et al, "Soil-like Deposits Observed by Sojourner, the Pathfinder Rover," *Journal of Geophysical Research*, Vol. 104, No. E4, pp. 8729-8746, April 1999.
11. Norris, J.S. et al, "PTEP: The Parallel Telemetry Processor," *IEEE Aerospace Conference*, Big Sky, MT, March 2001.
12. Schenker, P.S. et al, "FIDO: a Field Integrated Design & Operations Rover for Mars Surface Exploration," 6th *International Symposium on Artificial Intelligence, Robotics and Automation in Space*, Montreal Canada, Paper No. AM012, June 2001.
13. Schenker P.S. et al, "Planetary Rover Developments Supporting Mars Science, Sample Return and Future Human-Robotic Colonization," *Proc. IEEE 10th Intl. Conference on Advanced Robotics*, Budapest, Hungary, August 2001.
14. Tunstel, E. et al, "FIDO Rover Field Trials as Rehearsal for the 2003 Mars Exploration Rover Mission," submitted to 9th Intl. Symp. on Robotics and Applications, WAC, Orlando, FL June 2002.

Intelligent Autonomous Systems 7
M. Gini et al. (Eds.)
IOS Press, 2002

An Autonomous Inspection Robot System for Runways Using an Ultraviolet Image Sensor

Hiroshi WAJIMA[1][3]   Tatsuo MAKINO[1]   Yuzuru TAKAHAMA[1]
Takahiro SUGIYAMA[2]   Keiichi ABE[2][3]

[1] *Engineering Division, Aerospace Company, Kawasaki Heavy Industries, Ltd.*
[2] *Faculty of Information, Shizuoka University*
[3] *Graduate School of Science and Engineering, Shizuoka University*
E-mail : wajima_h@khi.co.jp

Abstract. This paper describes an autonomous mobile robot which detects foreign objects (FOs) on airport runways. FOs are one of the major causes of damage to airplanes, and may result in serious accidents. At present, visual detection of FOs is performed by humans. However, the job is hard and dull; moreover, detection of FOs is sometimes difficult, depending on their colors and the light condition.

We propose an autonomous mobile robot which has an ultraviolet image sensor in addition to a laser radar to solve these problems and to develop an automatic inspection system. In the ultraviolet range, the sky radiation component always makes a strong highlight on a ground object. This fact enables us to design a system which detects FOs robustly, regardless of the direction of the sun, the weather, and colors of FOs. The conditions for detecting FOs are studied in terms of reflectance, and their validity is shown by experiments.

1. Introduction

Though air traffic accidents do not happen often, once it occurs it brings a disaster. Therefore, efforts for various safety services are made to not only design and maintenance of airplanes but also of ground support equipment. Besides typical causes of accidents such as human mistakes and hardware troubles, foreign objects (FOs) on runways can be one of inducements to the hardware troubles not to be expected in airports. A piece of metal, paper and plastic, pebbles, and others are typical kinds of FOs. Among these FOs, a piece of metal has high-strength and toughness, and may cause the most serious FO damage. A typical FO damage by a piece of metal is the engine damage caused by an inhalation of FOs into the jet engine, a puncture of a tire, and a collision to a blown-up FO at the time of takeoff or landing. One such unfortunate accident was the case of the Concorde occurred in July 2000. An FO on the runway injured the airplane and brought the fire and fall. For preventing such an accident inspection staff patrol runways and apron zones by driving low speed vehicles or walking. However, there are many limitations to such a visual inspection; for instance, the inspection can be carried out in early morning before arriving and leaving flights start. In such circumstances, FOs are sometimes overlooked. Actually, depending on the direction of the sunlight, FOs may not show highlights and if the colors of FOs are close to that of the runway, they are not easy to find in a short time of watching. Development of a system which can detect FOs more easily in a short time of inspection is strongly required in such circumstances.

In this study, for the realization of such an automatic inspection system we propose an approach which adds an image sensor of an ultraviolet range as a detection sensor system of FOs to our autonomous mobile robot[1][2] using a line-scan laser radar. At present, though the sensors in the ultraviolet range are applied in various industries, most applications are employed in indoor environments and applications in outdoor environments are mostly limited to the field of remote sensing. In this field, ultraviolet sensors are used for measuring the ozone layer [3], volcanic explosion detection [4], aurora observation on the earth [5], and so on. In our study, we utilize the fact that the luminance of the sky radiation component is greater than that of the direct sunlight component in the ultraviolet range and that for this reason the sky ultraviolet radiation component always makes a strong highlight on a ground object. It enables us to design a system which detects the FOs robustly, regardless of the reflection angles of the sun beam. In the visible range, there are many researches such as wet conditions detection of the road surface, distinction of the materials (metal and plastic) of an object, and so on [6][7][8]. Most researches are based on polarization characteristics. In the field of intelligent transport systems, though some researches deal with the detection of FOs on a road, their target FOs are rather big and high [9][10]. Our sensing method uses the diffuse reflection component (sky light) made by ultraviolet radiation under the natural environment, and is essentially different from other researches. Furthermore, existing applications using polarization are affected by the angle between the camera and the light source, but our method does not require the consideration of the angle dependence on the camera and the light source.

In this way, we develop a method which can detect FOs robustly against the weather, the sun and the camera positions, amount of the sunlight, and colors of FOs. The sky radiation component is larger than the direct sunlight component in the ultraviolet range, and does not depend on the position of the camera and the sun. The FOs with the property of specular reflection make always strong highlights reflection because the large sky radiation component comes from every direction of the hemisphere, while asphalt of runway in the background has weaker reflection because of its diffuse reflection property. As a result, FOs makes higher intensity than asphalt in the background in ultraviolet images.

In Section 2, we introduce our autonomous mobile robot system for outdoor ground checking. To enforce its faculty for automatic runway inspection, the condition for detecting flat FOs in any sunlight circumstances is investigated in Section 3. We show the availability of this approach by the experiments described in Sections 4 and 5.

2. Autonomous Inspection System

We developed an autonomous mobile robot in an outdoor environment, and produced a test robot which can "Track a route set by an operator" and "Find and avoid obstacles" [1]. This robot has sensors for these fundamental faculties: a line-scan laser radar for recognition of the ground-surface shapes and detection of small obstacles, and a differential global positioning system (D-GPS), encoders, and so on for estimating its own positions. The specifications and performance of the robot are presented in Table 1 [2]. According to these faculties, the robot can go around automatically on the ground and detect FOs of 10cm high or higher using the laser radar, as shown Figure 1. In other words, the laser radar can not detect flat FOs of less than 10cm high, e.g., boards, planks, or panels. We had to add another imaging sensor to the robot for detecting which are not high FOs. In the following, we confine our purpose to detection of such flat FOs. We found out that an ultraviolet imaging sensor is effective as a sensor for this purpose. The effectiveness of an ultraviolet sensor is discussed in the following sections.

Table 1. Specifications and performance of the robot

Length x width x height	2.37 x 1.10 x 2.20 m
Max operating time	about 2 hour
Max speed - at avoid obstacle maneuvering - at road tracking maneuvering	about 4km/h about 15km/h
Minimum detectable obstacle size	10 cm high
Obstacle detecting range	about 3.5 m
Own-position estimation precision	standard div. 1.0 m

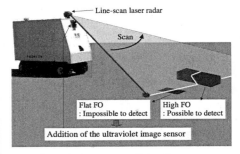

Figure 1. Autonomous Robot System
for Ground Checking

3. Theoretical Considerations

In an outdoor environment there are two kind of light sources: the direct light L_{direct} of the sun and the sky light L_{sky}, scattered sunlight in the atmosphere. Let us assume that a Lambertian surface and a mirror surface are placed in an outdoor environment and irradiated by these light sources, as shown in Figure 2. In this application, runway surfaces can be considered Lambertian surfaces because they are paved with asphalt, and FOs are can be considered mirror surfaces because they are metal parts of airplane, that is, the polish processing to enlarge a strength/weight ratio is given to these metal parts for reducing weight of the airplane. Furthermore, even if an FO has surfaces such as mat painting, a sheen phenomenon occurs and the mat surfaces behave as mirror surfaces, because the angle of dip of the camera is set up shallowly in order to take pictures of as wide area as possible in one image. On the Lambertian surface, both the direct light and the sky light are reflected diffusely. The mirror surface reflects a part of the sky light, and the direct light only when the positions of the sun and the camera makes the exact configuration for the specular reflection. These relations are shown by the following equations, where L_{dif} and L_{spc} denote the luminance of the reflections from the Lambertian surface and from the mirror surface, respectively, observed by the camera.

$$L_{dif} = f_{direct}\,(\,L_{direct},\,R_{dif}\,) + f_{sky}\,(\,L_{sky},\,R_{dif}\,) \qquad (1)$$
$$L_{spc} = g_{direct}(L_{direct},\,R_{spc},\,p(P_{sun},\,P_{cam})) + g_{sky}\,(L_{sky},\,R_{spc}) \qquad (2)$$

where $f_{direct}(\ )$ and $g_{direct}(\ )$ denote functions determining reflection components by the direct light on the Lambertian surface and the mirror surface, respectively; $f_{sky}(\ )$ and $g_{sky}(\ )$ are functions determining reflection components by the sky light on the Lambertian surface and the mirror surface, respectively. L_{direct} and L_{sky} stand for the luminances of the direct light and the sky light, respectively; R_{dif} and R_{spc} denote the reflectance of Lambertian surface and the mirror surface, $p(\ )$ is a function of the solar position P_{sun} and the camera position P_{cam}.

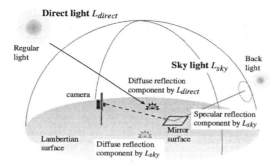

Figure 2. Two light sources and the reflection models in the outdoor environment

When the direct light of the luminance L_0 irradiates an ideal Lambertian surface, the luminance L observed by the camera is as follows [11] .

$$L = 1/\pi \cdot L_0$$

In addition to this reflection luminance, there is also the luminance of the sky observed by the camera under the uniform sky light from every direction. Thus Eq.(1) can be rewritten as follows.

$$L_{dif} = 1/\pi \cdot L_{direct} \cdot R_{dif} + L_{sky} \cdot R_{dif} \qquad (3)$$

Next, regarding the reflection of the mirror surface, the sun and the camera positions must be in the relation of the specular reflection so that a direct light can be observed by the camera. As for the sky light, its reflection is always observed because it comes from every direction. Eq.(2) can be rewritten as follows.

$$L_{spc} = L_{direct} \cdot R_{spc} \cdot p(\,P_{sun},\,P_{cam}\,) + L_{sky} \cdot R_{spc} \qquad (4)$$

where, if the sun and the camera positions are in the specular reflection relation, then $p(P_{sun},P_{cam})=1$, otherwise $p(P_{sun},P_{cam})=0$. We assume that the camera is in a general position and thus $p(P_{sun},P_{cam})=0$ hereafter.

Now, we define L_{obj} be the luminance of an object which has a mirror surface with the reflectance R_{obj}, and L_{back} be the luminance of the background which has a Lambertian surface with the reflectance R_{back}, in the visible range and the ultraviolet range. Here, we consider the luminance ratio D of L_{obj} to L_{back} according to Eqs.(3) and (4). The luminance ratio D is shown in Eq.(5).

$$D = L_{obj} / L_{back} = L_{spc} / L_{dif} \qquad (5)$$

It means if D is larger than 1, then the object is seen brighter than the background; otherwise the object is seen darker than the background. We assume that the object should be imaged always brighter than the background for the sake of an easy and robust implementation of the inspection system. If an object makes a brighter image in one situation and a darker image in another, then there should be some very difficult situation for detection in between.

Using Eq.(5), we would like to derive the conditions of detection possibility of an object from the background in the visible range and in the ultraviolet range. For this purpose we have to know the luminances of the direct light and the sky light in the visible range and in the ultraviolet range. They depend on the weather and are shown in Table 2 [12]. Though the absolute values L_{direct} and L_{sky} in the visible range and L_{direct} and L_{sky} in the ultraviolet range are different, we need only the ratio L_{direct}/L_{sky} because L_{obj}/L_{back} are discussed in the visible range and in the ultraviolet range separately.

Table 2. The ratios of the direct light and the sky light due to the weather

	Clear	Cloudy
Visible range	$L_{direct} = 9 \cdot L_{sky}$	$L_{direct} = 0$
Ultraviolet range	$L_{direct} = 2/3 \cdot L_{sky}$	$L_{direct} = 0$

By substituting Eqs.(3) and (4), and the values in Table 2 into Eq.(5), we obtain the condition for detecting an object with a mirror surface from the background with a Lambertian surface as a brighter region in the image, in the visible range and in the ultraviolet range. The results are shown in Table 3.

Table 3. The conditions for detection ($D > 1$) in the visible range and in the ultraviolet range

		Visible range	Ultraviolet range
clear	Regular light	$\times : r > 4$	$\bigcirc : r > 1.2$
	Backlight	$\circledcirc : r > 0.4$	$\circledcirc : r > 0.7$
Cloudy		$\bigcirc : r > 1.0$	$\bigcirc : r > 1.0$

$r = R_{obj}/R_{back}$

$\circledcirc$: The objects can be seen brightly even if R_{obj} is smaller than R_{back}. $\bigcirc$: The objects can be seen brightly if R_{obj} is larger than or close to R_{back}. $\times$: The objects can not be seen brightly unless R_{obj} is much larger than R_{back}.

From the Table 3, in the case of clear weather in the visible range, the influence of the light source whether it is a regular light or a backlight is very big. There is only a little influence in the ultraviolet range, however. In the visible range with a regular light of clear weather, even if the reflectance of the object is 1.0, the reflectance of the background must be less than 0.25. On the other hand, in the ultraviolet range, the reflectance of the background must be less than 0.83. These facts show that in the ultraviolet range an object gives stronger reflection than the background even if the background has a high reflectance. Therefore, construction of a robust system becomes easier in the following points by using the ultraviolet range rather than the visible range.

- It is more robust against the changes of environment such as the sun and the camera positions or the weather like clear, fine, or cloudy.
- The range of reflectance of the applicable objects is much wider for a fixed reflectance of the background.

4. Experimental Results

Now we want to prove our method proposed in this paper to be practically applicable. We use a multi-spectrum camera whose specifications are shown in Table 4.

Table 4. Specifications of the multi-spectrum camera used

Items	Specifications
Sensor	SITe Back-Illuminated CCD
Angle of view	8.3° □
Camera	Apogee Instruments Inc. (USA) :AP-7p S.W.S. 200 to 1100nm
Lens	P.W.R. 220 to 1000nm
UVB filter	P.W.R. 295 to 315nm
UVAB filter	P.W.R. 280 to 370nm
Visible filter	P.W.R. 390 to 680nm

Notes: S.W.S. is the Sensor Wavelength Sensitivity, and P.W.R. is the Permeation Wavelength Range.

When the weather is clear or cloudy, the experimental images obtained in the condition of a regular light and a backlight are shown for the visible range, UVAB, and UVB in Figure 3. The test piece of an aluminum board (L: 300 mm, W: 100 mm) painted in gray frosting was put on asphalt. The paint color is similar to the color of asphalt, so that distinction of two colors is sometimes difficult by human eyes.

A subjective judgement on the detectability of the object in the images in Figure 3 is summarized in Table 5.

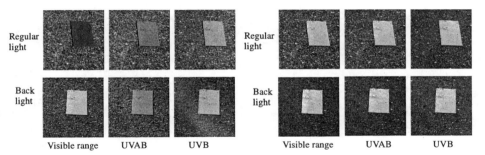

	Visible range	UVAB	UVB		Visible range	UVAB	UVB
Regular light				Regular light			
Back light				Back light			

Figure 3. (a) Weather clear :　　　　　　Figure 3. (b) Weather cloudy :
Visible range, UVAB, and UVB images in a regular light and a backlight.

Table 5. Summary of the experiment results of the object

Environment		Visible range	UVAB	UVB
Clear	Regular light	✕	△	○
	Backlight	◎	○	◎
Cloudy	Regular light	◎	◎	◎
	Backlight	◎	◎	◎

◎: much brighter than the background,　○: brighter than the background,
△ : a little brighter than the background,　✕ : nearly equal to the background

The object shown in Figure 3 has almost the same luminance all over the surface because it is board-shaped and painted uniformly. One might consider that the object and the asphalt can be discriminated by using the difference of the texture. However, real FOs on runways often have irregular texture by non-uniform painting, scratches, and cracks. Therefore, the object detection can not be based on the difference of the texture, and should rely on only the difference of the brightness.

From Table 5 the following can be concluded.
- In (clear, backlight), (cloudy, regular light), and (cloudy, backlight) cases the visible range camera and the UVB camera behave as well. However, in (clear, regular light) case the visible range works poor in object detection, while the ultraviolet behaves much better.
- The UVAB range camera behaves intermediately between the visible range and the UVB.
- In order to detect the object robustly in any environment of the weather and the relative positions of the sun, the object, and the camera, use of the UVB range is the best. This is because in the UVB range the sky light is relatively stronger than the direct sunlight, and the asphalt in the background has a Lambertian surface which makes a weaker reflection than the object.

Next, we studied on the influence of painting colors of the object from the practical viewpoint. For this purpose the spectral reflectances were investigated for various paints under the sunlight. These painting colors are used for actual airplanes. In this experiment, the standard white color board is made by applying white reflection coating (sulfuric acid barium melted with ethyl alcohol, Munsell corporation, USA) to an aluminum board. The spectral reflectance of this coating is almost 1 from ultraviolet to near infrared. Therefore, we measured the ratio of reflectances of each color board to this standard white color board and assumed that it is the reflectance of the color board. The results of spectral reflectances of paints for airplanes measured are shown in Figure 4. The following are seen from this result.
- The reflectances of various colors in the ultraviolet range spread narrower than the reflectances in the visible range and the near infrared range do.
- In the ultraviolet range, the difference of the reflectances among the colors locates between the reflectances of white and black (including fluorescent orange).
- The reflectances in UVB are larger than the reflectances in UVA.

Table 6 shows the condition for the reflectance of objects to be detected brighter, referring to Table 3, where the background of asphalt has the reflectance 0.09 in UVB and the reflectance 0.12 in visible range.

Consequently, in order to detect an object on asphalt under any environmental condition the requirement for the reflectance becomes the following.

Visible range : higher than the reflectance 0.48
Ultraviolet range: higher than the reflectance 0.11

Here, both limitations are put in the clear and regular light situation.

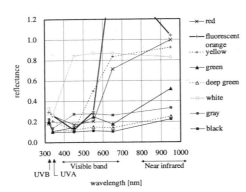

Table 6. The reflectance conditions for detecting an object as brighter on asphalt in the visible and ultraviolet ranges.

		Visible range	UVB range
clear	Regular light	$R_{obj} > 0.48$	$R_{obj} > 0.11$
	back-light	$R_{obj} > 0.05$	$R_{obj} > 0.06$
Cloudy		$R_{obj} > 0.12$	$R_{obj} > 0.09$

Figure 4. The measurement results of the spectral reflectances of various paints for airplanes

The reflectances of the objects are now to be considered. As noted in Section 1, critical damages of airplanes are caused by metal pieces whose strength and stiffness are high. Aluminum is most popular material of airplanes. The reflectance of bare aluminum is almost uniform and high from ultraviolet to near infrared. Though it depends on grinding condition, reflectance is 0.8 or less. Therefore, as for the bare aluminum, from the above consideration, detection is possible in all environmental conditions both in the visible range and in the ultraviolet range. The painted metal pieces have much lower reflectance, but they are about 0.2 to 0.3 in UVB as shown in Figure 4. Accordingly, in UVB all painted objects satisfy the reflectance conditions for detection shown in Table 6. On the other hand, robust detection is not possible in the visible range for painted metal pieces. This is because the condition for detection in Table 6 forces reflectance larger than 0.48 in the worst case while the reflectances of color pieces ranges from 0.09 to 0.9 (except for fluorescent orange).

In UVB range the worst case is clear weather and regular light and the object color black from Table 6 and Figure 4. In this case Table 6 says "An object can be detected brighter than the background if $R_{obj} > 0.11$", and the reflectance of the test piece painted in black is 0.19. Therefore, this test piece should be detectable. The UVB image and its cross section profile depicted in Figure 5 indicates that this object is brighter than the background in average, showing that the experimental result is consistent with the theoretical one.

We also tested the case that the background is concrete whose reflectance is different from asphalt. The reflectance of concrete in UVB is 0.2. This value makes the condition for detectability $R_{obj} > 0.24$ as $R_{back}=0.2$. For the reflectance of the black object is $R_{obj}=0.19$, it does not satisfy the above condition. The experimental result is shown in Figure 6 and this object looks a little darker than the background. This implies that an object in some color near black is very difficult to be detected from the concrete background.

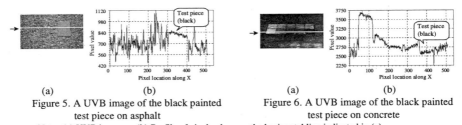

(a)	(b)	(a)	(b)

Figure 5. A UVB image of the black painted test piece on asphalt

Figure 6. A UVB image of the black painted test piece on concrete

Notes (a):UVB image   (b):Profile of pixel values on the horizontal line indicated in (a)

From these experiments we can conclude the following.
- In UVB range metal pieces of any color give a detectably brighter image than the asphalt background; in visible range this does not hold.
- If the background is concrete, in UVB range a black metal piece gives a slightly darker image than the background and objects in some colors are supposed to be difficult to discriminate from the concrete.

5. Image Processing method

We tried to extract painted objects on asphalt from UV images by image processing. Based on the proposed method, the processing is essentially only a binarization. If a simple binarization process suffices, the detection system can achieve high speed as well as low cost. These factors are important for practical use. Here is a problem, however, how to set up the threshold for binarization.

In the case of this application, the threshold must be variable corresponding to the change of the reflection from asphalt depending on the amount of sunlight at the time of the day, the direction of the sun, and the weather as well as the shutter speed or the amount of iris. In this study the p-tile method is used for setting up threshold, under the assumption that the area of an FO in a taken image is small compared to the area of the background. Now, the distribution of the pixel values of the asphalt image is supposed almost normal, and the part of the strong reflection caused by an FO is in a small percentage. Then the threshold is to be set between the part of the strong reflection by the FO and the rest: the appropriate threshold may be $m + 2\sigma$ (about 2% of the upside probability) to $m + 1.6\sigma$ (about 5% of the upside probability), where m is the average of the pixel values in the image and σ is the standard deviation of them. In the following experiment, we used the threshold value set at $m + 2\sigma$.

Table 7 Variable thresholds in various environments

Condition of environment		UVB (background)		$m + 2\sigma =$ threshold
		mean m	standard deviation σ	
clear	regular light	94.5	39.7	174
	backlight	70.9	35.6	142
cloudy	regular light	79.5	30.7	141
	backlight	77.5	28.0	134

Using the threshold determined by the calculation in Table 7, extraction experiments were made using the image shown in Figure 3. Note that the object is painted in a similar color to that of asphalt. The processing results are shown in Figure 7. Though there is small noise in these results, the FOs are easily detectable if their areas are taken into account. Accordingly, the algorithm of the detection system of FOs on runways can be achieved by the block diagram in Figure 8, though there needs a broader field test further.

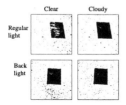

Figure 7. Results of image processing. Figure 8. Detection system of the foreign objects on runways

6. Conclusions and discussions

A significant point of this study is to provide a way of realizing an automatic system, using an autonomous mobile robot already developed, for the inspection work on runways which has been done by humans so far. We found out that an ultraviolet image sensor is effective for detecting flat FOs which a laser radar was not able to detect. Under the natural environment, our method of detecting FOs is based on the property that the sky radiation component is larger than the direct light component in ultraviolet range. Therefore, FOs always have highlights even when they are observed from any direction, independent of colors of FOs. As a result, the probability of overlooking FOs can be decreased. The conditions for detecting FOs on runways were analyzed theoretically and then validated experimentally. Finally, an image processing algorithm for detecting FOs on an asphalt runway was proposed, and its effectiveness was shown.

There are a few problems for practical use of the system. One is restriction on the speed of the mobile robot. If the speed of the robot is increased without changing the present performance, it will not be able to stop or avoid the detected FOs. Another is a requirement of setting the sensor at an angle of a shallow dip, so that FOs can be detected from an observation point as far as possible. However, the measurement error of the laser radar will grow larger due to vibrations of the vehicle.

We discuss here approaches to resolve these problems. As for the first problem, the prototype robot can be improved by decreasing its weight and enhancing the brake performance. Then the robot will be able to run at a speed of about 30 km / h, because it has already been achieved with another research of Autonomous Ground Vehicle(AGV) of an equal size[13]. As for the second, the characteristics of the two sensors used in the present system are summarized in Table 8. However, to use only a UV imaging sensor without a laser radar has advantages of practical use in costs, resolution, and speedup of the vehicle. For this solution it is necessary to develop an image processing algorithm which can detect both high and flat FOs only using a UV sensor. This is a future research subject.

Table 8. Characteristics of Laser Radar and UV Imaging Sensor

Items	Laser Radar	UV Imaging Sensor
Detection of FO	Higher than 10cm	Flat objects, less than 10cm high
Costs	expensive	inexpensive
Scanning	Line scan (Covering narrow area)	Solid-state image sensor :CCD (Covering wide area)
Refresh rate	26ms/line	33ms/image
Resolution	0.5deg.	Below 0.04deg.

Another solution for this problem is a laser radar mounted on a gimbal to isolate it from vibrations of the vehicle. Using a gimbal, even a low-priced gimbal of the present goods on the market, can almost eliminate the problem about vibrations. That is to say, the gimbal can get a space stability of about 50 micro radian, then an error is only about 4mm within a limit of distance measurement of 80m with a laser radar (equivalent to the laser protection class level 1) by our consideration. By taking such an approach, compared to the present speed 5-10km/h by inspection staffs, it becomes possible to make the speed up to about 30km/h with a future automatic inspection system.

As another assignment on using a UV sensor, we should consider cases where mirror surfaces of FOs are spoilt by rough flaw, dirt, and so on. We made an analysis using the dichromatic reflection model with a diffusion component considered. By this analysis, even when surfaces of FOs changed from mirror surfaces to diffusion surfaces, there is only a little difference in the reflectance conditions for detecting FOs in the UV range. This analytic result is based on strong sky radiation in the UV range. Therefore, it is confirmed that the difference of surface conditions of FOs hardly influences propriety of the detection.

If we want to apply this system to roads rather than runways, a decline of the detectability performance is expected due to the decrease of the sky radiation, hidden by buildings. Subsidiary light sources and combination with other methods will be necessary in such cases.

Acknowledgement The author would like to thank Yasuyuki KITAMURA of Kawasaki Heavy Industries, Ltd. in Japan.

7.   Reference

[1] Y. Mitani, N. Awaki, et al., "A Study of an Autonomous Mobile Robot in an outdoor environment", Advanced Robotics Vol.15, No.3, pp. 344-350(2001)

[2] Y. Takahama, Y. Mitani, et al., "Autonomous Ground Experimental Vehicle in an outdoor environment", Proc. The 28th AUVSI's Unmanned Systems Symposium & Exhibition (2001)

[3] J. Joanna, B. Pawan, et al., "Remote sensing applications of continuous spectral backscatter ultraviolet measurements", Proc. SPIE Vol.2831, pp.94-103 (1996)

[4] McPeters Richard, "Satellite measurements of the back-scattered ultraviolet to determine ozone trends, volcanic SO_2, and nitric oxide", Proc. SPIE Vol.1715, pp.522-526 (1993)

[5] Torr M. R., Torr, D. G., "Imaging of the terrestrial aurora in the vacuum ultraviolet", Proc. SPIE Vol. 2008, Instrumentation for Magnetospheric Imagery II, pp. 172-172 (1993)

[6] K. Ueda, I. Horiba, et al., "A detecting method of wet condition on road using image processing", Trans. Information Processing Society Japan, Vol.35, No.6, pp.1072-1080 (1994) (in Japanese)

[7] M. Yamada, I. Horiba, N. Sugie, et al., "Discrimination of the road condition based on image information", Trans. Information Processing Society Japan, Vol. 38, No.12, pp.2648-2655 (1997) (in Japanese)

[8] Wolff, L. B., "Palarization-based material classification from specular reflection", IEEE Trans., Vol.PAMI-12, No.11, pp.1059-1071 (1990)

[9] R. Bin, L. Henry, et al., "Vision-based object detection and recognition system for intelligent vehicles", Proc. SPIE Vol. 3525, Mobile Robots XIII and Intelligent Transportation Systems, pp.326-337 (1999)

[10] S. Baten, "Autonomous road and contour following with a tracked vehicle", Proc. SPIE Vol. 4024, Unmanned Ground Vehicle Technology II, pp.32-42 (2000)

[11] Berthold K. P. Horn, Robot Vison, The MIT Press, Cambridge, Massachusetts USA(1982).

[12] Edited by Illuminating Engineering Institute Japan, UV and Creature Industry, Yokendo Co., Ltd., Tokyo Japan (1998) (in Japanese)

[13] Rankin, A.L., Owens, K., Matthies, L., and Litwin, T., " Terrain Adaptive Gaze and Velocity Control for UGV Obstacle Detection", Proc. The 25th AUVSI's Unmanned Systems Symposium & Exhibition (1998)

Intelligent Autonomous Systems 7
M. Gini et al. (Eds.)
IOS Press, 2002

An Architecture for Task and Behavior Object Transmission Among Multiple Autonomous Robots

ZHIDONG WANG*, TAKAYUKI TAKAHASHI**, TAKAFUMI NITSUMA**,
TAKASHI NINJOJI***, TAKUJI MATSUKAWA* and EIJI NAKANO**
* Graduate School of Engineering   ** Graduate School of Information Sciences,
Tohoku University, Aoba-ku, Sendai 980-8579, JAPAN
***NTT Information Sharing Platform Lab., NTT. Co., JAPAN
{ wang,taka,nitsuma,takuji,nakano} @robotics.is.tohoku.ac.jp, ninjouji@slab.ntt.jp

Abstract

This paper proposes a new communication and control architecture which improves the capability and the flexibility of multiple autonomous robots system in performing a complicated task and coping with unpredictable situation. This system treats robot's information as "Behavior Element Object: (B.E.O.)" and "Task Object: (T.O.)" in terms of Object Oriented paradigm. Both BEO and TO can be serialized, so it can be communicated among robots and behavior server system in the network. Action manager module, device module and some checking mechanisms are also designed in executing new TO or BEO sent from other robots or a server system. A simulation and a basic experiments is presented for a situation of robots' relief for an emergency purpose.

1 Introduction

In performing a task by multiple autonomous robot system the robot should have the whole information about the task. Furthermore, it will also be desirable that a robot system has the flexibility on coping with various tasks by combining and organizing the necessary actions together, which the robot system already has, according to the tasks information. In this case, one of the countermeasure is to build and incorporate *all* tasks information and *all* possible decisions to different situations in the robot controller beforehand. To construct a robot system, which can acquire the task information and perform it in real time, can be thought of another solution. The information can either be acquired from other robots which works already on the same task or from a server system on a network which has a database of the

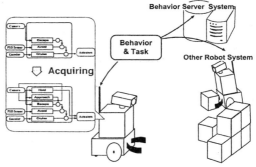

Figure 1: By adding some new behavior elements, robot is able to change to perform a new task

different behavior objects and related information on task performing. The latter have many advantages, it will lead to an autonomous robot increase its ability to perform new tasks and to cope with change of environment, the control system can be constructed easily in increment style, the size of the control system of a robot can be small since it only needs to contain some basic control functions. A smaller control system is really profitable on designing a multiple robots system, specially when the number of robots increases. Also the latter is realizable using the current computer network technology.

In the case of inter-robot communication, the methods of socket based communication and remote procedures are frequently used for transmitting parameters among robots or between robot and host. These are easily found in the researches of master-slave remote control system and the researches of teleoperation system[1][2] which are coping with large time delay caused by network in the control loop. It can be said that algorithm of a robot's behavior or subset of the algorithm is still not included in the contents of the inter-robot information transmission in these researches currently. Also some other researches[3] which are using JAVA RMI also demonstrated that some off-line planning path or command sequence are trans-fered to a teleoperated mobile robot from Web based interface. But the whole process is still un-der operator's control strictly rather than performed by a robot itself. Some researchers[4][5] proposed a remote brain architecture and a BetNet archi-tecture on robot control and implemented some plug-in mechanisms to let the robot be able to up-load a new behavior in runtime. But these are in the view point of increasing the efficiency of devel-opment of robot control system. Recently on de-signing autonomous robot system, some research groups[6][7] proposed some methods to carry some task and environment data by using the portable *Intelligent Data Carrier System* or visual marks with memory storage. In addressing fault toler-ance problem, a research[8] also demonstrated an autonomous robots' relief on cooperative box push-ing by two robots with inter-robot communication.

In the software engineering area, the plugin tech-nology and software auto-update system are widely used in recent years and shows the advantages of dynamic reconfigurable system. But these meth-ods and ideas are not enough. The controller of an autonomous robot system involves many sensor devices and actuator devices which interact with the real physical world and has some characteris-tic that devices in a computer system do not have. Sometimes, a robot controller has to manipulate the redundancy of sensor information according to their physical model. Also some status of an actu-ator will be critical on task performing. For exam-ple, you can not simply reset a gripper's input to zero when it is holding an object. Also for the sta-bility on control, human operator cannot interact with the robot's low level feedback control loop, but the software application usually ask the oper-ator to make a decision by some dialog windows in these case. Specially, to an autonomous robot sys-tem, some particular mechanisms which are able to manage behaviors or tasks should be considered.

In this paper, we propose a system which rep-resents both the task information and behavioral information as object. Then the proposed system allows both task and behavioral information to be shared among autonomous robots which are work-ing for the same tasks. Also a robot only need

to maintain a small size on-board control system and get the necessary task and behavior informa-tion from other robots or server system for some unpredictable situation. Additionally, it is easy to modify the robot's control system in proposed architecture since both the task and behavior in-formation can be renewed while robot is running. For achieving the purpose on realizing this system, we constructed a package, Little Object-Oriented Grounded User Environment for robot operation: LOGUE, on Java language. We also designed a ba-sic architecture which realizes task-behavior trans-ferring, checking and executing while a robot is running.

As an illustrative demonstration, we design a robot system to cope with the behavior and task modification in an emergency situation. The pro-posed LOGUE package is implemented in this robot's control system. In this paper, we name the robot control system to "LOGUE", the same name of the package. A simulation and basic experiment which verify the system "LOGUE" is shown in the end of this paper.

2 Object Representation of Behavioral Element and Task

From conventional robot programming researches to recent teleoperation researches, it has shown that function modules or high level command se-quence is useful in controlling a robot to perform some complicated tasks. By including some strate-gies, a behavior-based system[9] can also achieve some dynamic tasks, such as manipulating an ob-ject cooperatively by multiple autonomous robots[10]. The object-oriented representation shows a lot of advantages to an autonomous robots control[11].

In this research, the action decision of the robot was realized by a behavior-based architecture since it makes information for a task to achieve a good abstraction and easily reusable. Then information for transmitting and receiving among robots could be classified into two types: information of behav-ior and information of task.

2.1 Behavioral Element Object

The information and contents of a behavior, an action or a motion of the robot is represented by a *Behavioral Element Object* (BEO, Fig.2) in the research. Functions of the most BEO are tightly related to the robot's sensors, actuators and its control mechanism.

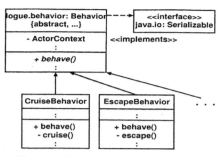

Figure 2: Behavioral Element Object(BEO)

In the LOGUE, the BEO is generated by inheriting from an abstract class *Behavior* which includes an abstract method called *behave()*. By implementing this method, a BEO has a particular meaning in controlling the robot. In the LOGUE system, each BEO has a unique name which is used as the key for searching and in checking its existence in a behavior object database. Each BEO has a list of native devices which include sensors and actuators involved by this behavior. A method *checkNativeDevices()* is working on checking the existence of those devices when this BEO is transferred to a robot. A method named *isActive()* is designed for showing the status of current behavior when the LOGUE system needs to stop a task or update a behavior. Furthermore, a BEO can also include other one or more BEOs.

2.2 Task Object

On the other hand, the task information contains all information of what the robot behaves, and includes starting and ending condition, etc. It is represented by *Task Object*(TO, Fig.3) in the LOGUE. Similar with the BEO described previously, the TO inherits from an abstract class *Task* which includes an abstract method called *perform()*. A TO is generated by implementing its *perform()* method according to the contents of the task. In each TO, starting condition, ending condition and at least one BEO should be included. Furthermore, a module representing the network among BEOs is contained on controlling and making the BEO to be functioning. When a TO is functioning, BEOs included in the TO will become active or inactive respectively according to the status of the robot and the situation of the task performing. The function of the BEO which is in active state will be expressed as the robot's behavioral attribution. Additionally, each TO has a priority

index which is used for deciding the running order of tasks, and the methods, *setPriority()* and *getPriority()*, are implemented for this purpose.

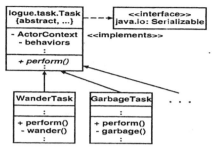

Figure 3: Task Object(TO)

3 Transmitting and Managing Objects

The whole system LOGUE is shown in Fig.4 which can be run on both autonomous robots and behavior server systems. In the system LOGUE, three components are included. The Communication Module are working on TO and BEO transmission. The Action Management Module includes Behavior Manager Module and Task Manager Module. It works on reconstructing objects transfered from the network, checking executive ability, storing them and running the task objects, as the key component of the LOGUE system. The Devices Module is designed for providing a common interface from robot's native devices such as sensors, actuators and some peculiar parameters for the robot control.

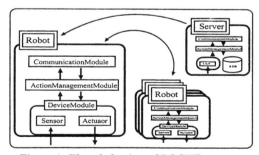

Figure 4: The whole view of LOGUE system

The LOGUE is implemented on the Java(tm) JDK1.2 which is running on Linux 2.0 on experimental robot and is implemented on JDK1.3 which is running on Linux 2.2 on simulation and behavior server system. For objects transmission among robots and behavior server, the distributed object environment java.rmi package is employed(Fig.5). Excepting in the behavior database and GUI interface included in behavior server system, the implementation of the LOGUE in each autonomous robot and the server system are the same.

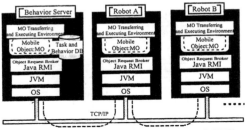

Figure 5: Layers for object passing in LOGUE

3.1 Communication Module

A communication module functions on invoking and transmitting objects among robots and servers in the LOGUE. In the JavaRMI, a remote object which the Remote interface is implemented can be invoked by sending arguments from the local virtual machine to a remote one. The results are marshaled from the remote machine and sent back to the local virtual machine in the same style with the result by running a local object. Also an object, which includes some arguments or results are not invokable remotely, will be copied to the remote virtual machine. This is because the invocation of an object should be done in the same virtual machine. Therefore, to guarantee consistency of a behavior-based controller, the whole object of BEOs or TOs should be transferred to the virtual machine of current robot. And both types of objects are implemented to be serializable.

For the object transmission in the LOGUE, both the sending robot and the receiving robot should have the object class. But only a class running on a virtual machine can not be transferred to other JAVA virtual machine directly. The file of the class should be sent first and the class loader in the receiver system makes it available in its system. In the LOGUE, the object class transmission is realized by preparing an object *Message*, which reads the class file and converts it to a binary stream.

3.2 Action Management Module

In action management module of LOGUE, all behavioral element objects are managed by an object called *BehaviorManager*. Similarly, all task objects are managed by an object called *TaskManager*(Fig.6). The behavior manager object contains a BEO list to record all BEOs on the current LOGUE. In this object, a method called *requestBehavior()* sends a request message about necessary BEOs to other robots or behavior server through the communication module. Also a *checkDevices()* method and a *checkBehaviors()* method are implemented for checking if the required devices and behaviors exist in the current LOGUE when a new BEO is received. The new BEO can not be run on the current robot if there is a devices checking error. Then the behavior manager object will inform the sender that the current BEO is infeasible on the current robot, and the sender will send the BEO or related TO to other robots. But if behaviors checking find one or more behaviors do not exist, the *requestBehavior()* will be called to obtain them. After passing both checks, *registerBehavior()* method registers the new BEO to the BEO list in behavior manager object.

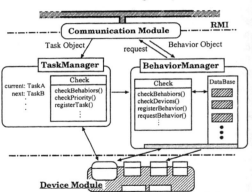

Figure 6: Action management module

Similarly, a task manager object also contains a priority based TO list for all TOs on current robot system. Three methods, *checkBehaviors()*, *checkPriority()* and *registerTask()*, are implemented for managing the TO list. When a new TO are received from the communication module, the *checkBehaviors()* will pass the list of BEOs in the TO to current robot's behavior manager object for checking and completing a set of the necessary BEOs. After the checking of behaviors, the new TO will be added to the TO list according to its priority. Additionally, two methods *checkActives()* and *changePriority()* are installed for managing the task switching and behavior switching in the task manager object. Also the behavior manager object and the task manager object include a method to delete objects which is unused in a certain time for keeping the controller in small size.

3.3 Device module

In the system LOGUE, device module is designed for two purposes. The first one provides the abstracted interface of sensors and actuators to BEO and TO for a good portability. The second purpose is to maintain inner status information of the robot such as current positions, orientation or camera heading, etc.

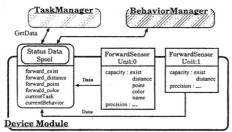

Figure 7: An example of sensor units

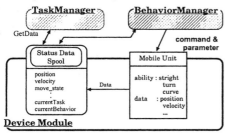

Figure 8: An example of mobile unit

For realizing these purposes, a concept of abstracted device module and *Status Data Pool* is used. Abstracted sensor devices(e.g. for sensing obstacle, Fig.7) and actuator devices(e.g. for mobile platform, Fig.8) are designed for convenience of devices checking by action management module. The *Status Data Pool*, a blackboard architecture, is designed for storing the whole sensor modules output and robot's inner status data. It also handles the redundant information in robot's sensor system. For example, in our experimental system, both PSD range sensor and on-board camera can detect obstacle in front of a robot but only *forward_exist* and *forward_distance* will be shown to BEO of system LOGUE from *Status Data Pool*. In the device module, the abstracted sensor or device module is written in Java and manipulates the devices written in C by each native method in JNI.

This makes the system LOGUE handle the real time sensing and control loop with various sampling frequency.

3.4 Task and Behavior Switching

A TO will be run when the starting conditions are satisfied and it has the highest priority in the TO list of the task manager object. A TO which is running can be stopped and abandoned in the following two cases, when all ending conditions to the task represented in the TO is satisfied and when a new TO which is to perform some emergency tasks is registered. When the running task is stopped normally, the next task object is evaluated and started by the task manager object. Also in the case of updating a BEO from other robot or behavior server, two cases can be thought: replacing an inactive BEO and replacing an active one. Replacing an inactive BEO is done by updating the BEO class file and BEO list in the behavior manager object.

In situation where a task should be interrupted in case of emergency or a BEO should be replaced even the former one is active, some special treatments which not only save the current status into a stack, like most software systems do, should be done. This is because a task or a behavior running on a robot may involve some physical interaction with the real world, such as holding an object or moving in some velocity. Without considering these physical interaction, a change of a task or behavior may cause some problems, such as lost grasping of the object or running out of the safety area. These problem depends on native devices of the robot and should be done deliberatively. A basic mechanism called *dialog-mode* is designed in current version of action manager module of LOGUE which automatically sets velocity and angular velocity of all actuators to 0 when it resumes the status of the robot. Currently, some basic simulation have been done on verifying the function of *dialog-mode* in dynamic task-behavior change in the LOGUE.

4 Simulation Result

Here, we show a scenario of a simulation for transmitting the task object and its class. Two robots, RockHopper and Adelie, are set in simulation environment. Robot RockHopper is working in a task "Garbage Collection(GarbageTask)". This defined as its highest priority command and

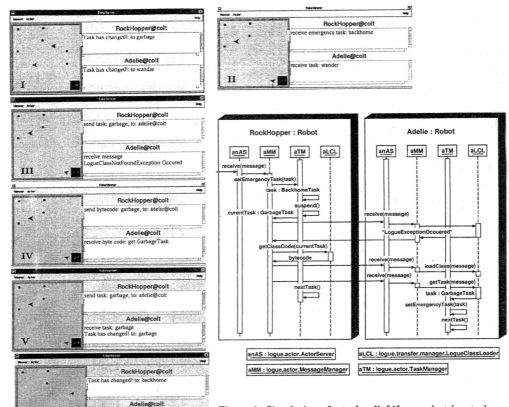

Figure 9: Simulation of a task relief if one robot has to leave and the sequence of transmitting task

should be achieved as soon as possible. In this moment, the robot Adelie is running on "Wandering in the Environment(WanderTask)" and does not own the object GarbageTask(Fig.3.3-I).

For an emergent requirement, the RockHopper should leave from the working area(BackhomeTask) (Fig.3.3-II) even though it interrupts its current job. Then, Adelie needs to perform the duty of the Garbage Collection task. In this moment, some actions occurred between the RockHopper and the Adelie, the system LOGUE succeeded on transmitting the GarbageTask to Adelie(Fig.3.3-V).

5   Experiment: Behavior Element Object Transmitting

A two wheels driven mobile robot, which includes a PSD range sensor and a camera with 2DOF heading control, is used to verify the experiment of the system LOGUE(Fig.10). Also encoder on each wheel is used for velocity measurement and odometry. Two computer systems are installed on the robot. The system LOGUE is running on a PC-AT machine with a wireless Ethernet module and an image processing board. A MC68332 controller board, which realizes all sensors(except the camera) signal processing and realtime control of the mobile platform and the camera heading, are connected to the main computer via RS232C.

The basic task is moving to a goal position which is located at the left conner of a fixed obstacle(Fig.11 top). Three behaviors CruiseRightBehavior, Cruise-LeftBehavior and ExcapeBehvior are used for this task. The robot was moving to the right side of the fixed obstacle(for easy to understand motion of robot, a plate is used as the fixed obstacle) in the beginning. When the robot's realtime vision

Figure 10: The experiment robot system

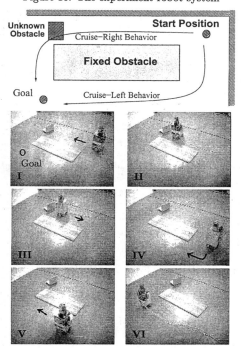

Figure 11: Experiment 1: Ordinary actions

system[12] and PSD range sensor detected an unknown obstacle in front of it, the *CruiseLeftBehavior* became active and lead the robot to the goal position(Fig.11-I~VI).

In the next experiment, another unknown obstacle appeared in the path behind the robot(Fig.12-II) and the *EscapeBehavior* also prevented the *CruiseLeftBehavior* to work to the final goal. The control of the robot was switching between *CruiseLeftBehavior* and *CruiseRightBehavior*, and the robot was blocked. A server system with a human operator[1] has knowledge on that the left small ob-

---

[1]It should be emphasized that the human operator just

stacle can be pushed away from the path and an emergency requirement need the robot to escape from the deadlock even there is a risk on making wheels slip by pushing an obstacle, which will cause large error on robot's odometry. The server system sent a new task including a BEO *pushObstacle* which the robot did not have. After replacing the new task, the robot asked the server for the *pushObstacle* (Fig.12-V). By integrating the new behavior, the robot removed the small obstacle in the left and reached the final goal position(Fig.12-VII~IX). The total processing time of task and behavior object transmitting and invoking here is not more than 1s which is including the time delay of network communication. It can be thought that this is a feasible value on task decision level of the autonomous robot control.

6 Conclusion

In this paper, the LOGUE, an architecture of communication and management system for object transmitting among autonomous robots or between a robot and behavior server system is proposed and the representation of the robot's behavioral elements and the task object are described. Also a structure of communication, action management and devices module are proposed. Problems on behavior checking and task switching, which are different from software auto-update system, are also discussed. A simulation and a basic experiment shown to verify processes of the task and the behavior object transmitting and switching by the LOGUE system in cases where the control system is running but replaced behavior is not active.

Automatically acquiring and running tasks or behaviors from other robot or a server system via network will make the robot maintain its control system in relatively small size. Also this makes it possible that, beyond its own sensing ability and storage, an autonomous robot can get proper information, tasks and behaviors via the network. It is meaningful on using some wide range of information detected from sensors which are connected to the server or installed on other robots since limited sensing ability on an autonomous robot is not enough for most unpredictable or emergency situ-

---

provides the necessary knowledge for emergency action and is not in charge any procedure on obtaining the necessary BEO or TO. If the knowledge: "pushing obstacle is a feasible plan" is installed in the system, the human operator is unnecessary exactly.

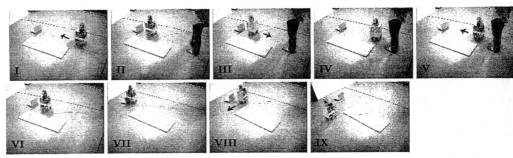

Figure 12: Experiment 2: Emergency action including an obstacle pushing behavior to the robot

ation. With this, the LOGUE system and inter-robot network extend a single robot to an autonomous multiple robots system. Also it may be possible to apply the proposed architecture to a learning based autonomous robots system as a component for sharing a behavior to other robots from one robot who got better learning results on the behavior.

It should be emphases that the proposed system is high scalabe on the number of robots. Because the BEO and the TO transmitting is based on JAVA RMI, network computer communication technology overcomes all basic technical difficulties for a system with a large number of robots. Furthermore, the LOGUE package provides simple but useful standard on management of the BEO and the TO. From these two points, it can be considered that robots construct a distributed database of the BEO and the TO in the LOGUE environment.

Detailed implementation of switching process of active task and active behavior, advanced task-behavior object verification and exception management in LOGUE system are our future work.

Acknowledgment

This work has been supported by Kayamori Foundation of Informational Science Advancement, and the Scientific Grants #11780243, Ministry of Education and Science, Japan.

References

[1] J.Kikuchi, K.Takeo and K.Kosuge, Teleoperation System Via Computer Network for Dynamic Environment, *IEEE Int. Conf. on Robotics and Automation*, pp.3534-3539, 1998.

[2] I.Elhajj, J.Tan, N.Xi,*et.al* Multi-Site Internet-Based Cooperative Control of Robotic Operations, *IEEE Int. Conf. on Intelligent Robots and Systems*, pp.826-831, 2000.

[3] P.G.Backes, K.S.Tso,*et.al*, Internet-based operations for the Mars Polar Lander mission, *IEEE Int. Conf. on Robotics and Automation*,pp.984-991, 2000.

[4] T.Oka, M.Inaba, H.Inoue, Describing a modular motion system based on a real time process network model, *IEEE/RSJ Int. Conf. on Intelligent Robots and Systems*, pp.821 -827, 1997.

[5] S.Kagami, F.Kanehiro, *et.al*, Design and implementation of brain real-time part for remote-brained robot approach, *IEEE Int. Conf. on Robotics and Automation*, pp.828 -835, 1997.

[6] T.Fujii, H.Asama, *et.al*, Intelligent data carrier system for cooperative behaviors emerged among collective robots, *IEEE Int. Conf. on Systems, Man and Cybernetics*, pp.299-304, 1997.

[7] J.Ota, M.Yamamoto, *et.al*, Environmental support method for mobile robots using visual marks with memory storage, *IEEE Int. Conf. on Robotics and Automation*, pp.2976-2981, 1999.

[8] L.E.Parker, ALLIANCE: an architecture for fault tolerant multirobot cooperation, *IEEE Tran. on Robotics and Automation*, 14-2 pp.220-240, 1998.

[9] R.C.Arkin, T.Balch, and E.Nitz, Communication of Behavioral State in Multi-agent Retrieval Tasks, *IEEE Int. Conf. on Robotics and Automation*, pp.588-594, 1993.

[10] Z.D.Wang, M.N.Ahmadabadi, E.Nakano and T.Takahashi, A Multiple Robot System for Cooperative Object Transportation with Various Requirements on Task Performing, *IEEE Int. Conf. on Robotics and Automation*, pp.1226-1233, 1999.

[11] T.Fujita and H.Kimura, Tight Cooperative Working System by Multiple Robots, *IEEE/RSJ Int. Conf. on Intelligent Robots and Systems*, pp.1405-1410,1998.

[12] J.M.Maja, T.Takahashi, Z.D.Wang and E.Nakano, Real-time Obs tacle Avoidance Algorithm for Visual Navigation, *EEE/RSJ Int. Conf. on Intelligent Robots and Systems*, pp.925-930, 2000.

Intelligent Autonomous Systems 7
M. Gini et al. (Eds.)
IOS Press, 2002

Maintaining a Common Co-ordinate System for a Group of Robots based on Vision

Dennis Wildermuth and Frank E. Schneider
Research Establishment for Applied Sciences (FGAN), Wachtberg, Germany
E-mail: dennis@fgan.de, schneid1@fgan.de

This paper presents a novel approach to the problem of establishing and maintaining a common co-ordinate system for a group of robots. A camera system mounted on top of a robot and vision algorithms are used to calculate the relative position of each surrounding robot. The watched movement of each robot is compared to the reported movement which is sent over some communication link. From this comparison a co-ordinate transformation is calculated. The algorithm was tested in simulation and is at the moment being implemented on a real robot system.

Keywords: common co-ordinate system, multi-robot, vision

1   Introduction

In many applications for multi-robot systems, the existence of a common co-ordinate system is favourable. An important precondition for the generation and maintenance of such a shared co-ordinate system is a precise localisation method. Most localisation techniques are based on global strategies that make use of special landmarks or other kinds of a priori knowledge about the environment of the robots.

This paper introduces a method for establishing a common co-ordinate system with reference only to the robots themselves. (We will call this a 'relative' common co-ordinate system throughout the rest of this manuscript.) Since it is not possible to map such a relative co-ordinate system to any system of global world co-ordinates, it is of course not useful for all multi robot applications. But for special problems it is sufficient, for example moving in formation, which is an important sub-topic in a demining application.

In order to build the common co-ordinate system the robots are equipped with cameras that are continuously taking pictures of the surrounding robots. Based on these images the relative positions of all visible robots can be calculated. Whenever a robot moves and therefore its position inside the camera image changes the movement with respect to the watching robot's co-ordinate system can be computed. By comparing this movement with the one the moving robot itself reports it is possible to calculate the transformation matrix between the co-ordinate system of that robot and some arbitrarily chosen reference co-ordinate system.

Some authors worked on similar concepts in order to reduce the odometry error of a single robot system. Murray [4] and Braithwaite [1] introduced a movable stereo camera system to follow remarkable points in the surrounding environment. Using the measured distances and rotation angles of the camera, they calculate the actual movement of the robot.

In several approaches, these results are transferred from a single to a multi robot system [5, 6]. Because in these works the aim is to generate and maintain a global co-ordinate system a great accuracy is needed. Just one robot moves at any given period of time while the others are standing still, thereby functioning as 'temporary' landmarks. In their so-called Cooperative Positioning System (CPS) Kurazume and Hirose extend this method by moving several robots in parallel [3]. Suzuki and Yamashita [7] present an approach to building a common co-ordinate system in which all robots may move simultaneously, but they use a simulated and somewhat idealised robot system. In their simulation, for example, every robot has a full 360 degree view and is capable of error-free measurement of the relative positions of the other robots.

The method presented in this paper is based on Suzuki's and Yamashita's ideas is aiming at real robot systems. Because the relative positions retrieved from a real vision system are distorted we developed an additional error model in order to weight and correct each measurement. By these means we improve the resulting co-ordinate system and make it possible to maintain it over longer periods.

2   Approach

A two step approach is used to establish the common co-ordinate system and share it between a group of robots. The first step consists of a vision process in which one or more camera equipped robots continuously grab images of the surrounding robots and hence calculate their relative positions. In the second step these position information in combination with the robots' movement data is used to establish and share the so called 'relative' common co-ordinate system.

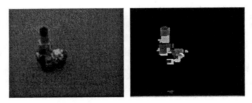

Figure 1: Example image segmentation

2.1 Acquisition of Robot Positions

The images necessary for the mutual localisation are continuously acquired by the mounted camera systems and then analysed and segmented in order to extract the different objects visible in the picture. The criteria relevant for the segmentation are:
- brightness of neighbouring pixels,
- size and shape of objects,
- special movement parameters.

Figure 1 shows an example of such an image segmentation. On the left side an image containing one of our robots can be seen, the right image presents the complete segmentation generated from the original. All robots have special distinct markers mounted on top which can be found as easily distinguishable areas in the segmented image as well.

The segmentation process generates a list of objects within the segmented image. For each object in this list its size, its position and its YUV colour coding is stored and compared to the characteristics of the each robot. If one or more objects found in an image segmentation can be identified as part of a robot or – to be more precise – as belonging to the mounted marker their relative positions are computed. We use a standard photogrammetric projection algorithm as, for example, described in [2] to calculate the robot's relative co-ordinates based on its position inside the image.

For each image taken by one of the robots' cameras first its projection centre (which roughly means the position of the camera lens) is computed. Afterwards the actual position of the image with reference to the robot's local co-ordinate system has to be calculated. Since the camera is mounted on a pan/tilt system its actual viewing direction, i.e. the rotation angles ω, φ, and κ around the x-, y-, and z-axis are used to calculate a spatial rotation matrix R. As $\varphi = 0$ because the camera is mounted exactly upright it is

$$R_{\omega,\kappa} = \begin{pmatrix} \cos\kappa & \sin\kappa & 0 \\ -\cos\omega\sin\kappa & \cos\omega\cos\kappa & \sin\omega \\ \sin\omega\sin\kappa & -\sin\omega\cos\kappa & \cos\omega \end{pmatrix}. \tag{1}$$

Let X_O, Y_O, and Z_O denote the co-ordinates of the formerly computed projection center, ξ_0 and η_0 the co-ordinates of the image center point, ξ and η the co-ordinates of a relevant image segment and c the 'camera constant', then the corresponding object co-ordinates X, Y, and Z can be derived from the two equations (2a) and (2b) (see [2]):

$$X = X_O + (Z - Z_O)\frac{r_{11}(\xi - \xi_0) + r_{12}(\eta - \eta_0) - r_{13}c}{r_{31}(\xi - \xi_0) + r_{32}(\eta - \eta_0) - r_{33}c} \tag{2a}$$

and

$$Y = Y_O + (Z - Z_O)\frac{r_{21}(\xi - \xi_0) + r_{22}(\eta - \eta_0) - r_{23}c}{r_{31}(\xi - \xi_0) + r_{32}(\eta - \eta_0) - r_{33}c}, \tag{2b}$$

where r_{ij} represent the components of matrix $R_{\omega,\kappa}$.

As the height of each robot and thereby the height of the relevant image segments belonging to the top mounted markers is well known Z has a constant value. Using this additional knowledge it is possible to assign a unique point in the robot's local co-ordinate system to each relevant image segment and – as a result – compute the relative position of every visible robot.

2.2 Establishing the Relative Common Co-ordinate System

Using the other robots' relative positions which are computed by the vision process described in the last chapter it is possible the establish a so called 'relative' common co-ordinate system for all visible robots. In this context the attribute 'relative' means that there is no reference to any fixed global co-ordinate system but instead the robots just share a common co-ordinate system between each other. Figure 2 schematically describes the process leading to this relative co-ordinate system.

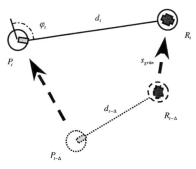

Figure 2: Camera equipped robot watching another robot while both are moving

One camera equipped robot observes another robot and calculates its relative positions while he is moving from position $R_{t-\Delta}$ to R_t during a time interval Δ. In the same interval the observer itself might be moving from point $P_{t-\Delta}$ to point P_t. In order to establish a common co-ordinate system for these two robots the watched robot functions as some kind of moving landmark.

First, it is considered that the camera equipped robots is not moving but stands still at a given position throughout the observation process. Whenever a new position for another robot is accepted by the vision and evaluation process the total movement during a specified period of time is computed based an the last few measurements.

As figure 3 illustrates the starting point A_l and the end point A_l' of this movement can be expressed as points in a global reference co-ordinate system. For the same period of time, the movement of the robot measured by its odometry sensors is transferred over a communication link. A_r and A_r' represent this route with respect to the local co-ordinate system of that robot. Using these four points a (2-dimensional) translation matrix can be computed which transfers the 'communicated' vector into the 'watched' one and thus transfers the local co-ordinate system of one robot into the reference co-ordinate system.

In a next step the pre-condition of the camera equipped robot standing still at one position is dropped. If the observing robot itself has moved while watching the other robot this additional translation and rotation has to be considered when calculating the vector from point A_l to point A_l'. The other steps of the calculation remain unchanged but the resulting transformation matrix of course suffers from the additional odometry error which enters the calculation process.

Obviously the whole process does not need any information about the environment (apart from the knowledge which robot uses which marker). The resulting reference co-ordinate system therefore is not 'global' in the sense that it has a fixed reference to world co-ordinates. We call it a 'relative' common co-ordinate system because it is just shared among the participants of the robot group but can diverge from world co-ordinates over time. Once the common co-ordinate system is established the robots can start working on their task. During the work of the robot group the whole process must be repeated regularly in order to maintain and correct the transformations between the robots' local co-ordinate systems and the reference co-ordinate system.

Since the used camera system, of course, only has a limited opening angle it was mounted on a pan/tilt platform. Therefore, it is not necessary to move the whole robot in order to take images of the surrounding robots. Simply the camera is moved towards the direction in which a robot is expected to be found. A simple prediction algorithm was implemented to find a robot's probable course from its last viewed positions. With the help of this prediction algorithm the sequence of measured robots as well as the direction the camera has to be moved to in order to

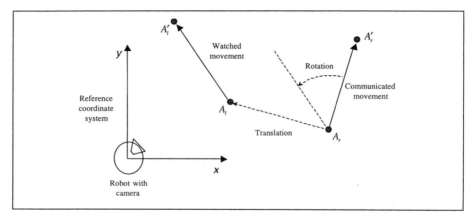

Figure 3: Transformation from robot's local co-ordinate system into reference co-ordinate system.

point towards a special robot can be optimised. As a result the frequency of measurements can be increased and thus the mean accuracy of the resulting relative co-ordinate system can be improved.

If a robot is not visible over a longer period of time because he works behind some obstacle or he is too far away, it might become necessary to use an additional, globally referenced localisation method. For such a task a vision-based approach is probably not the best choice. The last chapter will shortly describe the overall application our institute is working on and for which this relative localisation method is sufficient.

3 Validation Studies

To verify and validate the presented approach two studies were carried out. First, as a supplement for our multi-robot simulation environment a very simple camera simulation was added. With this rough model of the vision the rest of the algorithm could already be implemented and tested. Second the vision system was installed in order to undertake first real world experiments.

3.1 Simulation Results

A simple simulation component for our multi-robot simulation environment was added which delivers relative positions of the robots inside a virtual camera field of view. In order to emulate the inherent inaccuracy of this vision based process a simple randomly distributed error offset is added to each position information. Additionally the multi-robot simulation models the odometry error for the robots thus providing a rather realistic environment for testing complex higher level algorithms for multi-robot systems.

Figure 4 demonstrates an example run of the complete co-ordination algorithm inside the simulation environment. The robot group in this example consists of four robots, of which one is equipped with a top mounted and movable camera system. This robot takes the role of a group leader throughout this run. The task for the group is to set up a common co-ordinate system and build a line formation behind the leading robot. This formation is generated by simply sending each robot a target position with reference to the shared co-ordinate system.

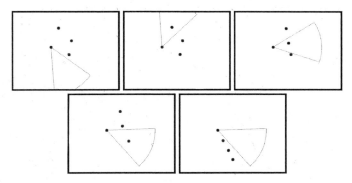

Figure 4: Example simulation run – one camera equipped robot co-ordinates three other group members. The pictures are sorted upper left to lower right and are taken at a fixed time interval.

The five pictures in figure 4 are sorted upper left to lower right and shall describe the complete procedure. As a first step the group leader turns his camera until all participants are found. Then he successively commands each robot to move forward about one meter. Pictures 2 till 4 present these steps of the algorithm. As described in chapter 2 the observer generates a common relative co-ordinate system by comparing the travelled path he observed with the path reported by each robot.

Each resulting transformation matrix is sent to the corresponding robot. With this matrix the robots are able to transform movement commands expressed in terms of the common co-ordinate system into locally referenced steering commands. The last picture presents the result for the given task – all robots line up behind the group leader. Because of the imprecise relative positioning and the inherent odometry errors this is, of course, not a perfect line but is somewhat deformed.

3.2 Experiments with Real Robots

Since the results from simulation looked rather promising we started to implement the approach on our real robot system. It consists of one B21 robot by RWI Inc. which is equipped with a camera system mounted on top of a pan/tilt platform and four of RWI's Pioneer I robots. Special two-coloured markers were mounted on top of the Pioneers which make them distinguishable for the vision process. After having implemented the basic vision algorithms for locating and identifying a robot inside a camera picture as described in chapter 2 we started to evaluate the quality of the delivered relative position information.

The first result was a large number of wrong and deviating segmentations nearly for every measurement. The main reason for the variation of the segmentation results is that, depending on the lighting of the scene, a single red coloured area is often segmented into more then one partitions. In order to use all these segments and to distinguish them from simple erroneous segments we developed a simple selection algorithm: The circle (of the robot's diameter) in which the most correctly coloured segments are found is considered to be the position of the robot. Of course, this method still is error prone but – as figure 5 shows – produces already good results. One Pioneer robot was positioned at different positions spread over the camera's field of view (marked by the small circles inside the chart). Then five pictures were taken and the robot's relative position was calculated (marked by the small crosses). The average computed from these five measurements is marked by the small diamond shaped figures inside the chart.

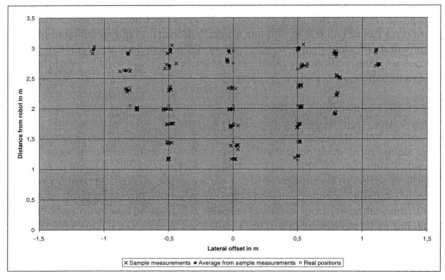

Figure 5: Measured positions of a specially marked robot using a real camera system

Figure 6: From left to right: a moving robot passing the observer's field of view

Figure 6 shows the first experimental set-up with three pictures ordered from left to right representing one example experimental run. A Pioneer I robot passes the observing B21 robot. Each picture was augmented with a little overlay representing the maximal field of view of the camera mounted on top of the robot. The Pioneer robot moved very slow at a speed of about 15 cm per second in order to get several measurements while passing the camera's view. Table 1 lists the relative positions of the vehicle as they were computed by the vision process during the experiment. Thereafter several runs were performed during which the camera equipped B21 also moved at a rather slow speed of about 10 cm per second. The Pioneer I again passes the field of view of the camera at about 15 cm per second. The relative positions generated by the vision process were recorded and are also presented in table 1.

The values listed in the right column of table 1 which belong to this experiments show a gap of several seconds in which the observed robot could not be detected at all. This effect also occurs when the watching robot does not move but the Pioneer moves rather fast. The reason for this problem is the grabbing software that gets the raw camera image from the frame grabber hardware and probably the camera as well. Whenever an image contains moving objects – and when the camera itself moves, of course, all objects are moving – the image which enters the segmentation process is strongly distorted. As a result the segmented image consists of lots of small parts which are not large enough to be interpreted as part of a robot. We are currently evaluating the possibility of changing the grabber software and/or the camera hardware in order to avoid this effect.

Table 1: Relative positions of a marked robot measured during two example runs, first with a still standing observer, second with a moving one

Time (in sec.)	Example 1: Observer standing (x/y position in m)	Example 2: Observer moving (x/y position in m)
0	-/-	-/-
1.5	0.58/1.67	-0.95/2.89
3	0.26/1.71	-0.78/2.87
4.5	-0.12/1.78	-/-
6	-0.45/1.94	-/-
7.5	-0.75/2.05	0.14/1.94
9	-/-	0.51/1.80
10.5	-/-	-/-

As long as all participants are moving slow enough the vision process works rather well and produces steady position data. The frequency of measurements can be higher than the here presented one result every 1.5 seconds because a pause of one second was added after each position estimation to decrease the amount of data listed in the result table.

Although the resulting positions seem to be rather steady the probability of completely wrong measurements increases when more than one robot is visible inside the same image. In order to reduce the probability of such errors an additional testing algorithm was implemented.

With every measurement the last positions computed for a robot are compared to its actual position. If the new position is impossible, e.g. because the robot's maximum speed would have been exceeded, or unlikely, e.g. because it differs strongly from the direction the robot had before, a repetition of the last measurement is initiated.

4   Conclusions and Future Work

This paper presents a new approach to the problem of establishing and maintaining a common co-ordinate system for a group of robots. A camera system mounted on top of a robot and vision algorithms are used to calculate the relative position of each surrounding robot. The watched movement of each robot is compared to the reported movement that is sent over some communication link. From this comparison the co-ordinate transformation is calculated.

The method was tested inside a multi-robot simulation environment which fully models the odometry errors of the robot vehicles. Additionally first validating experiments with real robots and a real camera system were performed and produced promising results. The vision process generates consistent position information for other visible robots.

Several problems occurred throughout the implementation process. For example, the segmentation process ceases to work properly when the watched vehicles and especially the camera equipped robot move too fast because of a decreasing image quality. The range in which a robot position can be calculated from a camera image is at the moment limited to about three meters. Different camera hardware has to be tested to increase this range to a more applicable distance. Additional vision systems must be mounted on participating robots and their separate results must be fused in order to achieve greater precision for the resulting coordinate system.

As a major improvement it is planned to combine this solely vision based approach with the distance data coming from each robot's laser rangefinder. After a robot is identified and its initial position is measured its ongoing movement is tracked using laser data. Whenever the error probability of this tracking process rises beyond some upper limit a new vision based identification is performed. With this enhancement the overall number of necessary camera measurements can be greatly reduced.

5   References

[1] R. N. Braithwaite and B. Bhanu. *Robust Guidance of a Conventionally Steered Vehicle Using Destination Bearing*. Proceedings of the 1993 International Symposium on Intelligent Control, September 1993.

[2] K. Kraus. *Photogrammetrie – Band 1: Grundlagen und Standardverfahren*. Dümmlers Verlag, Bonn, 1994.

[3] R. Kurazume and S. Hirose. *An Experimental Study of a Cooperative Positioning System*. Autonomous Robots 8 (1), pp. 43-52, Kluwer Academic Publishers, Boston, 2000.

[4] D. W. Murray, I. D. Reid and A. J. Davison. *Steering and Navigation Behaviours using Fixation*. British Machine Vision Conference, 1996.

[5] I. M. Rekleitis, G. Dudek and E. E. Milios. *Multi-Robot Exploration of an Unknown Environment, Efficiently Reducing the Odometry Error*. Proceedings of the International Joint Conference in Artificial Intelligence. Nagoya, August 1997.

[6] I. M. Rekleitis, G. Dudek and E. E. Milios. *Reducing odometry error through cooperating robots during the exploration of an unknown world*. Proceedings of the 5th IASTED International Conference on Robotics and Manufacturing. Cancun, Mai 1997.

[7] I. Suzuki and M. Yamashita. *Agreement on a Common X-Y Coordinate System by a Group of Mobile Robots*. Proceedings of the Dagstuhl Seminar on Modeling and Planing for Sensor-Based Intelligent Robot Systems. Dagstuhl, September 1996.

Intelligent Autonomous Systems 7
M. Gini et al. (Eds.)
IOS Press, 2002

Industrial robot manipulator with integrated automatic collision-free path planner

Heinz Wörn   Björn Hein
woern@ira.uka.de   hein@ira.uka.de
Computer Science Department, Universität Karlsruhe (TH)
Institute for Process Control and Robotics (IPR)
Kaiserstrasse 12, D-76128 Karlsruhe, Germany

Abstract. In this paper the integration of an automatic path planner in a robot controller for industrial applications is presented. A given position will be reached collision-free and time-optimized. The application allows the input of the workcell with the robot and offers an interface to simulation systems. A command interpreter controls the programm execution. The planning method is based on the A*-search algorithm and needs no essential off-line computations and is therefor on-line capable.

1 Introduction

When working with robot manipulators and handling objects (e.g. Pick & Place, spotwelding) there is the problem of finding fast collision-free movements between two locations. Normally it is to the programmer and his experience to teach manually a collision-free path. When the environment changes (e.g. new obstacles) a new collision-free path has to be taught. It would be fine to let the robot controller do this task and just to update the environment data or the target location. Due to the increased performance of common CPUs, the robot controllers can handle now a days more than just the controlling of the manipulator. Therefor we integrated an automatic collision-free path planner into a robot controller, computing on-line a collision free movement between two given locations.

To explain this system and its possibilities this paper is organized as follow: Section 2 gives an overview of the robot controller, manipulator and the integrated application of the planning system. In Section 3 a tool for generating the environment data is proposed. Section 4 give some information about the planning algorithm. This paper finishes with a conclusion and an outlook.

2 Overview

This Section gives information about the robot system, that is used for testing. After that the application, that allows the automatic collision-free path planning will be presented.

2.1 Controller, input device and robot

The robot system consists of three components: The robot controller, the input device and the robot.

(a)   (b)   (c)

Figure 1: (a) The robot controller, (b) KCP - control panel, (c) a KR15 robot

The used robot controller is the KRC-1 from KUKA Roboter GmbH. It is currently based on a 400Mhz Celeron processor, the robot control algorithms are running on the realtime operating system VxWorks. Microsoft Windows95 is running as a task of VxWorks offering an GUI and API to the controller.

For programming and moving the robot it exists an input device : the KCP (KUKA Control Panel). This input device consists of a TFT-display for graphical output, a keyboard for entering program data and special buttons for faster access to menus, robot control and other often used functionality.

KRC-1 can control different kind of robots. For testing we use a KR15, a robot with six degrees of freedom and a maximum payload of 15 kg.

2.2 Planning application

The planning application is realized in Visual Basic offering the GUI and is using the planning library programmed in C/C++ (s. Section 4). The layout is specially designed for the use with the KCP and it's special buttons. There are two main program modes: layout and planning mode.

2.2.1 Layout mode

This mode allows the handling of the environment data: loading, creating, modifying and storing. For describing objects in the actual environment boxes, spheres and cylinders are supported (s. Figure 4). For instance the application offers some simple ways of entering the data: alphanumerical using the keyboard or moving the robot and marking the extreme points of an obstacle or using the both input methods mixed. In the first case the programmer must enter the position of an obstacle by filling the necessary parameters (e.g. box: height, width, length and position; sphere: diameter and position), in the second case the robot and an attached tool is moved to mark points on the object and use this information for computing the dimension of the obstacle, some kind of simple "digitizing" process. The robot programmer is guided through the teach-in process, and is been told, which point should be entered next. If a position cannot be reached with the robot, there are some strategies to enter it manually (s. Figure 2 and Figure 3). Figure 5 illustrates the graphical output on the KCP giving a visual feedback to the programmer.

Entering the environment data the way above is time consuming and needs some

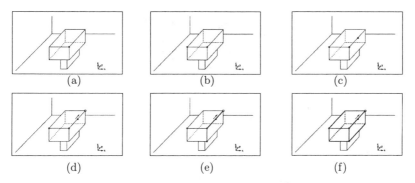

Figure 2: First sample for teaching in an obstacle with the robot: (a) first point, (b) second point, (c) only a point on the edge can be reached with the robot, (d) manually entered length of the edge gives information of the position of the third point, (f) fourth point is entered, (g) a box is created by this four points.

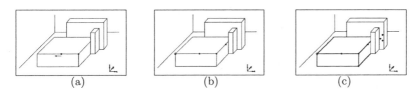

Figure 3: Second sample for teaching an obstacle with the robot: (a) mixed input of second point (s. Figure 2(c)+(d), (b) again only a point on the edge can reached, but instead of entering the data manually, the user can define a bounding surface (with three points); after entering the last point, a box is created.

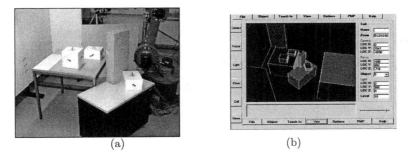

Figure 4: (a) Sample workcell (a table, some boxes and a roll-boy). (b) Cell entered with by using the robot as input device.

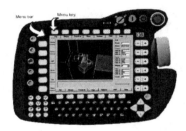

Figure 5: Visualization of the worcell in Figure 4(a) displayed on the KCP.

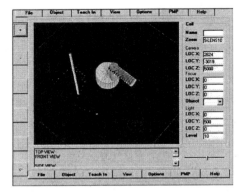

Figure 6: Sample of planning: Robot has to move around staff. After planing the movement of the TCP (tool center point) is visualized with points.

exercise. Therefor it is possible to import the environment data from a simulation system (s. Section 3).

2.2.2 Planning Mode

The planning mode allows the configuration of the path planner and of the parameters of the planning algorithm. An interpreter allows command sequences concerning the movement of the robot (moveTo <location>, moveHome) and command sequences changing the environment (grab <object>, release <object>, addObstacle <object>, removeObstacle <name>), the latter automatically updates the data of the distance computation.

After planning, the path can be visualized before execution is done with the robot (s. Figure 6).

3 Interface to simulation system

As mentioned above, it is not easy to teach a workcell using only the keyboard and a robot. The obstacles are in the majority of cases more complex than boxes or cylinders. Therefor it is possible to layout the workcell with professional robot simulation tools like eM-Workplace(Tecnomatix) or IGRIP (Delmia) (s. Figure 7), store this information and use it later in the planning application. A further advantage of this interface is the

Figure 7: Export application (ROBCAD): This application allows the generation of environment data for the planning application out a commercial simulation system.

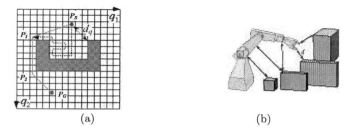

(a)   (b)

Figure 8: (a) A 2D illustration of the path search in the implicit C-space from the start P_s to the goal P_G using the obstacle distance d_q for collision detection, d_q is the conversion from the Cartesian to C-space distance., (b) Collision detection in the explicit workspace by computing the minimum distance d between robot and obstacles

huge variety of objects (torches, fixtures, tools) already available as libraries included with the simulation systems.

4 Planning library

The Planning system consists of two parts: a motion planner and optimization algorithm based on rounding. The following subsection will give an short overview of the both components, for detailed information please refer to referenced sources.

4.1 Collision-free motion planner

The motion planner [6] is based on the A*-algorithm [1]. The search takes place in the discrete configuration space (C-space), spawn by the joints of the robot (Fig. 8(a)).

One of the main features of this planner is that obstacles are not transformed from the 3D-workspace to the C-space. That means obstacles are not really existing in C-space, they are represented by the distance of a fast, hierarchical distance calculation done in 3D-workspace [3] during the search (Fig. 8(b)). Therefor we call it implicit representation of obstacles in C-space, or short implicit C-space. This allows the planner to be used in dynamic environments.

As result of the planning process we get a collision free path - P_0 .. P_i .. P_n - consisting of PTP-movements (Point-To-Point) in C-Space (PTP = lines in C-space,

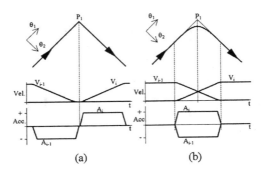

Figure 9: Velocity and acceleration of robot axis without (a) and with (b) rounding

Fig. 8).

4.2 Path optimation

As mentioned above, the output of the motion planner is a path consisting of PTP-movements. There are different possibilities to optimize this movement. One interesting method is to use spline movements [4][5], but most of the standard robot controllers do not support this motion command. Therefor it would be necessary to modify the robot controller software and integrate a new motion command. This is in most cases not possible. Due to this our approach uses for optimation a robot command available on every standard robot controller: rounding.

Figure 9(a) shows the velocity and acceleration profile of two PTP-movements without rounding. For easier understanding the first movement only changes joint 1 (Θ_1) and the second movement changes joint 2 (Θ_2). This will result in a trajectory in C-space having an angle of 90 in Point P. Without rounding the velocity in P is 0. With rounding the two movement will overlap as seen in Figure 9(b). The velocity in P is not 0 anymore, execution time is reduced, but P is not reached any longer. This is the main problem when using rounding. The resulting trajectory cannot be guaranteed to be collision-free anymore. Our method [2] modifies the path in such a way, that all obstacles are avoided in spite of using rounded movements (s. Figure 10) .

5 Conclusion and outlook

In this paper we introduced a automatic collision-free path planning application integrated on a robot controller. This allows the collision-free movement between location in changing environments. The application offers the possibility to layout the workcell using the KCP and the robot, or using the data from commercial robot simulation tools. An command interpreter controls the execution of the planning task. This allows the usage of a robot manipulator for industrial purpose in changing environments and simplifies the programming process.

The proposed path planning library finds and optimizes a collision-free path based on the environment data of the planning application.

Following features will be added to the application in the next future:

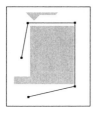      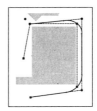

(a)   (b)   (c)

Figure 10: Sample for the optimation algorithm using rounded movements.(a) Path after planning process, all PTP-movements are collision-free, but the robot has to stop at every point. (b) Three "virtual" points are created and the new path using rounding is tested for collision (third movements collides and cannot replaced by a rounded movement). (c) Optimizes path. Rounding is used as much as possible.

- Using a navigation system instead of the robot for input of the environment data.

- Add more functionality to the command interpreter.

- Adding functionality for calibrating the imported environment data from the simulation system with the robot or the navigation system.

For the improvement of the optimation process the dynamic constraints of the robot will be taken into account (mass, torque forces, max. velocity or acceleration).

References

[1] Raphael B. Hart P.E., Nilsson N.J. A formal basis for the heuristic determination of minmum cost paths. In *IEEE Trans. Syst. Sci. Cybern*, pages 100–107, 1968.

[2] B. Hein, Marcos Salonia, and Heinz Wörn. Automated generated collision-free time optimized robot movements in industrial environments based on rounding. In *ISATP, International Symposium on Assembly and Task Planning*, 2001.

[3] Suei Chen Marcos Salonia, Sander Karl. Disco, fast hierarchical distance computation. wwwipr.ira.uka.de/~iirob, 2001.

[4] H. Ozaki and C.J. Lin. Optimal B-spline joint trajectory generation for collision-free movements of a manipulator under dynamic constraints. In *Proceedings of IEEE Internation Conference on Robotics and Automation*, pages 3592–3597, 1996.

[5] M.A. Norris S.Dubowsky and Z.Shiller. Time optimal trajectory planning for robotic manipulators with obstacle avoidance: A cad approach. In *Proceedings of IEEE International Conference on Robotics and Automation*, pages 1906–1912, 1986.

[6] C. Wurll. *Mehrziel-Bahnplanung für Industrieroboter*. PhD thesis, Insitut for Process Control and Automation, Universität Karlsruhe (TH), 2000.

Intelligent Autonomous Systems 7
M. Gini et al. (Eds.)
IOS Press, 2002

Collision Avoidance and Handling for a Mobile Manipulator[*]

Thomas Wösch and Werner Neubauer
Corporate Technology, Information and Communications
Siemens AG
81730 Munich, Germany
{thomas.woesch,werner.neubauer}@mchp.siemens.de

Abstract

In general service robots move autonomously in partially known environments. This paper proposes a collision avoidance and collision handling scheme to move through such working areas. Both schemes are implemented in a single paradigm: artificial and measured forces are combined, resulting in force/torque vectors. The force/torque vectors are superimposed and the robot's joint velocities are determined. We present results from our experimental system consisting of a 8 degrees of freedom manipulator arm mounted on a mobile platform. In the illustrated example, a human enforces collisions by touching the robot. The robot evades the human's touch without robot self-collisions.

1 Introduction

We propose a collision avoidance and collision handling scheme to move safety in partially known environments. The implementation is done in a single paradigm, see Figure 1. The collision avoidance is used to prevent collisions between the robot and the known environment. Self-collisions are also avoided. The collision handling system detects collisions, that have not been prevented by the collision avoidance system. This may occur, when the environment model is incomplete, e.g. because of high-dynamically scenes. Dependent on the system mode the environment model can be updated by the tactile sensor information of the collision handling system. The implemented collision handling scheme can also be used as a interaction interface, e.g. the user can guide the robot by touching it.

In literature there can be found various sensors used to detect collisions on robot systems, ranging from infrared sensors [1] to capacitive sensors [2]. A disadvantage of most of these sensor systems is, that their shock absorbency is low.

To control a robot system in an environment of average complexity a lot of work was done during recent years. Very popular have been artificial forces [3, 4] resulting in pure reactive control schemes. Because generally these methods suffer under local minima several planning techniques have been proposed [5]. Different frameworks as [6] combine the advantages of reactive controllers (fastness) and the advantages of planners (find a solution whenever one exists).

[*]This work was partially funded under contract *01IN601A2* and *01IL902DO* by BMBF, Morpha project.

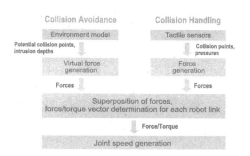

Figure 1: Collision handling and collision avoidance scheme overview

Figure 2: Experimental system

2 Experimental Setup

Our experimental setup consists of a manipulator arm with 8 degrees of freedom (DOF). The arm is mounted on a non-holonomic platform. For moving through an environment and determining the robots position in the world we use the navigation system described in [7]. Figure 2 illustrates our robot. The laser-scanner on top of our experimental setup is used to get a 3D surface model of a current scene. Tactile sensors [†] are mounted on the manipulator arm and mobile platform. With these tactile sensors, the robot can detect collisions. When pressure is exerted on a tactile sensor segment, the sensor generates information about the amount and location of pressure. A more detailed description of our system is given in [8].

3 Force Generation

The collision avoidance scheme determines virtual forces based on an environment model to avoid collisions before they occur. The environment model is built by the geometry of the robot itself, pre-knowledge of the environment and sensor information. When an obstacle gets closer to a link than its safety distance s, see Figure 3, the minimum distance $d = \|\underline{h} - \underline{c}\|$ between the obstacle and the link is determined. According to the minimum distance a virtual force can be evaluated by

$$\underline{F}_v = \begin{cases} -K_f(s-d)\frac{\underline{h}-\underline{c}}{d} & d \le s \\ 0 & d > s \end{cases} \tag{1}$$

The force is acting on the collision point $\underline{c}$. The size of a links safety distance, depends on the task behavior and the state of the current task. E.g. when we grasp a obstacle the safety distances of the links have to be reduced.

Collisions that have not been avoided by the collision avoidance scheme or that have been enforced, e.g. by a human, are detected and controlled by the collision handling scheme using the physical forces determined by tactile sensors mounted on our robot. How the system responds to a collision depends on the current system mode. In the simplest case the system evades the collision, by generating appropriate forces attached to the collision points. According to the pressure p examined by the tactile sensors a force acting on the collision point $\underline{c}$ is given by

$$\underline{F}_p = -K_f p \underline{e}_f \tag{2}$$

[†]Siemens patent, cat. no. 19959703.0

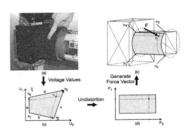

Figure 3: Virtual force generation   Figure 4: Physical force evaluation

where $\underline{e}_f$ describes the evaluated direction of the force, see Figure 4.

4 Superposition of Collision Avoidance and Handling

For L forces (generated by the collision avoidance or handling scheme), acting on the collision points $\underline{c}_l$ of a robot link, the resulting force $\underline{F}_{\text{link}} = \sum_{l=1}^{L} \underline{F}_l$ acting on the control point $\underline{p} = \sum_{i=1}^{L} \underline{c}_i$ is determined. Using the distance vectors $\underline{r}_i = \underline{c}_i - \underline{p}$ a torque on the link $\underline{M}_{\text{link}} = \sum_{i=1}^{L} \underline{r}_i \times \underline{F}_i$ is evaluated. The resulting force and torque is summarized in a force/torque vector δ for each robot link. Transforming this vector into joint space by

$$\underline{\tau} = \left(\underline{\underline{J}}_p\right)^T \underline{\delta} \tag{3}$$

yields appropriate joint torques for moving the control point $\underline{p}$.

5 Results

For testing the proposed schemes we have set the robot in *Push* mode. This means, that the collision handling system evades a collision. In Figure 5 the collisions are forced by a human who touches the tactile sensors. The environment model of the robot consists of its self geometry and the floor where the robot is positioned. The implementation of the proposed method allows the user to move the robot links, by touching the tactile sensors. The user is able to freely interact with the robot. Thus the user is moving the robot, the user must not take care of robot self-collisions. The introduced collision avoidance scheme is running in the background and detects potential collisions before the arm moves. Appropriate motions according the users touch and the collision avoidance system are generated.

6 Summary, Conclusion and Future Work

The paper introduces the combination of a collision avoidance and collision detection system. Both systems are implemented in a single paradigm: Virtual forces according to an environment model and forces proportional to the pressure detected on tactile sensors are combined with torques, resulting into force-torque vectors. These vectors determine the joint velocities. Experimental results show a human, enforcing collisions by touching the robot. The collision handling system evades the human touch. While the collision avoidance system prevents self-collisions.

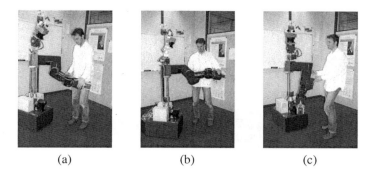

(a)   (b)   (c)

Figure 5: Example for the collision handling and avoidance system.

In this paper the collision handling system only evades collisions. More sophisticated modes than this simple *Push* mode will be implemented. A *Tactile Exploration* mode might be of interest, where the robot will actively (though carefully) seek contact with the environment and update the environment model. Other tactile interactions will be investigated, like the definition of a *Tactile Alphabet* to communicate with the robot *via* the tactile interface. The redundant DOF in our setup will be utilized for collision avoidance and object manipulation at the same time.

References

[1] D.Um, B.Stankovic, K. Giles, T. Hammond, and V. Lumelsky, "A modularized sensitive skin for motion planning in uncertain environments," in *International Conference on Robotics and Automation*, Leuven, May 1998.

[2] H. Shinoda and H. Oasa, "Passive wireless sensing element for sensitive skin," in *International Conference on Intelligent Robots and Systems*, 2000, pp. 1516–1521.

[3] O. Khatib, "Real-time obstacle avoidance for manipulators and mobile robots," *International Journal of Robotics Research*, vol. 5, no. 1, pp. 90–98, 1986.

[4] H. P. Xie, R. V. Patel, S. Kalaycioglu, and H. Asmer, "Real–time collision avoidance for a redundant manipulator in an unstructured environment," in *International Conference on Intelligent Robots and Systems*, October 1998, p. 1925.

[5] C. Nissoux, T. Simeon, and J. P. Laumond, "Visibility based probabilistic roadmaps," in *Intelligent Robots and Systems*, 1999.

[6] O. Brock and O. Khatib, "Elastic strips: A framework for integrated planning and execution," in *Preprints, 6th International Symposium on Experimental Robotics (ISER'99)*, Sydney, 1999.

[7] Gisbert Lawitzky, "Das Navigationssystem SINAS," in *Proceedings Robotik 2000, VDI-Berichte 1582*, Düsseldorf, June 2000, pp. 77–82, VDI-Verlag.

[8] G. v. Wichert, Th. Wösch, St. Gutmann, and G. Lawitzky, "Mobman – Ein mobiler Manipulator für Alltagsumgebungen," in *Autonome Mobile Systeme (AMS'00)*, R. Dillmann, H. Wörn, and M. v. Ehr, Eds. 2000, Informatik aktuell, pp. 55–62, Springer Verlag, Heidelberg.

Intelligent Autonomous Systems 7
M. Gini et al. (Eds.)
IOS Press, 2002

Intelligent Mobile Robot System Theory by Use of Network Thermodynamics

Hideo YUASA*,** and Tamio ARAI*

**Graduate School of Engineering, The University of Tokyo, Japan*
{yuasa,arai}@prince.pe.u-tokyo.ac.jp
***Bio-mimetic Control research center, RIKEN, Japan*

Abstract. The intelligence of a mobile robot is determined by the connection with its sensor input and its actuator output. If we interpret the information flow as energy one, we can consider them the basis for the definition of an autopoietic loop, in almost the same fashion as it happens in living systems. We propose a new structure of dynamic connection with input and output, which has duality between input recognition and output generation.

1 Introduction

It is said that a mobile robot is embodied within their environment. This means that a mobile robot is situated, it acquires information about its environment only through its own sensors and it moves by use of its own actuators. Then, the intelligence of such a mobile robot is determined how to connect with its sensor input and its actuator output. One of the basic research about such intelligence is "Braitenberg's Vehicles[1]." He started to consider it from the simplest case of the connection between single sensor input and single actuator output (Vehicle I). Then, the consideration of "Vehicles" step up to more complex input, output and connections between them (Vehicle II~), and their behavior becomes more complex. Unfortunately, Braitenberg was not interested in the quantitative analysis of the relation between its behavior and its input-output connection. But this is the key point to construct some desired intelligence of an autonomous mobile robot.

One of the promising research about such an intelligent mobile robot is named "Roboticle[2]." The basic idea of Roboticle is to think about *information flow*, which flow in robot through its sensors, and next through the conservative and dissipative process, finally out the environment through its actuators. If we interpret this process as energy flows, we can consider them a basis for the definition of an autopoietic loop, in the same fashion as it happens in living systems. Then, we want to expand "Roboticle" to be able to treat n-dimensional input and output, moreover to treat a multi-robot system which organizes their behaviour themselves suitable for their environment.

This short paper is organized as follows. In the section 2, we start to consider actual self-organizing systems. To use self-organizing property for a mobile robot system, common properties of self-organizing system are discussed. In section 3, it is considered the duality between input recognition and output generation. Section 4 provides the new structure of dynamic intelligence for an autonomous mobile robot.

2 Spatiotemporal dynamics of Self-organizing Systems

A multi-robot system consists of many autonomous mobile robots (subsystems) which are interacted only within their sensor range (neighbors). This situation can be expressed by a graph. The intelligence should have a property that gets a model of environment (internal model principle). Therefore, the graph expression of dynamic input-output connection is promising method to realize the intelligence of autonomous robots.

An autonomous robot usually senses the same physical quantity, which is usually distance from other robots and/or obstacles. And it is continuous against space and time. Almost all self-organizing systems are generally modeled by spatiotemporal evolution equations. It is shown that such a self-organizing system can be analyzed by the time evolution of spatiotemporal evolution equations in function space[3]. In [3], self-organized spatial pattern was expressed by function of position which defines the value of the physical quantity at the position. Then, the spatiotemporal evolution equation determines the trajectories of pattern in function space. This is regarded as expressing some self-organizing process.

One of the simplest dynamics is generated by a gradient system whose velocity vector field is proportional to the gradient vector field of a potential function. In [3], a potential functional whose minima correspond to some desired patterns is designed, then the gradient system according to this potential functional organizes one of the desired pattern. At this time, the micro dynamics expressed by the evolution equation is derived by differentiating the potential functional with respect to pattern. These subsystems whose behaviors obey the micro dynamics cooperate with their neighbors and generate one of the desired orders. This is derived not only on usual continuous media, but also on a graph which is suitable for expressing many kinds of artificial systems, especially for a multi-robot system. That is autonomous mobile robots are expressed by nodes of graph, and their interactions are expressed by edges of it.

3 Duality between input recognition and output generation

Unfortunately, gradient system can express a dissipative process only. General system usually consists of some dissipative part and conservative one. Conservative part is usually expressed by "Hamilton System." In there, symplecticity or duality of a system play a very important rôle for keeping Hamiltonian.

In Network Thermodynamics[4], this duality is expressed by dual KCL and KVL variables. It is shown that multiplying the incidence matrix A_a or its transpose A_a^T correspond to operating exterior derivation, and that multiplying the mesh matrix M or its transpose M^T do that. For example, the gradient 1-form field of node function $f \in C(V)$ calculate as $df = -A_a^T f \in C(E)$ where $C(V)$ is functions set on a set of nodes V and $C(E)$ is that on a set of edges E. It is well known that the exterior derivative of df vanishes, that is $d \cdot df = 0$, which correspond to $MA_a^T f = 0$. It is also said on dual graph that $d^*g = M^{T*}g \in C(^*E)$ and $d \cdot d^*g = A_a M^{T*}g = 0$. Figure 1 shows this relationship.

Here, we want to use this duality to treat intelligence of an autonomous mobile robot. To treat the input-output duality, actuator output is expressed by a vector field while sensor input is expressed by a 1-form field. In other words, actuator output is expressed by KVL variables while sensor input is expressed by KCL variables. Then, the intelligence of a mobile robot is determined by the connection with them.

$$\begin{array}{ccccc}
\text{vertex} & & \text{edge} & & \text{mesh} \\
f & \to A_a^T \to & df & \to M \to & d \cdot df = 0 \\
& & \uparrow \text{constitutive relation} \downarrow & & \\
d \cdot d^* g = 0 & \leftarrow A_a \leftarrow & d^* g & \leftarrow M^T \leftarrow & {}^* g
\end{array}$$

Figure 1: Duality of KCL and KVL variables on graph and dual graph

4   New Dynamic Structure of Intelligent Mobile Robot

To consider the dual system, it is suitable for treating a function on a planar graph $G(V, E)$ because it has the dual one G^*. The codifferentiation operator $d : C(V) \to C(E)$ can be regarded as exterior differentiation on the graph. Note that this gradient 1-form field $df \in C(E)$ is *not* a 1-form function on V, but a function on E. This means that the information flow only exists on graph (nodes and edges) and there is no reason to think about it in another place.

According to network thermodynamics, vector field on graph $g : E \to \Re$ is decomposed into two parts, one is KCL variable g_f and another is KVL one g_e. KCL variable g_f is satisfied with Kirchhoff Current Law:

$$\forall u \in V, \quad \sum_{e \in E(u)} g_f(e) = 0 \tag{1}$$

then the degree of freedom of KCL variable g_f is $|E| - |V| + 1$, where $|E|$ is the number of edges and $|V|$ is that of nodes. On the other hand, KVL variable is satisfied with Kirchhoff Voltage Law:

$$\forall l \in \text{mesh}, \sum_{e \in l} g_e(e) = 0 \tag{2}$$

then the degree of freedom of KVL variable g_e is $|V| - 1$. When we consider that on the dual graph G^*, this KVL condition is transformed into dual KCL* one.

$$\forall u^* \in V^*, \quad \sum_{e^* \in E^*(u^*)} g_e(e^*) = 0 \tag{3}$$

The relation between KCL variable $g_f(e)$ and KVL one $g_e(e)$ on each edge e is called "constitutive relation." In linear electric circuit theory, this constitutive relation is expressed by impedance or admittance. Note that this constitutive relation is dynamic one because this usually includes time differential operators. This constitutive relation constrains this system the $|E|$ degree of freedom, then totally this network dynamic is determined uniquely. This formulation also says that dynamic property in the network system is *not* determined at each node, *but* at each edge. In a multi-robot system, we should design the dynamics of interaction to construct some intelligence.

To use node analysis, we have the following node equation.

$$A_a D_y A_a^T (\hat{v}_1, \cdots, \hat{v}_u, \cdots, \hat{v}_\rho)^T = 0 \tag{4}$$

Where $D_y = \text{diag}[Y_1, \cdots, Y_e, \cdots, Y_b]$ is $|E| \times |E|$ admittance diagonal matrix. Note that equation (4) is a Laplace equation on graph whose metric tensor is D_y. On the other hand to use mesh analysis, we also have the following mesh equation.

$$M D_z M^T (\hat{i}_1, \cdots, \hat{i}_l, \cdots, \hat{i}_\mu)^T = 0 \tag{5}$$

Where $D_z = \text{diag}[Z_1, \cdots, Z_e, \cdots, Z_b]$ is $|E| \times |E|$ impedance diagonal matrix. Note that equation (5) is also a Laplace equation on the dual graph whose metric tensor is D_z.

This relation can use the realization of intelligence for an autonomous mobile robot shown in figure 2.

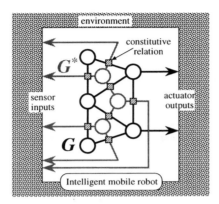

Figure 2: **New structure of intelligent mobile robot**

This figure can be interpreted that G^* is an input recognition graph and G is an output generation graph. These constitutive relations play an important rôle for dynamic input-output connection, which determine the intelligence of this autonomous robot.

5 Conclusion

In this short paper, the basic representation of an intelligent mobile robot system is considered. Focused on duality, the new structure of intelligence for an autonomous mobile robot is proposed. The duality of a planar graph corresponds to that between input recognition and output generation, and dynamic connection between them is determined by the intelligence of it. Some simulations and/or experiments of this intelligence are necessary.

References

[1] V. Breitenberg: Vehicles, experiments in synthetic psychology, MIT Press, New York, 1996

[2] A. D'Angelo, F. Montesello and E. Pagello: Building Autonomy within Self-Organizing Dynamical Agents, Intelligent Autonomous Systems 6 (IAS-6), pp.43-50(2000)

[3] H. Yuasa and M. Ito: Internal Observation Systems and a Theory of Reaction-Diffusion Equation on a Graph, Proc. of 1998 IEEE International Conference on Systems, Man, and Cybernetics(SMC'98), U.S.A., 3669/3673 (1998)

[4] G. F. Oster, A. S. Perelson and A. Katchalsky: Network Thermodynamics: Dynamic Modelling of Biophysical Systems, Quarterly Reviews of Biophysics, Vol. 6, No. 1, Cambridge University Press (1973)

Intelligent Autonomous Systems 7
M. Gini et al. (Eds.)
IOS Press, 2002

Algorithms for Utility-based Role Exchange

Xin Zhang and Henry Hexmoor

Computer Science & Computer Engineering Department, Engineering Hall, Room 313,
Fayetteville, AR 72701
{xxz03, hexmoor}@uark.edu

Abstract. We present three algorithms for utility-based role exchange that are inspired by game theory. We introduce some methods for comparison of algorithms. While all three algorithms provide similar gains, the extent to which utilities of other agents are considered in role exchange, the number of exchanges is minimized.

1 Introduction

In multiagent systems, agent roles have been studied in regards to plans and distribution of tasks [2][3]. Little attention is paid to an agent's preference over its roles, and reasoning about swapping roles with other agents. In this paper we will consider agents operating based on the social welfare of their group. As such role exchange between agents will be to benefit the collective. Therefore, some agents need to sacrifice their individual utility for the sake of their community [1].

Careful selection of a pattern of roles for adoption is a key point for improving the group utility. [12] presented the concept of *individual utility*, which measures each agent's utility in a specific role and uses the *total utility*, which represents the team utility in the formation of roles. [12] also presented the concept of *role exchange values (REV)*, which measures each agent's utility gain during the hypothetical role exchange. We considered role exchange based on the individual agent's utility gain and the algorithm allowed role exchange between any pair of agents who may experience net positive gain from the exchange. This *any role exchange algorithm* is not optimal for the consideration of the total numbers of role exchange times before reaching the maximum total utility.

In this paper we will show two other variants of the algorithm and compare their performance. At first we will introduce the related works in the field of formation-based roles in section 2. Then we will provide some assumptions, formula, and the conditions for role exchange. Based on the formula and conditions, we will introduce the *any role exchange algorithm* in section 4.1, *Individual Optimal Role Exchange Algorithm* in section 4.2, and *Group Optimal Role Exchange Algorithm* in section 4.3. The implementation and comparison of the three algorithms will be presented in section 5 with the results. We have not tested our algorithms on real data with real agents. Our results are based on simulated data. In section 6, we will provide some concluding remarks.

2   Related Work

Formation as basis of collaboration among agents is introduced in [4] and [8]. Formations are commonly found in the game of soccer or Robocup [11]. Agents adopt an initial assignment of roles, but this assignment may need to be revised as the situation changes. Therefore, re-assignment of roles or some role exchanges become necessary [6]. Formations are dynamic and flexible, and each formation is a team structure to divide the task into a set of roles. There is an initial formation, but there are also the run-time triggers for dynamic changes of the formations [5]. Formation-based role assignment ensures flexibility of role exchange, which was beneficial to the FC Portugal team [7].

Various formations as role models are discussed in [5]. A role defines a position and a set of responsibilities within a role model, and roles are assigned to agents in an application. Role models can have various forms of dynamic behavior, which ensure the modeling mobility and adaptive behavior. Agent organizations can take on various formations, and role assignments can be assigned dynamically [5].

Consider a group of distinct n agents, $A = \{a_1, a_2, ..., a_n\}$, where a_i is the ith agent and a set of distinct n roles, $R = \{r_1, r_2, ..., r_n\}$, where r_i is the ith role, such that $i \neq j \Rightarrow r_i \neq r_j$. For any agent a_i and role r_j, there is a utiltiy $U(a_i, r_j) = u_{ij}$, where u_{ij} stands for the utility of adopting role j by agent i and function $U(a_i, r_j)$ is to get u_{ij}. Any formation F is a set of $\{< a_i, r_j, u_{ij} > | where\ i, j \in [1..n]$, for any pair of $< a_i, r_j, u_{ij} >$ and $< a_k, r_l, u_{kl} >$, $a_i \neq a_j \Rightarrow r_k \neq r_l\}$, and each agent is assigned a single distinct role[12].

3   The Assumptions and Properties for Role Exchange

Assume the agents we discuss in this paper obey Pareto-optimality. We make the following assumptions.

1. With N agent and N roles, there is a one to one assignment of roles to agents.
2. Each agent has a unique utility value per role. That is., $V(A, R)$ is agent A's unique utility in role R.
3. An agent's adoption of a role will not affect the utility of another agent adopting another role.
4. The *total utility* of a number of agents is equal to the sum utilities from each of these agents. I.e. *Total Utility* $= \sum_{i=1}^{i=n} V(A_i, R_i)$
5. The role exchange process takes place only between a pair of roles at one time.
6. If the margin of gain from a hypothetical role exchange is positive for a pair of agents, they are obliged to exchange. This is due to the Pareto-optimality cooperative agents.
7. The time consumption or utility loss due to the process of role exchange is assumed to be negligible and will not be considered.

The concepts of *Role Exchange Value* (REV) and *Individual Utility Gain* (IUG) for each agent are computed when considering role exchange in an agent pair [12]. REV involves a pair of agents and a pair of roles, i.e., agent A, agent B, role R1 and role R2. We have the following concepts related with role exchange:

1. $V(A, R1)$ represents the unique utility of agent A taking role R1.
2. $V(A, R1, B, R2)$ represents the sum utility of agent A taking role R1 and agent B taking role R2. I.e., based on assumption 3 and 4, $V(A, R1, B, R2) = V(A, R1) + V(B, R2)$.
3. $REV_A(A, R1, B, R2)$ represents the role exchange value of agent A in the formation that agent A taking role R1 while agent B taking role R2.

In role exchange for agent pair (A, B), the REV of agent A in the formation that agent A taking role R2 while agent B taking R2 as equation (1):

$$REV_A(A, R1, B, R2) = 0.5 * \{V(A, R1) + V(A, R2, B, R1) - V(B, R2)\} \quad (1).$$

The *individual utility gain* for agent A in this role exchange formation is as equation (2):

$$IUG_A(A, R1, B, R2) = REV_A(A, R1, B, R2) - V(A, R1) \quad (2).$$

Based on equations (1) and (2), we use the following 3 conditions in Figure 1 to check if role exchange is applicable. Section 5 of this paper gives an example that uses equation (1) and (2).

---

1. If $IUG_A(A, R1, B, R2) < 0$, role exchange will degrade to total utility for the entire group and original role formation is better.

2. If $IUG_A(A, R1, B, R2) = 0$, role exchange is not necessary. There is no difference between before and after role exchange.

3. If $IUG_A(A, R1, B, R2) > 0$, role exchange will be beneficial to the entire group.

Figure 1. Conditions for Role Exchange

4 The Algorithms for Role Exchange

The algorithms for utility-based role exchange take the following steps. In our algorithms, t is the time index. For instance, $t = 0$ is the time before any role exchange. $t = 1$ is the time at the 1st exchange. Function $add(< x, y, z >, S)$ adds the triple $< x, y, z >$ to set S. Function $delete(< x, y, z >, S)$ deletes the triple $< x, y, z >$ from set S. Function $IsMaxForAll(i, k, j, l)$ is true if $IUG_i(i, k, j, l)$ is the maximum among all the formations. Function $IsMaxFori(i, k, j, l)$ is true if $IUG_i(i, k, j, l)$ is the maximum among the formations, which correspond to role exchange with agent i. $stop$ stands for termination of the algorithm. Predicate "formation" picks out a specific formation, e.g., F.

We discussed an algorithm for utility-based role exchange in [12]. In this paper we will present 2 other algorithms for role exchange and compare them to see what the advantage and disadvantage for each of them. The initial conditions are the following.

- Agent-role utility table has been set up, and the utility value is randomly generated.
- Each agent adopts an initial role randomly. We assume that the ith agent adopts ith role prior to any role exchange.
- Pairs of agents are selected in order to explore applicability of role exchange.

4.1 Any Role Exchange Algorithm

Role exchange happens whenever the agent pair's IUG>0. Since any role exchange will occur when it is applicable, we call it "any role exchange" algorithm. This algorithm was presented in [12] and is given below in Figure 2:

---

1. There is no role adoption for any agent at the very beginning: $t = 0 \Rightarrow F = \varnothing$.

2. When role adoption starts, each agent adopts a role randomly, which means that the agent may adopt any role at first: $t = 1 \Rightarrow \forall i \exists j, agent(i) \wedge role(j) \wedge U(i,j) = u_{ij} \wedge add(<i,j,u_{ij}>,F)$.

3. Search the agent pairs from the first agent for role exchange. If the IUG of the given pair of agents is positive, the agent pair will make role exchange; otherwise search the next agent pair for role exchange:

$\forall i, j \forall k, l \forall F, agent(i) \wedge agent(j) \wedge role(k) \wedge role(l) \wedge formation(F) \wedge <i,j,u_{ik}> \in F \wedge <i,j,u_{jl}> \in F \wedge$

$IUG_i(i,k,j,l) > 0 \Rightarrow add(<i,l,u_{il}>,F) \wedge add(<j,k,u_{jk}>,F) \wedge delete(<i,k,u_{ik}>,F) \wedge delete(<j,l,u_{jl}>,F)$

4. Role exchanges will stop when the utility gain of any agent pair is no more than zero:

$\forall i, j \forall k, l \, agent(i) \wedge agent(j) \wedge role(k) \wedge role(l) \wedge \sum_{i=1}^{n} IUG_i(i,k,j,l) \leq 0 \Rightarrow stop.$

Figure 2. Any Role Exchange Algorithm

4.2 Individual Optimal Role Exchange Algorithm

We start role exchange from the first agent. Search all other agents paired with first agent, find the pairing with the maximum IUG, perform the role exchange corresponding to the maximum IUG. Then repeat this step with other agents. If all agent pairings produce the IUGs less or equal to zero, no role exchange is needed. Since for every agent we find the best applicable role exchange and the agents will be checked in order (from the first one to the last one), we will call it "individual optimal role exchange" algorithm. It is shown in the following Figure 3.

---

1. There is no role adoption for any agent at the very beginning: $t = 0 \Rightarrow F = \varnothing$.

2. When role adoption starts, each agent adopts a role randomly, which means that the agent may adopt any role at first:
$t = 1 \Rightarrow \forall i \exists j, agent(i) \wedge role(j) \wedge U(i,j) = u_{ij} \wedge add(<i,j,u_{ij}>,F)$.

3. Search the agent pairs from the first agent for role exchange. Find the maximum IUG of the given agent pair related with the first agent. If the IUG of the given pair of agents is positive, the agent pair will make role exchange; otherwise search the next agent pair start with the second agent for role exchange. Repeat step 3:
$\forall i, j \forall k, l \forall F, agent(i) \wedge agent(j) \wedge role(k) \wedge role(l) \wedge formation(F) \wedge <i,j,u_{ik}> \in F \wedge <i,j,u_{jl}> \in F \wedge$

$IUG_i(i,k,j,l) > 0 \wedge IsMaxFori(i,k,j,l) \Rightarrow add(<i,l,u_{il}>,F) \wedge add(<j,k,u_{jk}>,F) \wedge delete(<i,k,u_{ik}>,F) \wedge$

$delete(<j,l,u_{jl}>,F)$

4. Role exchanges will stop when the utility gain of any agent pair is no more than zero: $\forall i, j \forall k, l \, agent(i) \wedge agent(j) \wedge role(k) \wedge role(l) \wedge \sum_{i=1}^{n} IUG_i(i,k,j,l) \leq 0 \Rightarrow stop.$

Figure 3. Individual Optimal Role Exchange Algorithm

4.3 Group Optimal Role Exchange Algorithm

Here agents share knowledge of IUGs and refrain from early role exchanges. IUGs are computed, the best one is performed, and this repeated until no more role exchange is applicable. For example, search all the agent pairs; find the maximum IUG pair; and do role exchange. Repeat this steps until the IUG for all the agent pairs are no more than 0. We can also call it "group optimal role exchange" algorithm in Figure 4.

---

1. There is no role adoption for any agent at the very beginning: $t = 0 \Rightarrow F = \varnothing$.

2. When role adoption starts, each agent adopts a role randomly, which means that the agent may adopt any role at first:

$t = 1 \Rightarrow \forall i \exists j, agent(i) \wedge role(j) \wedge U(i,j) = u_{ij} \wedge add(<i,j,u_{ij}>, F)$.

3. Search all the agent pairs from the first agent. Find the agent pair with maximum IUG, if the IUG of the given pair of agent is positive, do role exchange:

$\forall i, j \forall k, l \forall F, agent(i) \wedge agent(j) \wedge role(k) \wedge role(l) \wedge formation(F) \wedge <i,j,u_{ik}> \in F \wedge <i,j,u_{jl}> \in F \wedge$

$IUG_i(i,k,j,l) > 0 \wedge IsMaxForAll(i,k,j,l) \Rightarrow add(<i,l,u_{il}>, F) \wedge add(<j,k,u_{jk}>, F) \wedge delete(<i,k,u_{ik}>, F) \wedge$

$delete(<j,l,u_{jl}>, F)$

4. Role exchanges will stop when the utility gain of any agent pair is no more than

zero. $\forall i, j \forall k, l \ agent(i) \wedge agent(j) \wedge role(k) \wedge role(l) \wedge \sum_{i=1}^{n} IUG_i(i,k,j,l) \leq 0 \Rightarrow stop.$

Figure 4. Group Optimal Role Exchange Algorithm

5 Implementation and Comparison of Algorithms

Consider the following example involving role exchanges with 2 agents and their 2 roles. We can use a matrix to model this problem. Based on assumption 1 and 2, suppose there are N agents and N roles. We can use N*N matrix to represent the relationship between agents and roles. The rows represent agents such as agent A0, and columns represent roles such as role R0. The value at the intersection of an agent row and a role column, such as element (i, j), represents the utility that agent i adopting role j. In an implementation of this algorithm, we use a 10*10 matrix shown in Table 1, whose utilities are randomly generated. Here we used the Wichiman-Hill algorithm to generate uniform distributed random number in (0,1), and we repeated the implementation for several times.

5.1 Implementation

According to the algorithm we discussed above, no role has been adopted at first, so we may just assign each agent A_i with role R_i, as the entities highlighted in the table. Based on assumption 3 and 4, at this time, the initial total utility of the group is $\sum_{i=0}^{i=9} V(Ai, Ri) = 71$.

Then based on assumption 5, 6 and 7, we will check each agent pair to decide if role exchange is necessary based on conditions we discussed in 2.1 and the condition of the algorithms.

Table 1: Agent-Role Table Pairs

	R0	R1	R2	R3	R4	R5	R6	R7	R8	R9
A0	1	7	14	0	9	4	18	18	2	4
A1	5	5	1	7	1	11	15	2	7	16
A2	11	4	2	13	12	2	1	16	18	15
A3	7	6	11	1	8	9	12	7	19	15
A4	14	3	11	2	13	13	4	1	11	13
A5	8	7	4	2	17	17	19	3	1	9
A6	18	16	15	10	2	8	6	0	2	4
A7	8	6	5	10	9	10	10	6	1	13
A8	8	9	3	4	14	16	0	6	16	11
A9	8	4	19	6	3	17	18	18	2	9

5.2 Comparison of the 3 Algorithms

Figure 5 shows the results of comparison between 3 utility-based role exchange algorithms. Series 1 represents the role exchange using algorithm 1, Series 2 represents corresponds to algorithm 2, and Series 3 represents algorithm 3.

By comparing our 3 algorithms, we find that algorithm 1's overall computation time is the smallest, but it suggest the most number of role exchanges. Algorithm 2's overall computation time and the number of role exchanges are both at mod-level. Algorithm 3's overall computation time is the largest, but the number of role exchanges is the least. If we consider time consumption or utility loss due to the role exchange to be nontrivial, then algorithm 3 should be the best choice.

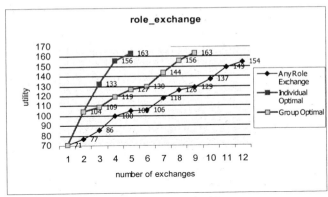

Figure 5. Role exchange with 3 algorithms

Since the utility values in the agent-role table are randomly chosen, it's hard to decide which algorithm will yield the best formation after role exchange in a specific situation. But, we know that for each agent-role table that there must be an optimum formation with regards to utility. Here, we define a term "optimization percentage" in equation (3).

$$\text{"optimization percentage"} = total\_utility/optimized\ utility \qquad (3).$$

Total_utility is the final utility of the group of agents after each role exchange algorithm stops. Optimized_utility is the utility yielded by the optimized formation, which yields the maximum total group utility.

We manually (not by either of our algorithms) calculate this optimized value by considering all possible formations which can be generated from the given agent-role table. We then compare this value with the output results from our 3 algorithms. For an empirical result, we repeated this 1000 times with different agent-role utility tables and accumulated the results. Figure 5 shows result of one of 1000 runs. Our results of optimization percentage for 3 algorithms are 97.7274%, 97.5173%, and 98.2839% respectively as in Table 2. These results suggest that the 3 algorithms' "optimization percentage" are statistically equivalent and does not provide an adequate reason for selection. We now define the term "unit_time_gain" in equation (4) to measure the average utility gain per role exchange:

$$\text{"unit\_time\_gain"} = total\_utility\_gain / role\_exchange\_times \quad (4).$$

Total_utility_gain is the utility gain for a run of the algorithm. Role_exchange_times is the number of role exchanges at the end of the algorithm.

From Figure 5, we see that the total utility of the whole group changes from 71 to 154 with algorithm 1, from 71 to 163 with algorithm 2, and from 71 to 163 with algorithm 3. The unit_time_gain for algorithm 1 is 6.91, for algorithm 2 it is 10.2, and for algorithm 3 it is 23.

The comparison results are shown in Table 2. We compare those 3 algorithms in three aspects, such as the total utility, the optimization percentage, and the unit time gain for each algorithm respectively. If, based on a number of experiments, we can assume that all 3 algorithms yield the "same" total utility, the algorithm with the highest "unit_time_gain" will be the best one. So, the "group optimal role exchange" algorithm is the best one among these 3 algorithms for utility-based role exchange.

Table 2: Comparison results of the Algorithms

	Total Utility	Optimization Percentage	Unit Time Gain
Algorithm 1	154	97.7274%	6.91
Algorithm 2	163	97.5173%	10.2
Algorithm 3	163	98.2839%	23

6 Conclusion

We presented three utility-based role exchange algorithms. Two of these algorithms had not been reported prior to this paper. We presented the concept of "optimization percentage" and "unit_time_gain" to measure the performance of the algorithms for utility-based role exchange. Then, we discussed the implementation of the algorithms and analyzed the advantage and disadvantage of each algorithm for utility-based role exchange. All three algorithms yield similar results. However, if we assume role exchange to involve non-trivial cost, a committee based choice of role exchange that we called "group optimal role exchange" algorithm produces the minimum exchanges and is preferred.

References

[1] Boman, M., Kummeneje, J., Lyback, D., Younes, H. L.: UBU Team. RoboCup-99 Team Descriptions Simulation League, (1999) pp 133-138.

[2] Cavedon, L., and Sonenberg, L., On Social Commitments, Roles, and Preferred Goals, In Proceedings of 3rd Int Confernce on Multiagent Systems, pp. 80-87, (1998).

[3] Castelfranchi, C. and Falcone, R., From Task Delegation to Role Delegation, In Proedings of AI-IA 1997: Advances in Artificial Intelligence Congress, LNAI 1321, pp. 278-289.

[4] Hexmoor, H., Zhang, X., Norms, Roles and Simulated RoboCup. 2nd workshop on norms and institutions in multiagent systems (Agents 2001), Montreal, CA, ACM press, (2001).

[5] Kendall, E., Role Models, Patterns of Agent System Analysis and Design. British Telecom Technical Journal Special Issue on Decentralized Business Systems, 1999.

[6] Marsella, S., Adibi, J., Al-Onaizau, Y., Kaminka, G., Musela, I., Tambe, M., Experiences acquired in the design of Robocup teams: A comparison of two fielded teams, Journal of Autonomous agents and Multi-agent Systems Journal of Autonomous agents and Multi-agent Systems (JAAMAS), special issue on "Best of Agents'99", (2000).

[7] Reis, L. P., Lau, N., FC Portugal Team Description: RoboCup 2000 Simulation League Champion. RoboCup 2000: Robot Soccer World Cup IV, Berlin, (2001).

[8] Stone, P., Veloso, M., Task Decomposition, Dynamic Role Assignment, and Low-Bandwidth communication for Real-Time Strategic Teamwork. Artificial Intelligence Journal (AIJ), Volume 100, number 2, June (1999).

[9] Tu, M.T., Griffel, F., Merz, M., Lamersdorf, W., A plug-In Architecture Providing Dynamic Negotiation Capabilities for Mobile Agents. Mobil Agents: Second International Workshop (MA'98), Stuttgart, Germany, Sep., 1998, pp 222-236.

[10] Veloso, M., Stone, P., Han, K., The CMUnited-97 Robotic Soccer Team: Perception and Multiagent Control. Agents'98, Minneapolis, May, 1998, pp. 78-86.

[11] Veloso, M., Pagello, E., Kitano, H., RoboCup-99: Robot Soccer World Cup III. RoboCup, (1999).

[12] Zhang, X., Hexmoor, H., Utility-based Role Exchange. In the Proceedings of Central and Eastern Europe on Multi-Agent Systems (CEEMAS-01), Poland. (2001), pp. 413-420.

404

Intelligent Autonomous Systems 7
M. Gini et al. (Eds.)
IOS Press, 2002

PQ–Learning: An Efficient Robot Learning Method for Intelligent Behavior Acquisition

Weiyu Zhu,   Stephen Levinson

Beckman Institute, 405 N. Mathews, Urbana, IL61801, USA
Dept. of Electrical and Computer Engineering, Univ. of Illinois at Urbana-Champaign
{weiyuzhu, sel}@ifp.uiuc.edu

Abstract This paper presents an efficient reinforcement learning method, called the *PQ*-learning, for intelligent behavior acquisition by an autonomous robot. This method uses a special action value propagation technique, named the spatial propagation and temporal propagation, to achieve fast learning convergence in large state spaces. Compared with the approaches in literature, the proposed method offers three benefits for robot learning. First, this is a general method, which should be applicable to most reinforcement learning tasks. Second, the learning is guaranteed to converge to the optimum with a much faster converging speed than the traditional Q and $Q(\lambda)$-learning methods. Third, it supports both self and teacher-directed learning, where the help from the teacher is directing the robot to explore, instead of explicitly offering labels or ground truths as in the supervised-learning regime. The proposed method had been tested with a simulated robot navigation-learning problem. The results show that this method significantly outperforms the $Q(\lambda)$-learning algorithm in terms of the learning speeds in both self and teacher-directed learning regimes.

1. Introduction

Intelligent behavior acquisition by a robot is a challenging topic in the field of Artificial Intelligence and Robotics. Traditional methods are usually based on planning or supervised learning, in which either complete task-specific knowledge from human beings is incorporated or extensive directions or feedbacks from an outside teacher are highly relied on. Although these methods have achieved impressive robotic intelligence in some fields, more often than not, the intelligent systems built in this way usually only work well on specific tasks and can hardly be transplanted from one to another. In addition, since these methods usually highly rely on humans' knowledge of the world, which might be incorrect sometimes, the acquired robotic systems may not be robust and adaptive in the real world.

In recent years, the Reinforcement Learning (RL) methodology has been paid increased attention for autonomous agent learning with little or no *a priori* knowledge of the environment [1][2][3]. An essential challenge of the current RL methods, such as the Watkins' Q-Learning algorithm [4], lies in the size of the learning state space. According to Watkins and Dayan [5], Q-Learning is guaranteed to converge to the optimum as long as all state-action pairs are continuously experienced and updated in learning. This requirement places great difficulties on many real learning tasks for two reasons. First, the size of the learning state spaces, which are usually obtained from the partitioned continuous feature spaces in the real world, could be very large in general. Secondly, learning with a real robot is practically time-consuming. It is usually unrealistic to let the robot exhaustively explore all situations in order for the learning to converge.

An intuitive scheme of coping with large state spaces is value function generalization. That is, the learning uses a function approximator, instead of a plain table, to represent value functions so that the robot is not necessary to explicitly visit all states in order to obtain a learned value function for each action. Some researchers have experimented with this scheme and some positive examples have been reported [6][7]. Nevertheless, in general this approach has been proved dangerous [8][9] because the estimate-based value updating cannot guarantee to reduce the errors of value functions at each step; consequently, the learning might not converge at all.

Adaptive resolution is another well-studied strategy for continuous feature space learning. In this scheme, feature spaces are partitioned adaptively according to a certain criterion so that different feature regions may receive different attention in learning. In [10], Moore proposed a PartiGame algorithm for solving puzzle problems, in which the agent learns to travel to a goal location in a cell graph by adaptively partitioning the graph to find the shortest feasible path. In [12], a decision tree based adaptive state partitioning method was proposed by Chapman and Kaelbling for solving learning problems in high dimensional binary spaces. In the real robot-learning domain, Yasutake *et al* [13] used local action models to partition feature spaces for robot soccer skill learning and Zhu *et al* have presented a hybrid state-partitioning strategy for autonomous robot navigation learning in [11]. Ideally, adaptive resolution approaches are capable of handling any state-partitioning learning problems. However, designing a clever adaptive strategy in general is not easy since it requires heuristic knowledge of each specific task. For instance, the Moore's PartiGame algorithm relies on the complete and known state transition models of the environment while Yasutake's local action model method assumes that the optimal actions of the physical states sharing the same action models must be the same, which might not hold in most other learning tasks.

In this paper, we proposed an efficient RL learning method called the *Propagated Q* (*PQ*)-Learning. This method is based on Watkins' Q-learning algorithm while action value functions are learned partially based on a recursive *spatial propagation* and *temporal propagation* process at each learning step, instead of the traditional one or multiple sequential step tracing scheme used in the Q or $Q(\lambda)$ learning. Compared with the current RL methods in literatures, this method possesses several advantages. a) It is a general model-free RL method. b) The learning is guaranteed to converge to the optimum with much less required number of learning episodes than the Q and $Q(\lambda)$ learning. c) It supports both self and teacher-directed learning, in which the correctness of the teacher, however, is not required. We have used the this method on a simulated navigation-learning problem (with 10,000 states defined), where it was proved to outperform the $Q(\lambda)$ learning in terms of the converging speeds under both self and teacher-directed learning scenarios.

The rest of this paper is organized as follows. Section 2 presents the details of the proposed *PQ*-learning algorithm. Section 3 is focused on the experiments carried out with the *PQ* and the $Q(\lambda)$ learning algorithms, followed by the conclusion in section 4.

2. Proposed Method

2.1 Q-learning

Before introducing the proposed *PQ*-learning method, we'd first give a brief overview to Watkins' Q-learning algorithm, on which this method is based. Q learning is a Temporal-

Difference (TD) based off-policy RL algorithm developed by Watkins in 1989 [4]. Its simplest form, the one-step-learning, is defined as

$$Q(s_t,a_t) \leftarrow Q(s_t,a_t) + \alpha[r_{r+1} + \gamma \max_{a'} Q(s',a') - Q(s,a)] \tag{1}$$

where $Q(s_t,a_t)$ is called the action-value function which indicates the goodness of taking
action a_t at state s_t. r_{t+1} is the instant reward received after the action is taken and α is a
constant step-size parameter. γ is called the *discount rate*, which reflects the influence of
future rewards on the current state.

The Q-learning algorithm updates only one action value at each time. Therefore, the
learning convergence could be very slow if the number of states is large since the number
of required learning steps may increase exponentially as the size of the state space increases. An improvement is called the $Q(\lambda)$-learning [4], in which a sequence of action values are traced and updated at each learning step according to the so-called *eligibility traces*
[4]. $Q(\lambda)$ algorithm learns much faster by updating multiple value functions at each step.
However, since the updating only takes place on the states experienced, the learning still
requires the explicit visits to all states in order for all value functions being updated.

2.2. Propagated Q-learning

The basic idea of the proposed *PQ*-learning method is to update value functions of multiple states simultaneously at the each step. However, in addition to feeding the updates to
the predecessors (temporal propagation), like what the $Q(\lambda)$-learning does, we also propagate the changes to the neighboring states meanwhile (spatial propagation) so that a state
could be learned even if it has not been visited once before.

General Assumptions
To simplify the analysis, the proposed method was developed based on a number of
general assumptions regarding the learning environment and task natures.
* Negative rewards are received only at the end of the failed learning episodes.
* For any non-terminal state (neither a failed nor a goal state), there exists at least one
  action for the robot to arrive at another non-failed state. That is, at least one path exists
  for the robot to reach its goal from any non-terminal state.
* State transitions and action rewarding policies are fixed, although the robot does not
  know them in advance.

Based on these assumptions, the learning scenario of the Q-learning algorithm can be
pictured like this: given all Q values being zero-initialized, these values would increase
monotonically towards the optimums in learning unless a negative reward is received, when
the corresponding Q value becomes negative. This analysis is helpful for the later on proof
of the convergence of the proposed *PQ*-learning method.

Spatial Propagation
The motivation of spatial propagation comes from the observation that, given the feature
space being "smooth" enough, the action values of two states close to each other are usually
very close too. Therefore, it might be reasonable to let the neighboring states "share" the
update of the current state so that more than one states could be learned simultaneously. To
achieve this idea, two issues should be considered in designing the value-propagation strategy in order to guarantee the correct convergence of the learning. First, the propagation
should not affect the learning of "critical" regions, i.e., the regions where optimal value
functions change dramatically, instead of smoothly as in most cases. Secondly, the propaga-

tion should not introduce oscillations in value function changes so that the updating would take place monotonically (or near monotonically) toward the optimums.

Figure 1 displays the propagation algorithm used in the proposed method, in which two facts are implied: 1) Propagations only "promote" the Q values of the neighboring states while the propagated updates never exceed the values of the current states. 2) Negative value functions of the neighboring states never receive promotions in propagations.

For each neighboring state s_i do:
    Compute the Euclidean distance d_i to the current state s
    Compute the *reducing factor* $\alpha_i = 1 - \text{MIN}(1, \beta d_i)$, where β is a positive constant factor
    Update $Q(s_i,a) \leftarrow \text{MAX}[Q(s_i,a), \alpha_i Q(s,a)]$ if $Q(s_i,a)$ is non-negative or $s_i = s$

Figure 1. Algorithm of spatial propagation of action value $Q(s,a)$

By choosing a proper distance factor β, the promoted Q values are guaranteed not to exceed the respective optimums. In addition, since negative value functions are never promoted, the updating introduced by propagations would be made monotonically toward the optimums for most states, except for the actions with negative Q values, when the promotions might be completely wrong. However, according to the algorithm details, the incorrectly promoted value functions still have chance to be adjusted if that state is visited in the future, when the learning of that value function is degraded to the Q-learning scheme.

The idea of the spatial propagation process is similar as that of the function approximation method. However, according to the above analysis, our algorithm guarantees the union of the two criteria placed before, i.e., value functions are updated monotonically toward the optimums while critical regions can still be correctly learned.

Temporal Propagation

Similar to the eligibility trace idea in $Q(\lambda)$-learning, the aim of temporal propagation is to exploit the updating information of the current state to reevaluate the previous experience. However, unlike the scheme in $Q(\lambda)$-learning, the proposed method propagates the updating information to, instead of the "eligible" states experienced in the current learning episode only, all state transitions in history so that the consistency of value functions among states are better maintained.

Figure 2 illustrates the temporal propagation process in the *PQ*-learning method. Once the value of a given state is updated, the new value function is used to "re-process" the learning of all predecessors (the triples of state, action and reward) by recursively "playing-back" the previous experience in mind. The "replay" process, which is similar to the "experience replay" technique used by Thrun [1] and Lin [14], does not affect the learning convergence in general.

If $(\Delta V(s) > \text{threshold})$
    For each predecessor triplet (s', a', r'), do:
        $Q(s',a') \leftarrow Q(s',a') + \alpha[r' + \gamma V(s) - Q(s',a')]$

where $V(s)$ is the value function of state s defined as $V(s) = \text{MAX}_a Q(s, a)$

Figure 2. Algorithm of temporal value propagation for state s

Summary of the proposed method

By combining the spatial and temporal propagation modules, a recursive value-updating scheme in *PQ*-learning is obtained, which is summarized in Figure 3.

Initialize $Q(s, a)$ to zeros, and do for each learning episode:
  Initiate an initial state s
  Repeat (for each step of episode):
    Choose action a for s using a certain policy (e.g., ε-greedy)
    Take action a, observe the instant reward r and the next state s'
    Initiate a queue with the state s
    Repeat
      Take the head element s from the queue
      Do spatial propagation (SP), including the state s itself
      Do temporal propagation (TP) for all states whose V values are updated in SP
      Insert all states whose Q values are updated in TP process into the queue
    Until the queue is empty
    $s \leftarrow s'$;
  until s is terminal

Figure 3. Summary of the PQ-learning algorithm

According to the analysis of the propagation properties, the PQ-learning method guarantees: a) The Q values of "non-failed" actions will increase monotonically toward the optimums and b) the values of "failed" actions could be learned and maintained thereafter if the corresponding state-action pair is explicitly experienced in learning.

3. Experiments

3.1. Simulation setup

The proposed learning method has been tested with a simulated robot navigation-learning task. The simulated robot started at a random location in a 5m×5m playing ground and learned to navigate toward a goal somewhere in the environment. The input to the robot consisted of the distance from the robot to the goal and the panning angle of its head, by moving which the robot tried to keep the goal in the middle of its viewing field. The outputs were four actions: going forward for 60 or 30 cm and turning left or right for 5°. Initially, the robot did not have any *a priori* knowledge of the environment except for knowing there were four actions to choose and also being aware of success and failure, when a learning episode terminated. A situation was said to be a success if the robot was within 30cm to the goal and the head angle was less than ±3°, when a positive reward +1000 was received. Similarly, the robot was said to have failed if it could not see the goal anymore (the head angle ≥ ±50°), when a negative reward −1000 was imposed. Learning states were obtained by uniformly partitioning the feature space with a resolution of 5cm and 1°, respectively. That is, the learning space consisted of a total 100×100 states.

The learning performance was evaluated in terms of decision accuracy and the speed of convergence to the ground truth, which was obtained by solving the following iterative Bellman equations [15] according to the known action and state-transition models.

$$\text{Policy Evaluation}: \quad V(s) \leftarrow r_{ss'}^{a^*} + \gamma V(s'), \quad \text{where } a^* \text{ is the current optimal action at } s \tag{2}$$
$$\text{Policy Improvement}: \quad a^* \leftarrow \arg\max_a \left[r_{ss'}^a + \gamma V(s') \right]$$

3.2. Self-learning without directions

The first experiment was to let the robot learn completely by itself without any external directions. The robot autonomously explored the environment and chose actions to perform according to the ε-greedy algorithm, with which the agent would always choose the current optimal action to perform with the probability of 1-ε while other actions are randomly selected with the remaining likelihood of ε. Initially all action values were set as zero, so the robot has to select actions randomly at the beginning.

Figure 4 shows the percentage of states learned with the *PQ* and *Q*(λ) (to be concise, *Q*(λ) is denoted as *Q* in all figures) algorithms in a varying number of learning episodes, where ε was set as 0.8 and all statistics were collected and averaged in five repeated experiments. A state s was said to be learned if its state value function $V(s)$ became positive, which means that the robot had already obtained a certain knowledge about action selection at state s to achieve its goal. The statistics in Figure 4 shows the superior efficiency of the *PQ*-learning method, where most states were learned after 100 episodes while only half states were learned with *Q*(λ)-learning in 1000 episodes.

Figure 5 depicts the averaged errors, which were defined as $\left| Q_{learnt} - Q_{truth} \right| / Q_{truth}$, of the learned value functions. Combined with Figure 4, one may observe that our method learned not only faster but more accurate than the *Q*(λ)-learning as well. In 5000 episodes, the value functions had almost (less than 8% in error) converged to the ground truth while the error of *Q*(λ)-learning was around 40% and there were still some states not learned yet. A visual representation of action values learned for each state in 5120 learning episodes is given in Figure 6 (a), where the values are linearly scaled into gray levels 0~255. The *PQ*-learning, as expected, offered a very good approximation to the ground truth.

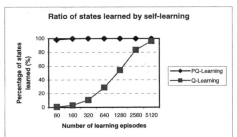

Figure 4. Learning efficiency by self-learning.   Figure 5. Averaged value errors by self-learning.

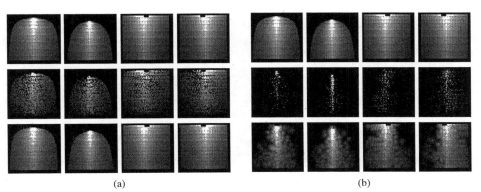

(a)   (b)

Figure 6. Visual representation of action values learned with *PQ* and *Q*(λ)-learning. (a) Self-learning in 5120 episodes. (b) Teacher-directed learning in 160 episodes. Top to bottom: ground truth, *Q*(λ)-learning and *PQ*-learning. Left to right: forward 30cm, forward 60cm, turn left and turn right. For each figure: x-axis: head angle (-50°~50° from left to right); y-axis: distance to the goal (0~500cm, from top to bottom).

The next test was using the learned value functions to predict actions. Since what we want is the robot being able to perform rationally to achieve its goal; therefore, in addition to optimal actions, we introduced a concept called the *semi-optimal action*, which was defined as the actions whose *Q* values exceeded 90% of that of the optimal actions. With this definition, decision accuracies could therefore be evaluated with respect to either optimal actions or semi-optimal actions. Figure 7 shows both prediction accuracies with the two algorithms. Again, the statistics were all collected from the learned state as in the previous experiment. Although the *PQ*-learning had to make more predictions (because more states were learned), it still outperformed the *Q*(λ)-learning in terms of both accuracies.

3.3. Learning under directions

In this experiment, the robot explored the environment by following the directions from a teacher at each step. The direction could be the optimal action or others according to the specified ε-greedy policy. However, the robot learned only from its own experience, which means that the correctness of the teacher was not required.

The similar learning statistics, as well as the visual representation of value functions, are given in Figure 6(b), 8, 9, 10, respectively, where ε=0.4. Encouragingly, remarkable accuracies in action prediction and value estimations were achieved with the *PQ*-learning in tens of learning episodes. According to the learning performance obtained, one may tell that, in addition to simulations, the proposed method is also desirable for real robot learning by interacting with a teacher.

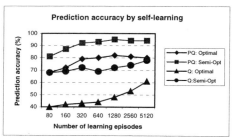

Figure 7. Prediction accuracy by self-learning.   Figure 8. Learning efficiency by directed-learning.

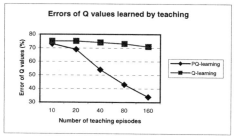
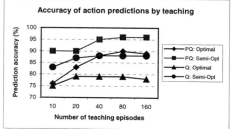

Figure 9. Averaged value errors by directed-learning.   Figure 10. Prediction accuracy by directed-learning.

4. Conclusion

An efficient reinforcement learning method, called the *PQ*-learning, is presented in this paper. The proposed method is based on Watkins' *Q*-learning while a recursive value propagation process, which consists of spatial propagation and temporal propagation, is introduced to boost the learning in large state spaces. The new algorithm offers several benefits for autonomous robot learning. First, it is a general delayed rewarding based RL method. Second, the learning is guaranteed to converge to the optimum with a much faster speed than the Q and $Q(\lambda)$-learning. Finally, this method supports both self and teacher-directed learning, in which the correctness of the teacher, however, is not required, which offers a relaxed and friendly interacting interface between the human and the robot.

In addition to the above benefits, the proposed method also has some limits in real applications. First, it assumes negative rewards only available when the robot fails the task, which might not be true in some cases. Secondly, this method does not consider uncertainties in state transitions and rewarding. Consequently, the learning might be noise sensitive since the influence of the noise would also be recursively propagated to all related states. How to ensure the robustness of the learning in real world is a crucial but still outstanding issue of this method.

Acknowledgement

This project is partly supported by the NSF Grant with the award number IIS-00-85980. The authors would like to acknowledge the anonymous reviewers for their invaluable comment and suggestion.

References

[1] S. Thrun, "An Approach to learning mobile robot navigation", *Robotics and Autonomous Systems*, vol. 15, pp.301-319, 1995.

[2] J. Millan and R. del, "Reinforcement learning of goal-directed obstacle-avoiding reaction strategies in an autonomous mobile robot", *Robotics & Autonomous Systems*, vol. 15, pp.275-99, 1995.

[3] J. Randlov and P. Alstrom, "Learning to drive a bicycle using reinforcement learning and shaping", *International Conference on Machine Learning*, pp.463-471, 1998.

[4] CJCH. Watkins, *Learning from delayed reward*, Ph.D. thesis, Cambridge University, 1989.

[5] CJCH. Watkins and P. Dayan, "Q-learning", *Machine Learning*, vol. 8, pp.279-92, 1992.

[6] G. Gordon, "Stable function approximation in dynamic programming", *Proc. of 12th Int'l Conf. on Machine Learning*, pp. 261-68, San Francisco, CA, 1995.

[7] L. Baird, "Residual algorithms: Reinforcement learning with function approximation", *Proc. of 12th Int'l Conf. on Machine Learning*, pp. 30-37, San Francisco, CA, 1995.

[8] S. Thrun and A. Schwartz, "Issues in using function approximation for reinforcement learning", *Proc. of 1993 Connectionist Models Summer School*, Hillsdale, NJ, 1993.

[9] J. Boyan and A. Moore, "Generalization in reinforcement learning: Safely approximating the value function", *Advances in Nerual Informatin Processing Systems*, MIT Press, 1995.

[10] A. Moore and C. Atkeson, "The Parti-game algorithm for variable resolution reinforcement learning in multidimensional state-spaces", *Machine Learning,* vol. 21, pp. 199-233, 1995.

[11] W. Zhu and S. Levinson, "Vision-based reinforcement learning for robot navigation", *Proc. of Int'l Joint Conf. on Neural Network,* vol. 2, pp. 1025-30, Washington DC, July, 2001.

[12] D. Chapman and L. Kaelbling, "Input generalization in delayed reinforcement learning: An algorithm and performance comparisons", *Proc. of Int'l Joint Conf. on Artificial Intelligence*, pp. 726-31, 1991.

[13] Y. Takahashi, M. Asada and K. Hosoda. "Reasonable Performance in Less Learning Time by Real Robot Based on Incremental State Space Segmentation", *Proc. of IEEE/RSJ Int'l Conf. on Intelligent Robots and Systems*, pp. 1518-24, 1996.

[14] L-J. Lin, *Self-supervised Learning by Reinforcement and Artificial Neural Networks*, Ph.D. Thesis, Carnegie Mellon University, Pittsburgh, PA, 1992.

[15] R. Bellman, *Dynamic Programming*, Princeton University Press, Princeton, 1957.

Author Index